MOLECULAR
BIOLOGY
INTELLIGENCE
UNIT

ORGANELLAR
PROTON-ATPASES

Nathan Nelson

Roche Institute of Molecular Biology
Roche Research Center
Nutley, New Jersey
U.S.A.

Springer-Verlag Berlin Heidelberg GmbH

MOLECULAR BIOLOGY INTELLIGENCE UNIT
ORGANELLAR PROTON-ATPASES

ISBN 978-3-662-22267-6   ISBN 978-3-662-22265-2 (eBook)
DOI 10.1007/978-3-662-22265-2

Library of Congress Cataloging-in-Publication Data

Nelson, Nathan
   Organellar Proton-ATPases / Nathan Nelson
     p. cm.—(Molecular biology intelligence unit)
   Includes bibliographical references and index.
   ISBN 978-3-662-22267-6
   1. Drug resistance in microorganisms. 2. Bacterial genetics.
   I. Title. II. Series.
     [DLNM: 1. Drug Resistance, Microbial—genetics. 2. Bacteria—genetics.
QW 52 A479o 1993]
QR177.A47 1993
616'.014—dc20
DNLM/DLC   93-37255
for Library of Congress   CIP

While the authors, editors and publisher believe that drug selection and dosage and the specifications and usage of equipment and devices, as set forth in this book, are in accord with current recommendations and practice at the time of publication, they make no warranty, expressed or implied, with respect to material described in this book. In view of the ongoing research, equipment development, changes in governmental regulations and the rapid accumulation of information relating to the biomedical sciences, the reader is urged to carefully review and evaluate the information provided herein.

Publisher's Note

R.G. Landes Company publishes five book series: *Medical Intelligence Unit, Molecular Biology Intelligence Unit, Neuroscience Intelligence Unit, Tissue Engineering Intelligence Unit* and *Biotechnology Intelligence Unit.* The authors of our books are acknowledged leaders in their fields and the topics are unique. Almost without exception, no other similar books exist on these topics.

Our goal is to publish books in important and rapidly changing areas of medicine for sophisticated researchers and clinicians. To achieve this goal, we have accelerated our publishing program to conform to the fast pace in which information grows in biomedical science. Most of our books are published within 90 to 120 days of receipt of the manuscript. We would like to thank our readers for their continuing interest and welcome any comments or suggestions they may have for future books.

Deborah Muir Molsberry
Publications Director
R.G. Landes Company

CONTENTS

CONTENTS

EDITOR

Nathan Nelson
Roche Institute of Molecular Biology
Roche Research Center
Nutley, New Jersey, U.S.A.
chapter 1

CONTRIBUTORS

Roland Baron
Department of Orthopedics
  and Cell Biology
Yale University School of Medicine
New Haven, Connecticut, U.S.A.
chapter 3

Marcjanna Bartkiewicz
Department of Orthopedics
  and Cell Biology
Yale University School of Medicine
New Haven, Connecticut, U.S.A.
chapter 3

Pe'er David
Department of Orthopedics
  and Cell Biology
Yale University School of Medicine
New Haven, Connecticut, U.S.A.
chapter 3

Julian A.T. Dow, M.A., Ph.D.
Department of Cell Biology
University of Glasgow
Scotland, United Kingdom
chapter 4

Stephen L. Gluck
Departments of Medicine, Cell
  Biology and Physiology
  and Pediatrics
Washington University School
  of Medicine
Seattle, Washington, USA
chapter 6

Sergio Grinstein
Division of Cell Biology
Hospital for Sick Children
Toronto, Canada
chapter 2

L. Shannon Holliday
Departments of Medicine, Cell
  Biology and Physiology
  and Pediatrics
Washington University School
  of Medicine
Seattle, Washington, USA
chapter 6

CONTRIBUTORS

Masahiro Iyori
Departments of Medicine, Cell
  Biology and Physiology
  and Pediatrics
Washington University School
  of Medicine
Seattle, Washington, USA
chapter 6

Howard T. Jacobs
Robertson Institute
  of Biotechnology
University of Glasgow
Scotland, United Kingdom
chapter 5

Beth S. M. Lee
Departments of Medicine, Cell
  Biology and Physiology
  and Pediatrics
Washington University School
  of Medicine
Seattle, Washington, USA
chapter 6

Gergely Lukacs
Division of Cell Biology
Hospital for Sick Children
  and Department of Surgery
Toronto General Hospital
Toronto, Canada
chapter 2

Raoul D. Nelson
Departments of Medicine, Cell
  Biology and Physiology
  and Pediatrics
Washington University School
  of Medicine
Seattle, Washington, USA
chapter 6

Ori D. Rotstein
Department of Surgery
Toronto General Hospital
Toronto, Canada
chapter 2

Natividad Hernando-Sobrino
Department of Orthopedics
  and Cell Biology
Yale University School of Medicine
New Haven, Connecticut, U.S.A.
chapter 3

THE WORLD
OF PROTON PUMPS

Nathan Nelson

Quite frequently an observer looks at a life phenomenon and asks himself why nature took this route out of several other available mechanisms. In order to understand this challenge he may first try to reconstruct the evolution of the process taking into account the main driving forces that were assumed to exist during the last 3.5 billion years on earth. Now we know that the electrochemical gradient of protons is the universal high-energy intermediate produced and utilized by every living cell in nature. This high energy intermediate is an expression of different concentrations of active protons in the two faces of biological membranes. Why then did nature elect to utilize a predominantly electrochemical gradient of protons and no other ion? The answer to this question may lie in the environment in which the first living creatures evolved. Some of them may have been challenged by an acidic environment that lowered their internal pH (by a proton leak through their membranes) to lethal levels. To counteract these incidents proton pumps evolved. Two independent systems were developed. One was a proton pump coupled to energetically downhill vectorial electron transport across membranes and the second was an ATP-dependent proton pump. Both of them pump protons outward from the cells generating an electrochemical gradient of protons. The more successful creatures learned to utilize this universal high energy intermediate to drive numerous secondary processes including ATP formation by the same proton pump coupled to the redox reactions that took place on the same membranes. Currently every living cell is utilizing this fundamental process.

The eukaryotic cell is defined by a separate nucleus containing most of the genetic material of the cell, semiautonomous organelles containing their own unique DNA and RNA molecules, and a vacuolar system composed of an internal network of membranes. The existence of numerous compartments inside the cell and the variable environment in which eukaryotic cells live requires a constant supply of energy for maintaining the precise concentration of solutes in the various cell compartments. Proton pumps play a major role in providing energy for several secondary uptake processes as well as maintaining the pH homeostasis required for every living cell. There are several kinds of proton pumps functioning in various life processes. They include electron transport reaction such as respiration, photosynthesis, nitrate reduction as well as numerous other redox reactions. These oxidation-reduction reactions are coupled to proton translocation across membranes. Essentially all of these systems involve proton pumps that derive their energy from light or compounds with negative redox potential. The electrochemical energy generated by these pumps can be utilized for ATP formation by the reversible H^+-ATPases; F-ATPases in eukaryotes, chloroplasts and mitochondria or V-ATPase in archaebacteria.[1] The electrochemical energy formed by proton pumps was termed protonmotive force (pmf), and according to the Mitchell hypothesis is the universal high energy intermediate across biological membranes.[2] The pmf can be utilized for driving numerous secondary transport systems that maintain the solute composition inside the vacuolar system different from that of the cytoplasm. The following equations express the thermodynamics of pmf:

$$ADP + P_i + nH^+_{out} \leftrightarrow ATP + nH^+_{in} \tag{1}$$

where n stands for the number of protons involved in a single phosphorylation or hydrolysis step. The relationship between the electrochemical gradient of protons and the formation of ATP is described by the relation,

$$\Delta G = n\Delta\mu H^+ + \Delta G_p \tag{2}$$

where ΔG is the Gibbs-free energy (free enthalpy), ΔG_p the free enthalpy of phosphorylation, and $\Delta\mu H^+$ the H^+ electrochemical proton gradient (pmf). When $\Delta G < 0$, ATP synthesis takes place, when $\Delta G > 0$, ATP hydrolysis takes place, and when $\Delta G = 0$, the reaction is at equilibrium. Protonmotive force is a combination of the proton gradient and the electric potential across the membrane. The relationship between the two is expressed as follows:

$$\Delta\mu H^+ = F\Delta\psi - 2.3\ RT\Delta pH \tag{3}$$

where $\Delta\psi$ represents the electrical potential across the membrane, ΔpH the difference in H^+ ion concentration on the two sides of the membrane, R the gas constant, T the absolute temperature, and F the

Faraday constant. Dividing equation 3 by F expresses the pmf in millivolts (mV) and a simple equation of pmf = $\Delta\psi$ + 58.8 ΔpH is thereby obtained (2.3 RT/F is equal to 58.8 mV at room temperature).

Eubacteria, chloroplasts and mitochondria contain an H^+-ATPase (F-ATPase) that utilize the pmf, generated by respiratory and photosynthetic electron transport chains, for phosphorylating ADP into ATP.[3] This enzyme operates at thermodynamic equilibrium ($\Delta G=0$) most of the time, and depending on the levels of ATP and pmf, may work in a mode of ATP-synthesis ($\Delta G<0$) or ATP-dependent proton pumping ($\Delta G>0$). On the other hand, V-ATPase functions in eukaryotic cells exclusively as a proton pump presumably at $\Delta G>0$ (see Equation 2). In archaebacteria V-ATPase functions in ATP-synthesis. The two families of proton pumps F- and V-ATPases have structural and mechanistic similarities and evolved from a common ancestral proton pump. In contrast to P-ATPases that function by forming a catalytic phospho-enzyme intermediate, F- and V-ATPases function without an apparent formation of such an intermediate.[1] Consequently the catalytic mechanism of F- and V-ATPases is distinct from that of P-ATPases, and the proton pumps among the P-ATPases have unique functions that do not include ATP formation.

In this book the current knowledge of F- and V-ATPases will be discussed and the function of P-ATPases will be mentioned only when relevant to the understanding of the former proton pumps. We will try to understand how and why the F- and V-ATPases function in the various organelles and cells and what the advantage is of employing such complicated enzymes as primary proton pumps. Why are one or both of these enzyme families present in every known living cell, and how can such a conserved enzyme as V-ATPase differentially function in specialized cells and organelles? The main function of F-ATPases is to synthesize ATP at the expense of the protonmotive force (pmf) generated by the photosynthetic and respiratory electron transport chains. However, under anaerobic conditions and certain metabolic circumstances F-ATPases may function as an H^+-pumping ATPase, generating pmf at the expense of ATP. A similar if not identical function is fulfilled by V-ATPases in archaebacteria. Since F-ATPases function in chloroplasts and mitochondria we assume that they are more efficient in an oxygenic environment than the V-ATPases. The reason for the presence of a V-ATPase in archaebacteria is not known and it may be that this is one of the reasons that they are not as successful as eubacteria.[3] On the other hand, V-ATPase may be advantageous under extreme conditions of low pH, high salinity and elevated temperatures. The vacuolar system of eukaryotic cells may have exploited the relative inefficiency of V-ATPases to gain better control over the extent of acidification of the various internal organelles.[3,4] We proposed that this control is governed by a proton slip resulting in the inability of these V-ATPases to synthesize ATP.[4] Consequently V-ATPase can function in a wide variety of organelles and membranes that have different requirements for the extent and composition of pmf. A special emphasis will be given to some of the organelles and tissues in which V-ATPases play a crucial role.

The first chapter will discuss the biochemistry and molecular biology of F-and V-ATPases. Most of the biochemical data came from studies of organelles from mammalian sources. They include neurotransmitter and hormone storing organelles such as chromaffin granules and synaptic vesicles as well as endosomes and specialized plasma membrane of kidney cells. Molecular biology of V-ATPases started by cloning cDNAs encoding subunits of the enzyme and developed into more sophisticated research by utilizing yeast genetics. The latter brought a wealth of information on the structure and function of the enzyme and helped to sort the fundamental subunits from the accessory polypeptides. Cloning of the genes encoding V-ATPase subunits also painted a beautiful and reliable scenario for the evolution of F- and V-ATPases.

Maintaining a neutral pH in the cytoplasm of eukaryotic cells and low pH inside several of their organelles is a life necessity. Dr. Grinstein and his colleagues review the processes involved in cytosolic pH regulation and the role of V-ATPases in this and related processes. Very small changes in the cytosolic pH have far reaching physiological consequences. Consequently, modulation of internal pH is being utilized for signal transduction, control of biochemical processes and a wide variety of other cellular events. These include receptor mediated processes as well as influencing protein sorting in the Golgi complex.

V-ATPase functions not only in the vacuolar system but also in the plasma membrane of the specialized cells. Dr. Gluck discusses the role of V-ATPase on the plasma membranes of specialized cells in the kidney. The kidney plays a vital role not only in cleaning the body of waste products but also in the acid-base balance of mammals. Hydrogen ion excretion involves several processes including bicarbonate reabsorption, carbonic anhydrase activity and regulated pumping of protons across the plasma membrane by V-ATPase. In epithelial cells of the proximal urinary tubule, V-ATPase is present in the apical membrane and functions in proton secretion. In the collecting duct V-ATPase may be found either in apical or basolateral membranes of specialized intercalated cells. These cells shuttle V-ATPase between intracellular vesicles and the plasma membrane in response to changes in the acid-base balance of the animal. It was shown that the distribution of V-ATPase, in apical or basolateral membranes of intercalated cells, is changing during adaptation to acidosis or alkalosis. The cells increase the number of V-ATPase enzymes in their apical membrane during acidosis and decrease their number during alkalosis. Therefore, V-ATPase plays a major role in maintaining pH homeostasis in mammals and other animals.

Dr. Baron and his colleagues elaborate on the involvement of V-ATPase in bone resorption. This process is necessary for bone growth, remodeling and repair. Osteoclasts are multinucleated and highly motile cells that migrate between the bone and bone marrow and function in bone resorption. They attach to the mineralized bone matrix forming a close space to which hydrolytic enzymes are secreted. These enzymes require low pH for their optimal activity and the low pH is provided by V-ATPase located in the part of the plasma membrane in contact with the bone. The principal bone mineral is hydroxyapatite, and protons are required for the release of each calcium ion from the mineral. The osteoclast V-ATPase provides all protons necessary for calcium resorption. As in kidney intercalated cells, osteoclasts shuttle V-ATPases between intracellular vesicles and the plasma membrane in response to attachment onto the bone matrix. This action renders an amoeba-like cell into a polar cell. The properties of the osteoclast V-ATPase is somewhat different from other V-ATPases in its greater sensitivity to vanadate as well as the possible presence of specialized subunit A in the enzyme. The pharmacological value of studying the osteoclast V-ATPase is apparent because a specific slow down in its activity may prevent the onset of osteoporosis. If indeed an enzyme with unique properties functions in osteoclasts, it would be more likely to discover a drug that will specifically inhibit the osteoclast V-ATPase.

Recent studies on the physiology of ion movements in insect organs revealed a pivotal function for V-ATPases in a wide variety of processes. Dr. Dow reviews this new exciting field that has a far reaching implication not only in the insect world but also in understanding the physiology of other organisms including mammals. Several organs energized in other creatures by P-ATPases utilize V-ATPases as a primary energy source for driving secondary uptake processes. This may be due to the sensitivity of P-ATPases to alkaloids which are abundant in the food of many insects. In contrast V-ATPases are not sensitive to alkaloids and may function well in their presence.

In the last 25 years the family of F-ATPases took center stage in the field of bioenergetics. Its discovery preceded that of V-ATPase by more than two decades and numerous articles, reviews and books were written on this family of proton pumps.[5] In this book Dr. Jacobs has the difficult task of reviewing recent discoveries in the field with an emphasis on diseases connected with mitochondrial F-ATPase.

REFERENCES

1. Nelson N, Taiz L. The evolution of H^+-ATPases. Trends Biochem. Sci. 1989; 14:113-116.
2. Mitchel P. Chemiosmotic Coupling and Energy Transduction. Bodmin, 1991.
3. Nelson N. Evolution of organellar proton-ATPases. Biochem Biophys Acta 1992; 1100:109-124.
4. Nelson N. Structure, molecular genetics and evolution of vacuolar H^+-AT-Pases. J. Bioenerg Biomembr 1989; 21:553-571.
5. Racker E. A New Look at Mechanisms in Bioenergetics. New York: Academic Press, Inc., 1976.

CHAPTER 1

MOLECULAR AND CELLULAR BIOLOGY OF F- AND V-ATPASES

Nathan Nelson

INTRODUCTION

There are two mechanistically distinct ATP-dependent proton pumps. One belongs to the family of P-ATPases that operates with a phospho-enzyme intermediate, and the second belongs to the families of F- and V-ATPases that operate without an apparent phospho-enzyme inter-mediate.[1,2] The P-type proton pumps are integral membrane proteins, having similar structure and mechanism of action to those of Na^+/K^+-ATPases and Ca^{++}-ATPases. The function of this proton pump is primarily in the plasma membrane of plant and fungal cells and in specialized mammalian cells such as parietal cells in the stomach.[3] F- and V-ATPases are more universal proton pumps and at least one of them is present in every living cell.[4] They share a common structure and mechanism of action and have a common evolutionary ancestry. In eukaryotic cells F-ATPases are confined to the semiautonomous organelles, chloroplasts and mitochondria that contain their own genes encoding some of the F-ATPase subunits.[5] F-ATPase is also vital for every known eubacterium acting in photosynthetic or respiratory ATP-formation and/or in gen-erating protonmotive force (pmf) by the reaction of ATP-dependent proton pumping.

V-ATPase functions as an ATP-dependent proton pump in the vacuolar system and plasma membrane of eukaryotic cells, and in ATP formation in the plasma membrane of archaebacteria.[6] The vacuolar system of eukaryotic cells consists of all of the internal membrane network of the cell excluding semiautonomous organelles such as mitochondria and chloroplasts.[7] V-ATPase pumps protons from the cytoplasm to the internal space of the organelles. Consequently, the pH inside the organelles is more acidic than that of the cytoplasm, a proton gradient across the membranes is established and a membrane potential positive inside is generated. Numerous organelles of the vacuolar system are energized by V-ATPases and each organelle has a special requirement for its internal pH and membrane potential. Thus endosomes operate with lower pH than lysosomes and synaptic vesicles accumulating glutamate maintain a larger membrane potential than chromaffin granules that take up catecholamines. Consequently, V-ATPase must be differentially regulated in the various organelles. The relations between F- and V-ATPases and their differential distribution and function is the subject of this chapter.

STRUCTURE AND FUNCTION OF F- AND V-ATPASES

FUNCTION

Respiratory and photosynthetic electron transport chains, as well as numerous other redox systems, generate electrochemical gradients of protons. The pmf generated by these proton pumps can be utilized for ATP formation by F- and V-ATPases. F-ATPase fulfills this function in eubacteria, mitochondria and chloroplasts and V-ATPase performs this reaction exclusively in archaebacteria. Therefore, one of the most notable distinctions between F- and V-ATPases is in their function in ATP formation. While the primary function of F-ATPases in eukaryotic cells is to form ATP at the expense of pmf, V-ATPases function exclusively as ATP-dependent proton pumps.[8,9] The coupling between a vectorial process of utilizing pmf and the scalar reaction of ATP formation took center stage in biochemical studies of the last 30 years. The thermodynamics of this reaction was discussed above but the big challenge of understanding the mechanism of this reaction is far from being settled. There is a general agreement that the coupling between the two reactions is mechanochemical and the mechanism of ADP phosphorylation, on the catalytic sector of these enzymes, involves conformational changes that function in releasing the product (ATP) from the enzyme.[8] We assume that a similar mechanism features ATP-dependent proton pumping by V-ATPases in eukaryotic cells. Yet these enzymes cannot function in pmf driven ATP formation and this property is probably due to alterations in the membrane sectors of these enzymes (see below). The pmf generated by V-ATPases in organelles of eukaryotic cells is utilized as a driving force for numerous secondary uptake processes. Several metabolic processes that take place in the internal membrane network of eukaryotic cells may be dependent or influenced by the function of V-ATPase. For example, V-ATPase plays an important role in the sperm during fertilization. Acidification of the interior of the acrosome is vital for the function of the sperm.[9] Upon binding of the sperm to the unfertilized egg an acrosome reaction takes place. The acrosome reaction results in the secretion of the acrosome contents as well as exposure of the inner acrosomal membrane to the egg zona pellucida. Consequently, a part of the acrosomal membrane rich in V-ATPases is exposed to a cleft between the membrane and the zona pellucida to which low pH-dependent hydrolytic enzymes were secreted. Thus the acrosomal V-ATPase acidifies the external cleft, providing a suitable environment for the activity of the hydrolytic enzymes. Therefore, the function of V-ATPase is vital for fertilization. Conversely, neutralization of the cleft's low pH, generated by the V-ATPase, prevents fertilization—an action that is as essential for the continuation of the human race as is fertilization.

The structure of F- and V-ATPases may disclose the clue to their mechanism of action. Both enzymes are multi-subunit protein complexes built from distinct catalytic and membrane sectors. The function of the catalytic sector is to provide the ATP-binding site and to catalyze the ATP-formation and/or ATPase activities of the enzymes. The main function of the membrane sectors is to conduct protons across the membrane. The energy coupling between these two processes is believed to be catalyzed via mechanochemical induced conformational changes.[9]

CATALYTIC SECTOR

Although very similar in their general structure the subunit structure of F- and V-ATPases contains several unique features. Figure 1.1 depicts a schematic presentation of the minimal subunit structure of F- and V-ATPases. The subunit structure of F-ATPase is drawn according to the *E. coli* enzyme that contains five polypeptides denoted as α, β, γ, δ and ϵ in the catalytic sector and three subunits denoted as a, b and c (proteolipid) in the membrane sector.[10] The membrane sector of *E. coli* F-ATPase contains 1a, 2b and 6-12 c subunits. The structure of V-ATPases was studied by electron microscopy.[11,12] The general structure of V-ATPases resembles that of mitochondrial and chloroplast

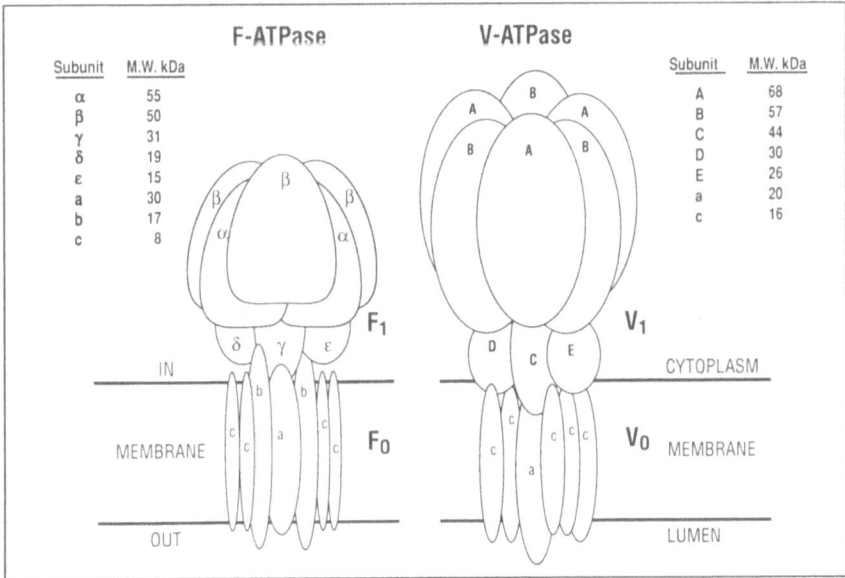

Fig. 1.1. Schematic structure of F- and V-ATPases.
The structure of F-ATPase is according to the subunit composition of the E. coli enzyme. The subunit structure of V-ATPase is a minimal depiction of identified subunits that may have analogous function to the F-ATPase subunits. More subunits are necessary for activity of the enzyme in eukaryotic cells.

F-ATPases with a notable exception of appendices protruding from the periphery of the catalytic sector. The subunit structure of V-ATPase is drawn from knowledge obtained from studies of enzymes from archaebacteria, yeast and mammalian sources. It contains five different polypeptides denoted as A, B, C, D and E in the catalytic sector, and subunits a (hypothetical) and c (proteolipid) in the membrane sector. The relative amounts of the subunits in the catalytic sector is 3:3:1:1:1 in both F- and V-ATPases.[6] The catalytic sector of F-ATPases can be readily dissociated from the membranes by incubating the membrane preparations in the presence of EDTA or subjecting them to sonication. The removal of the catalytic sector (F_1) results in the generation of a proton leak through the membrane sector. The catalytic sector of V-ATPases can also be removed by a mild treatment. Incubation of membranes on ice in the presence of MgATP and 0.2 M NaCl results in the removal of the catalytic sector but no proton leak is generated by this treatment.[13-15] This suggests that in one of the conformations of the catalytic cycle the contact between the catalytic and membrane sectors is hydrophobic in nature (hydrophobic interactions are weakened at low temperature). The exact organization of the catalytic sector is not known, but due to the homology and sequence conservation between F- and V-ATPases it is logical to assume that their general structure is similar. Moreover, it was demonstrated that the A and B subunit of V-ATPase from the archaebacterium *Sulfolobus acidocaldarius* are alternating similarly to the α and β subunits of F-ATPase from *E. coli*.[16]

The function of some of the individual subunits was studied in detail and a large wealth of information on their biochemistry is available. The β subunit of F-ATPases contains the catalytic ATP-binding site.[10] Similarly, subunit A (69 kDa) of V-ATPases contains the catalytic ATP-binding site of the enzyme. This was demonstrated by sequence homology between the two subunits as well as ATP binding to the A subunit by UV irradiation.[17-19] The amino acid sequence of this subunit contains a "glycine rich motif" that is common for ATP binding proteins. This motif contains, in the A subunit but not in β, two cysteine residues the modification of which causes inactivation of the enzyme.[20] Moreover, modification of a single cysteine on subunit A prevents dissociation of the catalytic sector from the membrane by cold treatment.[13] These and other observations leave little doubt that subunit A functions in the ATPase activity of V-ATPases by providing the catalytic ATP-binding site.

The α subunit of F-ATPases contains an ATP-binding site that may function in the regulation of the enzyme activities.[10] Subunit B of V-ATPases may contain a similar ATP-binding site. This subunit binds ATP analogs only under restricted conditions.[21] It does not bind ATP by UV irradiation.[17] Sequence analysis revealed an extensive homology to the α subunit of F-ATPases.[22] However, subunit B contains no glycine rich sequence, which is an indication of the existence of a nucleotide binding site. These and other observations suggest that subunit B may function in regulating the activity of V-ATPases, but may do so without the involvement of nucleotides.

The remaining subunits in the catalytic sector of V-ATPases have no homology to F-ATPase subunits. It is known that subunits γ, δ and ε are necessary for the function of F-ATPases but their precise role in the catalytic activity of the enzyme is not known. Similarly, there is no assigned function for subunits C, D,E and F of V-ATPases. Subunit C (41 kDa) is necessary for the assembly of V-ATPase in yeast cells.[23] Subunit D (34 kDa) was identified on SDS gels in preparations from a variety of sources from yeast to mammals. Immunological studies indicated that this polypeptide is copurifying with the enzyme and therefore is likely to be a subunit of V-ATPases.[6] Recently, we cloned a bovine cDNA and a yeast gene (VMA8) encoding subunit D of the V-ATPase.[91] We concluded that subunit D is an integral part of the catalytic sector of V-ATPase and is analogous to the γ subunit of F-ATPases. Subunit E (29 kDa) was one of the first to be identified in kidney V-ATPase by immunological studies, and the cDNA encoding this subunit was cloned and sequenced.[24] The function of this subunit is not known but it was implicated in controlling the biogenesis of the kidney V-ATPase.[25] The gene encoding this subunit in yeast cells was cloned, sequenced and interrupted. [26] It was shown that subunit E is necessary for the functional assembly of the enzyme. The rule of five subunits in the catalytic sector of F- and V-ATPases is violated in many instances. There are regulatory proteins such as F_1-inhibitor that modulate the activity of the mitochondrial enzyme. A new subunit was recently identified in insect, yeast or mammalian V-ATPases. It was first identified in V-ATPase from Tobacco Hornworm midgut. [27] It is a 14 kDa polypeptide that could be stripped from the membrane by the chaotropic agent KI, and may take part in the catalytic sector of the enzyme. Very recently we cloned a gene encoding highly homologous protein in yeast and found that a disruptant mutant of this gene fails to grow on medium buffered at pH 7.5.[28] This suggests that this polypeptide is a genuine subunit of V-ATPases and was denoted as subunit F of the catalytic sector. Figure 1.2 depicts the subunit structure of mammalian V-ATPase including some recently identified subunits. The function of the newly discovered subunit F is not known. Surprisingly, the antibody against this subunit inhibited ATPase activity of the insects enzyme. [27] This observation is quite odd because all the other antibodies including those that were raised against the catalytic subunits fail to inhibit the activities of the enzyme. It may be that this specific antibody removes the subunit from the enzyme and it was shown that the yeast enzyme is not functional without subunit F.[28]

The structure of the catalytic sector of F-ATPase from bovine heart mitochondria was recently determined at 6.5 A resolution.[8] It revealed an asymmetric structure with a 40 A stem which contains two α-helices in a coiled-coil arrangement. This structure presumably comes from the γ subunit and was implicated in the mechanochemical mechanism of ATP-dependent proton pumping and pmf-driven ATP formation. The ramifications of this structure are far reaching and they present a major step forward to understanding energy coupling in ATP-dependent proton pumps. There is an intriguing question of how much of this structure, and consequently the mechanism of action, will be preserved in V-ATPases. The stem in F-ATPases is part of the γ subunit that was implicated for a long time in energy coupling between the catalytic and membrane sectors. We propose that subunit D functions similarly to the γ subunit in the energy coupling between the catalytic and membrane sectors of V-ATPase.[91]

Membrane Sector

The function of the membrane sector is to conduct protons across the membrane and to couple this vectorial action with the scalar process of ATP formation or hydrolysis. While the *E. coli* membrane sector consists of three different subunits, that of mammalian mitochondria contains up to 10 different polypeptides. Similarly, the membrane

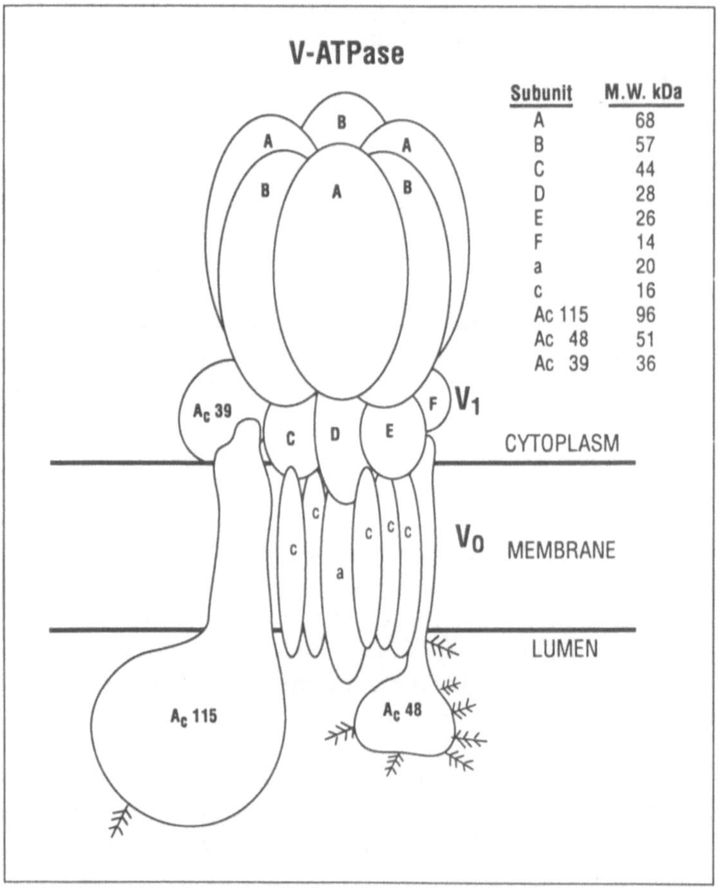

Fig. 1.2. Schematic subunit structure of V-ATPase of chromaffin granules. Subunits A, B, C, D and E of the catalytic sector were positively identified as subunits of the enzyme and cDNA's encoding these subunits were cloned and sequenced. Subunit D was identified by biochemical and immunological methods and its cDNA had recently been cloned. Subunit D is analogous to the γ subunit of F-ATPases. Subunit F was not identified in the enzyme from chromaffin granules but is present in enzymes from yeast and insects, and therefore we assume that it will comprise a genuine subunit in the mammalian enzyme. In the membrane sector, subunits Ac115, Ac39 and the proteolipid (subunit c) were positively identified by cDNA cloning. Subunit a is hypothetical and Ac48 is a specialized accessory polypeptide in chromaffin granules and other internal organelles but is not present in all mammalian enzymes. Potential glycosylation sites are indicated by (¥).

sector of V-ATPase from archaebacteria may be composed of only the proteolipid; and the membrane sector of the mammalian enzyme contains multiple subunits (see Figure 1.2). The proteolipid depicted in Figure 1.1 as subunit c is the principal subunit of the membrane sectors of both F- and V-ATPases. The identification of a DCCD binding site in proteolipids of F-ATPases led to an extensive study of their structure and function.[10] This protein of 8 kDa contains about 80 amino acids. It is highly hydrophobic and soluble in chloroform/methanol solution. Its detailed structure and orientation in the *E. coli* membrane is known.[29,30] It is built of two transmembrane helices with a hair pin facing the catalytic sector in the cytoplasm. In the middle of the second transmembrane segment there is a glutamyl or aspartyl residue that provides the binding site for DCCD. The binding of DCCD blocks proton conductance across the membrane and therefore inactivates the enzyme. The proteolipid of V-ATPase is also a highly hydrophobic protein that binds DCCD, but contains about 160 amino acids (16 kDa).[31] Binding of DCCD inactivates the proton pumping and ATPase activities of the enzyme.[32,33] cDNA and the gene encoding this subunit in mammals and yeast were cloned and sequenced.[31,34] The sequences revealed that the proteolipid evolved by gene duplication and fusion of an ancestral gene homologous to that present in F-ATPases. Disruption of the gene encoding the proteolipid in yeast cells showed that it is necessary for the assembly of all other subunits of the enzyme.[35] The proteolipid is likely to be involved in the process of proton translocation across the membrane. Subunit a of V-ATPase is still hypothetical, and except for the proteolipid there is no evidence of involvement of other proteins in proton conductance. In contrast, in F-ATPases it was clearly demonstrated that all three subunits (a, b and proteolipid) are required for proton conductance and/or proper assembly of the membrane sector.[36] While subunit a was implicated in the catalysis of proton conduction across the membrane, subunit b was suggested as the provider of the binding site for the catalytic sector. It also may act in the mechanochemical coupling between the ATPase and proton transport sites.

As depicted in Figure 1.2, mammalian V-ATPase may be composed of at least five different subunits. The genes or cDNA's encoding four of them were cloned and sequenced, and the cloning of the gene encoding the 20 kDa polypeptide (presumably subunit a) is on the way. Subunits Ac 115 and Ac 39 were identified, studied by immunological methods and the genes encoding these subunits were cloned and sequenced. Ac 115 is the most peculiar subunit of V-ATPases which has no homology nor analogy in F-ATPases. This subunit protrudes to the lumen of the vacuolar system and is glycosylated.[37] Its function is not known, but it is logical to assume that it may play a role in the recognition of V-ATPases at the cell exterior or in the lumen of the vacuolar system. This accessory subunit transverses the membrane several times and therefore is a bona fide membrane protein. The other accessory subunit is the Ac 39 associated with the membrane sector but contains no apparent transmembrane segments.[38] Its function is not known and it is located on the cytoplasmic side of the membrane.

Sequencing of the cDNA encoding this subunit revealed no sequence homology with any other protein and gave no clues to its function. However, recent cloning of the gene encoding the homologous subunit in yeast provided evidence that it is an integral part of the enzyme and is necessary for its assembly.[39]

A novel subunit was identified in V-ATPase from chromaffin granules.[40] This subunit was identified as a broad band on SDS polyacrylamide gels present between subunits B and C. Cloning of the cDNA encoding this polypeptide revealed seven potential glycosylation sites positioned in the luminal side of the membrane and a single transmembrane segment at the C-terminal part of the polypeptide. Immunological studies suggest that this polypeptide is restricted for internal organelles in mammalian cells with a potential function as an anchor for the enzyme in these organelles.

The fine structure of the membrane sectors of F- and V-ATPases is not known and it is unlikely to be resolved by crystallography in the near future. The function of this sector is not only to conduct protons across the membrane but also to couple the proton motive force with ATP synthesis and hydrolysis. Moreover, an additional function such as superseding the assembly of the enzyme as well as connecting the proton pump with other membrane proteins are the trait of the membrane sector.

MOLECULAR BIOLOGY OF V-ATPASES

The first step in studying the molecular biology of a system is cloning the relevant genes and cDNA's. It is not a secret that at this stage we advise our students "Do not think—clone". It was only since 1988 that cDNAs and genes encoding subunits of V-ATPases were cloned and sequenced.[18,19,22,24,31,41] The sequences revealed valuable information on the structure, function and evolution of the various subunits as well as the evolution of F- and V-ATPases.[2,6,42] It became apparent that subunits A and B of V-ATPases and subunits β and α of F-ATPases evolved from a common ancestral gene. The proteolipids of F- and V-ATPases also evolved from a common ancestral gene. All other subunits of F- and V-ATPases have no apparent homology and may have evolved independently of each other. In line with this evolutionary pattern subunits A, B and the proteolipid are the most conserved subunits in V-ATPases and are likely to be the most fundamental subunits of the enzyme. In the three conserved subunits there are sequences or amino acids indicative to their function. No such landmarks could be identified in all other subunits.

The cDNA and gene encoding subunit A were first to be cloned from plant,[19] fungi[18] and the archaebacterium *Sulfolobus acidocaldarius*.[43] It immediately became apparent that the enzyme that functions in ATP-synthesis in archaebacteria is V-ATPase and that subunit A is homologous to the β subunit of F-ATPases. It was also revealed that a yeast gene involved in trifluoperazine resistance, cloned the same year, encodes subunit A of V-ATPase.[44] This gene encodes a larger protein that undergoes protein splicing to give the mature subunit A.[45,46] Aligning the amino acid sequences of A and β subunits from various sources produced a wealth of information. The glycine-rich loop which contains the sequence GXXXXGKT/S is conserved in most nucleotide binding proteins. This sequence of amino motif was shown to bind ATP analogs and NBD-Cl in F-ATPases,[47] and NEM in V-ATPases.[20] The presence of ATP reduced the extent of chemical modification of this region, and site-directed mutagenesis in F-ATPase from *E. coli* showed high sensitivity to amino acid changes. A single NEM bound to C^{254} in the bovine subunit A is sufficient for inhibiting the activity of the enzyme. This and other cysteine residues render V-ATPases from eukaryotic cells sensitive to oxygen. It is interesting that archaebacteria exchanged the corresponding cysteine residue to serine and their V-ATPase is not sensitive to NEM.[43] The conserved glycine-rich loop in nucleotide binding proteins was implicated as a primordial common structure for nucleotide binding. All of these properties indicate that the A subunit of V-ATPases and the β subunit of F-ATPase are the catalytic subunits of these proton pumps.

Subunit B of V-ATPases is most homologous to the α subunit of F-ATPases. Significant homology in the amino acid sequences among the A and B subunits of V-ATPases and the β and α subunits of F-ATPases, left little doubt that all of these subunits evolved from a common ancestral gene.[42] Subunit B is the most conserved subunit in the catalytic sector of V-ATPases, and it was argued that the current subunit is the closest to the common ancestor of the others.[34] The function of subunit B is not clear but is likely to have a similar function in regulating the enzyme as the α subunit of F-ATPases. But unlike the α subunit, there are no wealth of experiments showing nucleotide binding to the B subunit. There is only one report showing the photoaffinity binding of the ATP analog 3-O-(4-benzoyl)benzoyladenosine 5'-triphosphate to the B subunit of plant V-ATPase.[48] Moreover, in contrast to the α subunit of F-ATPases that contains the consensus ATP-binding motif GXXXXGKT/S,[49] is not present in the B subunit of V-ATPases. Therefore it would not be a surprise to find that V-ATPases contain only three nucleotide binding sites, confined to subunit A, in variance from F-ATPases that contain six potential nucleotide binding sites, three on α and three on β subunits. Since the A subunit of archaebacterial V-ATPase also does not contain the ATP-binding motif, it is unlikely that its absence renders the ATPase to function exclusively as a proton pump. While yeast contains a single gene encoding the B subunit, bovine contains two genes encoding very similar proteins.[50] It is not known if each of these genes confers different properties to the enzyme.

Subunit C was first cloned from the bovine adrenal library.[51] The deduced amino acid sequence showed no homology with any other proteins in the GenBank. It was concluded that this subunit evolved independently of any F-ATPase subunit. The gene encoding subunit C in yeast exhibits only 37% identity with the bovine protein, and except for a small stretch of highly conserved amino acids, the remainder of the conserved residues are scattered all over the protein. A mutant was generated in which the gene encoding the C subunit was interrupted.[23] This mutant failed to grow in medium buffered at pH 7.5.

A chimeric gene comprised of the yeast and bovine sequences successfully replaced the yeast gene and permitted growth at pH 7.5. The amino acid sequence of bovine or yeast subunits C shows no sequence homology with the known V-ATPase subunits from archae-bacteria.[52] The function of subunit C is not known, but its requirement for the functional assembly of the yeast enzyme is well established. It may be analogous to the subunit of F-ATPases that is also poorly conserved and may function in energy coupling between the catalytic sector and the proton translocating membrane sector. The yeast gene encoding subunit D was cloned and interrupted.[91] The resulting mull mutation prevented the assembly and function of V-ATPase. Structural analysis suggested that subunit D is analogous to the γ subunit of F-ATPase.

The cDNA encoding subunit E of bovine kidney microsomes was cloned, sequenced and analyzed.[24,25] It exhibits no significant sequence homology with any of the subunits of F-ATPases. It also has no sequence homology with the predicted gene products of the archaebacterial operon encoding their V-ATPases. Studies with monoclonal antibodies, supported by partial DNA sequencing, revealed the existence of at least two isoforms of subunit E in the kidney. While V-ATPase isolated from kidney microsomes contains one form of subunit E, the enzyme from the kidney brash-border contains at least one additional form of subunit E. It was suggested that subunit E plays a role in the biogenesis of the enzyme and in regulating its polar assembly in two different types of intercalated cells.[25] The gene encoding subunit E in yeast cells was cloned, sequenced and interrupted.[26] The disruptant mutant exhibited a similar phenotype to all other V-ATPase disruptant mutants. While the proteolipid assembled into the membrane all subunits of the catalytic sector did not assemble. Consequently, the mutant was not able to grow in medium buffered at pH 7.5.[53] Molecular biology studies of the catalytic sector of V-ATPases are not complete. New potential gene products are emerging, some of which are likely to function in the assembly of the enzyme and others in regulating its activity.

The membrane sector of V-ATPases is dominated by a highly hydrophobic protein of 16 kDa called proteolipid. A cDNA encoding the proteolipid of V-ATPase was first cloned from the bovine adrenal library.[31] The predicted amino acid sequence revealed its relation to the proteolipids of F-ATPases and suggested that the proteolipid of eukaryotes evolved by gene duplication and fusion of an ancestral proteolipid gene. Subsequently, the gene encoding the yeast proteolipid was cloned and sequenced as well as genes and cDNA's encoding proteolipids from a variety of sources.[5] Alignment of the predicted amino acid sequences of proteolipids of various eukaryotes demonstrated that this protein is one of the most conserved hydrophobic proteins in nature. As shown in Figure 1.3, the third and fourth transmembrane helices were almost totally conserved in proteolipids from yeast to mammals. So far there is no other example in which the amino acid sequences of transmembrane helices are identical in yeast and mammalian proteins. Studies in yeast mutants showed that the proteolipid is necessary for the assembly of all other subunits.[35,54] The proteolipid, on the other hand, can assemble into the vacuolar membrane independently of the catalytic sector subunits. The fourth transmembrane helix contains the glutamyl residue that presumably binds DCCD; resulting in inactivation of the ATPase and proton transport activities of the enzyme. Numerous mutations, generated by site-directed mutagenesis, of the yeast proteolipid were reported.[54] Displacement of E^{137} by any amino acid except D inactivates the enzyme. Several other displacements rendered the enzyme inactive, and it was concluded that water molecules coordinated to amino acids in the transmembrane helices of the proteolipid may play a role in proton translocation across the membrane. Analysis of the second-site suppressor mutants in the yeast proteolipid suggests a very tight organization of the proteolipids in the vacuolar membrane.[55] More experiments are required to reveal the exact role of the proteolipid in proton translocation across the membrane.

Except for archaebacteria and presumably plants, a polypeptide of about 100 kDa is present in every preparation of V-ATPase. The cDNA encoding this subunit in bovine brain was cloned and sequenced.[37] It encodes a protein of 99 kDa with potential glycosylation sites and few transmembrane helices. A yeast gene encoding a protein with considerable similarity to the one from bovine was cloned and sequenced.[56] Disruption of this gene did not abolish the activity of V-ATPase, supporting the notion that this is an accessory polypeptide that does not directly participate in the ATPase or proton translocation activities of the enzyme.[6] However, inactivation of the gene interfered with the assembly of the enzyme, suggesting that this protein is a genuine subunit of V-ATPase. The function of this subunit is not known and it may have some role in sensing the environment in the interior of the vacuolar system. We still do not know whether this subunit is present in all V-ATPases in the wide variety of organelles in mammalian cells. There is an open question of what directs the V-ATPases to the different organelles and membranes and what factors confer the different

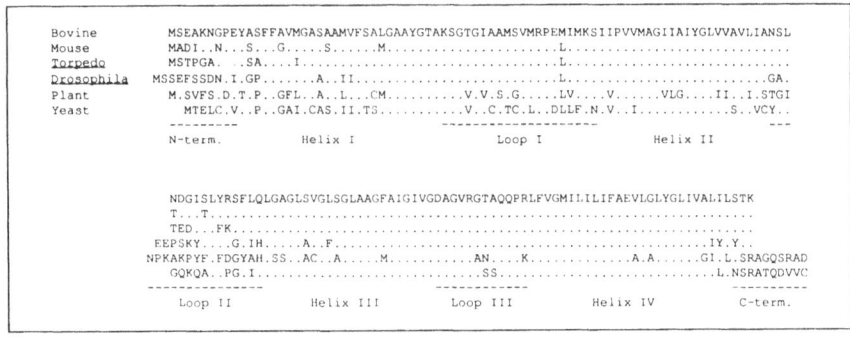

Fig. 1.3. Alignment of the amino acid sequences of proteolipid from various sources. The sequences were published previously and aligned manually. Bovine: proteolipid of V-ATPase from chromaffin granules;[31, Correction in GenBank] Mouse: mouse brain;[88] Torpedo: T. marmurata electric organ;[87] Drosophila: Drosophila melanogaster;[89] plant: oat;[90] Yeast: S. cerevisiae.[34]

activities of presumably identical enzymes in the numerous organelles having different requirements for internal pH and membrane potential. Very recently we cloned a cDNA encoding a potential subunit of 48 kDa in V-ATPase from chromaffin granules. The protein contains an N-terminal signal sequence, one potential transmembrane segment in its C-terminus and seven potential glycosylation sites in its luminal part.[40] This may represent one out of many specialized subunits that may confer specific properties into V-ATPases in different organelles.

A polypeptide (Ac 39) with an apparent molecular weight of 39 kDa is present in every preparation of V-ATPase from a mammalian source. The cDNA encoding this polypeptide was cloned from the bovine adrenal library.[38] It encodes a protein of about 36 kDa with no apparent transmembrane segments. Since it is fractionating with the membrane sector of V-ATPases, it is likely to be assembled as a complex with another membrane protein. The function of this polypeptide is not known and we termed this protein as an accessory subunit because of lack of evidence for its necessity. Recently a yeast gene encoding homologous protein to Ac 39 was cloned, sequenced and interrupted.[39] The resulting mutant failed to grow at high pH and the catalytic sector subunits did not assemble properly. These results promoted the subunit Ac 39 from an accessory polypeptide into a bona fide subunit of the enzyme.

The available phenotype of mutants lacking V-ATPase activity yielded several additional genes involved in the assembly or proper function of the enzyme. A yeast gene (*VMA11*) encoding a hydrophobic protein with 57% identical amino acids to the yeast proteolipid was cloned and studied by genetic analysis.[57] Disruption of this gene prevented the assembly of other subunits of the V-ATPase including the proteolipid. Attempts to identify the gene product of *VMA11* in yeast vacuoles were not successful. The function of the *VMA11* gene product may be in promoting the assembly of the membrane sector of the enzyme. The complexity of the yeast V-ATPase is further demonstrated by the emerging genes encoding proteins participating in the assembly or function of the enzyme. In yeast the genes denoted as *VMA 6, 12, 13, 21, 22* and *23* were identified to encode potential subunits or organizers of the V-ATPase.[58] In mammals potential specific inhibitors and enhancers were identified.[59] The involvement of more than 15 gene products in ATP-dependent pumping of the smallest ion in nature makes us keep wondering about its marvels.

EVOLUTION OF V-ATPASES — FROM ARCHAEBACTERIA TO HUMAN BRAIN

The fundamental function of F- and V-ATPases is a consequence of their early appearance in the most primitive life forms on earth. We assume that these primordial organisms evolved under the stress of high temperatures and low environmental pH.[4] Two independent proton pumps took shape in the plasma membrane of these primitive cells; one is an ATP-dependent pump (the ancestor of F- and V-ATPases), and the second is a redox potential-driven quinone pump (photo or chemosynthetic electron transport). Both pumped protons from the

from the cytoplasm to the acidic environment for maintaining the cytoplasmic pH neutral. While the reversibility of the electron transport-driven pump was not very useful, the reversibility of the ATP-driven pump was crucial to the evolution of more advanced life forms. When the environment became less acidic the two proton pumps started working in concert in ATP formation, where the electron transport chain provided the pmf and the ATPase utilized it for the phosphorylation of ADP to ATP. Presumably this was the way by which oxidation and photophosphorylation evolved.

It is assumed that the primordial atmosphere was anaerobic and the primitive organisms did not develop mechanisms for protection against oxidation damage. Amino acids such as cysteine and tryptophan were more abundant in the primordial proteins. The amino acid sequences of subunits A and B of V-ATPases subscribe to such proteins and therefore it was proposed that the catalytic sector of the primordial proton pump was closer to the current V-ATPases.[34] It was also proposed that F-ATPases evolved as a consequence of increasing oxygen tension concomitantly with the evolution of the oxygenic photosystem II in the photosynthetic electron transport chain. As a result of this event the cysteine and tryptophan residues are no longer present in the catalytically crucial positions of F-ATPases. The archaebacterial V-ATPase still contains much more tryptophan than F-ATPases, but the cysteine residue at the ATP-binding site in the A subunit was substituted to a serine residue. Today the V-ATPases of eukaryotic cells are protected against oxidation damage by their presence in the cytoplasm that maintains a relatively reduced environment. A schematic proposal for the main evolutionary events that progresses from the primordial enzyme to the current F- and V-ATPases is depicted in Figure 1.4. Apparently most of the fundamental events in the evolution of F- and V-ATPases took place over three billion years ago. It was proposed that these enzymes evolved from a single ancestral subunit proton pump containing both catalytic and membrane sectors, acting as a homohexamer.[4] Very early in the evolution of the system

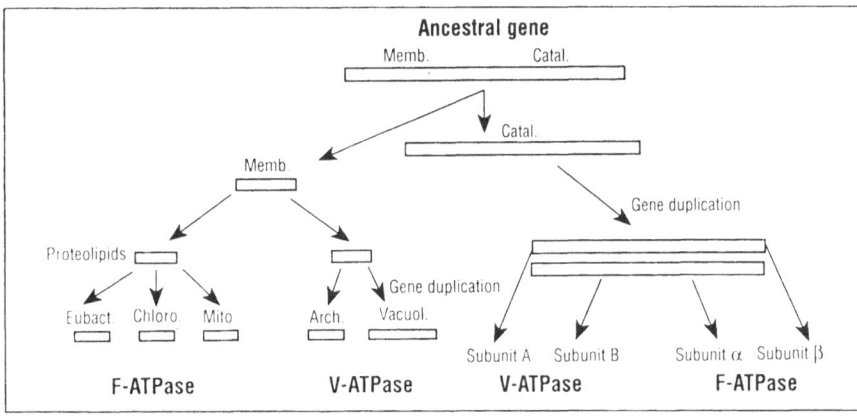

Fig. 1.4. Schematic proposal for early evolution of the catalytic and membrane sectors of F- and V-ATPases.

the gene encoding this enzyme was separated into two genes encoding the catalytic and membrane sectors. Soon after, the gene encoding the catalytic sector was duplicated to give a heterohexamer composed of three subunits A and three subunits B. Concomitantly with the introduction of oxygen to the atmosphere by oxygenic photosynthesis, the β subunit of F-ATPase evolved from the A subunit and the α subunit evolved from the B subunit. V-ATPases of archaebacteria and eukaryotes maintained the A and B subunits. In the remaining three billion years the F- and V-ATPases were busy adding more and more subunits to the enzyme, increasing its complexity and obtaining a highly specialized enzyme to act efficiently in diverse organelles, membranes and tissues.

The membrane sector underwent a separate evolution that was even more determinant to the specific function of F- and V-ATPases in different living creatures and in various cells and organelles. It is assumed that the ancestral membrane sector had three transmembrane segments, two of which evolved into the ancestral proteolipid.[4] The proteolipid maintained this structure in the two bacterial kingdoms as well as in chloroplasts and mitochondria. In all of these systems the main function of the F- or V-ATPase is to generate ATP at the expense of pmf. Very early on a gene duplication and fusion of the ancestral proteolipid took place. This form of proteolipid containing four transmembrane helices is present in all V-ATPases of eukaryotic cells. Concomitantly, or even consequently, these enzymes function exclusively as proton pumps and no longer can generate ATP at the expense of pmf. Thermodynamically this event may also be responsible for the difference in stoichiometry of H^+/ATP, which may be two in eukaryotic V-ATPases and three in F-ATPases.[6,60] The size of the proteolipid is also in correlation with the potential generation of proton leaks through the membrane sector in the absence of the catalytic sector. While most membrane sectors of F-ATPases and archaebacterial V-ATPases, when not gated by the catalytic sector, conduct protons down electrochemical gradients, the membrane sector of eukaryotic V-ATPases by itself is impermeable to protons.[61] All of these properties set up the proteolipids in the junction of energy transduction, and their evolution and conservation suggest a major role in the mechanism of ATP formation and proton pumping. The third transmembrane helix evolved into, or was displaced by, subunit a of F-ATPases and may be the 20 kDa polypeptide of V-ATPases.

As with the evolution of the catalytic sector, these events took place at the dawn of life on earth. In the subsequent three billion years more subunits were added to the various membrane sectors to optimize the function of these proton pumps in the numerous organelles and different membranes of the current eukaryotic cells.

FUNCTION OF V-ATPASES IN DIFFERENT ORGANELLES

SECRETORY GRANULES AND VESICLES

Neurotransmitters and hormones are accumulated and stored inside specialized organelles in presynaptic or secretory cells. In the brain these organelles are called synaptic vesicles or granules. The difference between the two neurotransmitter storage organelles is their content. While synaptic vesicles contain high concentrations of neurotransmitters and other low molecular weight substances such as ATP, granules contain in addition large amounts of proteins and enzymes. However, similar mechanisms underlay the uptake and storage of neurotransmitters into the two distinct organelles. The driving force for neurotransmitter uptake into these organelles is a protonmotive force generated exclusively by V-ATPase. Chromaffin granules are catecholamine storage vesicles of the adrenal medulla cells.[62] They contain up to 0.5M catecholamines (adrenaline, noradrenaline, etc.), about 120 mM ATP, 20 mM calcium, 5 mM magnesium, 20 mM ascorbate as well as several proteins and enzymes. Synaptic vesicles may contain very high concentrations of acetylcholine (up to 0.8 M) and ATP (up to 0.2M) or moderate concentrations of glutamate estimated to be lower than 50 mM.[63] Early studies of the energetization of catecholamines accumulation indicated that it is provided by an H^+-ATPase that generates pmf across the chromaffin granule membrane.[62] This was the first demonstration of the presence of V-ATPase in the vacuolar system of eukaryotic cells. Subsequently, the V-ATPase of chromaffin granules was isolated and reconstituted into phospholipid vesicles. Upon the addition of MgATP, proton uptake into the reconstituted vesicles could be observed.[64] The subunit structure of this enzyme was unraveled following the discovery of its cold lability.[13] It was demonstrated that incubation of chromaffin granules on ice in the presence of MgATP resulted in the dissociation of the catalytic sector from the membrane. The released catalytic sector contained five polypeptides denoted as subunits A to E in order of decreasing molecular weights. Similar cold inactivation was demonstrated with every membrane containing eukaryotic V-ATPase. The ready dissociation of the catalytic sector presented a bioenergetic enigma. It is well known that upon removal of the catalytic sector of F-ATPases their membrane sector conducts protons at high rates. Had similar phenomena occurred with the membrane sector of V-ATPase, chromaffin granules would lose all their catecholamines when even one of their enzymes lost its catalytic sector. Indeed it was not allowed by nature and was demonstrated that the membrane sector of V-ATPases is impermeable to protons.[61] All of the above properties make the chromaffin granules a nearly perfect bioenergetic machine. Their V-ATPase provides pmf by hydrolyzing ATP. The internal pH is about 5.5, and together with the ascorbate, provides an ideal environment to prevent oxidation of the catecholamines. The pH is maintained at 5.5 due to the proton slip that prevents overacidification of their interior.[65] The pH gradient generated by the V-ATPase is utilized for uptake and storage of the catecholamines. The catecholamines are taken up from the cytoplasm by a specific membrane protein that exchanges two protons for one catecholamine.[66] Figure 1.5 depicts the principal membrane proteins involved in this process. The cDNA encoding the catecholamines transporter was cloned and sequenced.[67] It is a hydrophobic protein containing 12 potential transmembrane helices. Similar vesicles storing catechola-

mines are present in various parts of the brain. While adrenaline secreted from the chromaffin granules acts on targets at long distance
from the gland, in the brain the catecholamines are secreted into synaptic clefts and interact as neurotransmitters with specialized receptors
on the post synaptic cells. The structure and properties of the brain
synaptic granule are similar to that of the chromaffin granules. This
makes the chromaffin granules an excellent model system for adrener-
gic granules in the brain.

   The neurotransmitter cycle is energized by two distinct ATPases.
While V-ATPase functions on the synaptic vesicles membrane, a
Na+/K+-ATPase functions in energizing the plasma membrane. Consequently the specific transporters for accumulation of neurotransmitters
into the synaptic vesicles are driven by an electrochemical gradient of
protons, and the specific transporters that function in the reuptake of
neurotransmitters from the synaptic cleft are driven by an electrochemical
gradient of sodium.[66] The subunits of the catalytic sector of V-ATPases
in synaptic vesicles are similar if not identical to those of chromaffin granules
and other organelles of the vacuolar system.[68] The subunit structure of

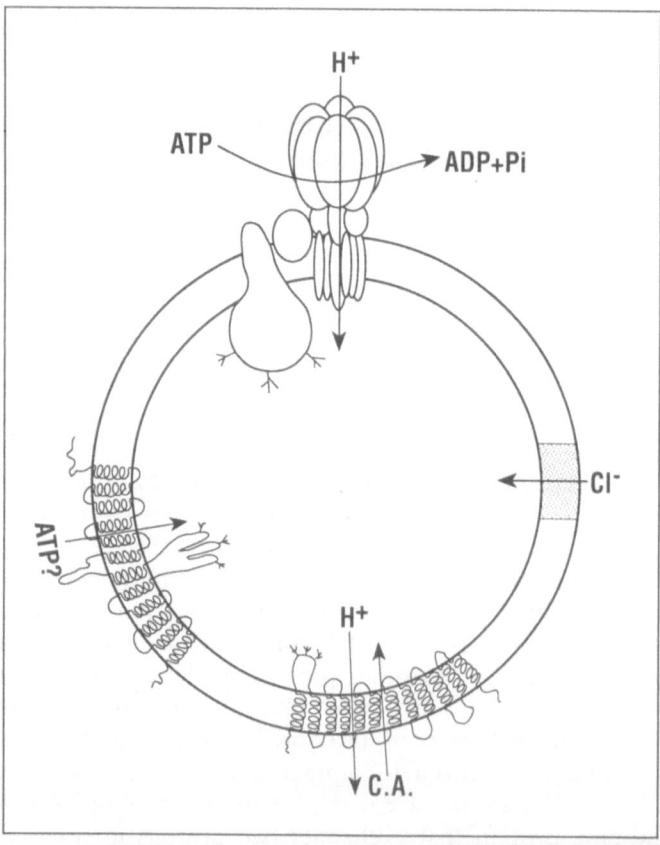

*Fig. 1.5. Schematic proposal for the principal proteins involved in ATP-dependent
catecholamine uptake into chromaffin granules. The V-ATPase provides the protonmotive
force that is accompanied by Cl and ATP transport into the granule. Catecholamines (C.A.)
are taken up by a vesicular transporter.*

the membrane sector was not studied but is also likely to be similar to other membrane sectors of V-ATPases. The accumulation of glutamate into the synaptic vesicles was studied in more detail than the other neurotransmitters, and it appears to utilize primarily the membrane potential provided by the V-ATPase rather than the ΔpH driving the catecholamines uptake.[68,69] The genes encoding glutamate, glycine and GABA transporters of the synaptic vesicles were not cloned as yet. The cDNA's encoding the sodium dependent neurotransmitter transporters of the plasma membrane were recently cloned and studied in detail.[70,71] The neurotransmission cycle is one of the fine examples of Mitchelian coupling between ATPases, the electrochemical gradient and a secondary process utilizing the electrochemical gradient.

THE ENDOCYTIC PATHWAY

Endo- and exocytosis are among the hallmarks of eukaryotic cells and are necessary processes for their life.[7] All the unique properties of eukaryots including highly organized tissues, nutrition, communication and signal transduction are intimately related to the function of the secretory pathway. Although there is no evidence of the direct involvement of V-ATPase in membrane traffic, there is ample evidence of its involvement in numerous secondary processes including modulation of the Golgi apparatus. As you can read in Dr. Grinstein's chapter, V-ATPase is crucial for the activity of several receptor-mediated endocytotic processes. In these processes a ligand binds to its receptor at neutral pH, then is taken up into clathrin-coated vesicles that are fused with secondary endosomes following clathrin decoating. These organelles contain V-ATPases that acidify their internal space and cause the release of the ligand from the receptor. In turn the ligand is taken up from the endosome into its site of action in the cytoplasm. Several receptor mediated processes do not require the release of the ligand from the receptor and following endocytosis are driven into the lysosomes for distraction.

Most catalytic activities inside lysosomes require a low pH. Therefore, lysosomes have an internal pH of about 5 which is two pH units below that of the cytoplasm.[72] The acidification of lysosomes is catalyzed by an ATP-dependent proton pump that was shown to be V-ATPase.[14,73,74] In addition, the V-ATPase provides most of the energy required for secondary transport systems in the lysosome membrane, some of which are vital for mammals and other higher eukaryotes.

The vacuoles of plants and fungi are analogous to lysosomes and play a major role in storage of metabolites, calcium homeostasis, their osmotic regulation and many other metabolic processes.[75,76] In most plant and fungal cells the vacuole occupies over 50% of the cell volume and the membrane contains numerous carriers, transporters, channels and enzymes. Here, too, the V-ATPase is the major energy provider and most secondary transport processes are driven by pmf generated by the enzyme. In plant vacuoles there exists a backup system of proton pyrophosphatase that also generates pmf by hydrolyzing pyrophosphate.[77] A pivotal function of plant and fungal vacuoles is their involvement in calcium homeostasis. In contrast to mammalian cells that contain Ca^{++}-ATPases and Ca^{++}/Na^+ exchangers in their plasma membrane, most calcium transport into and out of the cytoplasm of plant and fungal cells is mediated by the vacuolar membrane.[78] Therefore, calcium transport of plants and fungi is driven by pmf generated by V-ATPase or H^+- pyrophosphatase in plants. A Ca^{++}/H^+ exchanger functions in the vacuole membrane in Ca^{++} transport from the cytoplasm to the vacuole. Most studies on vacuolar calcium transport were conducted in yeast. It was shown that calcium is taken into isolated yeast vacuoles utilizing the ΔpH generated by their V-ATPase. Consequently, disruptant mutants in which the yeast V-ATPase was inactivated are sensitive to high calcium concentrations in the medium.[54,79] Though presumably the pmf in the vacuolar system is maintained in these mutants by fluid phase endocytosis,[35] it is not sufficient to counteract calcium flow into the cytoplasm at high external calcium concentrations (see Figure 1.6). These yeast mutants are also sensitive to low calcium obtained by including 25 mM EGTA in the medium.[54] This may result from higher than normal pH inside the vacuolar system not sufficient for dissociating the CaEGTA inside the vacuole (Figure 1.6). These studies demonstrated the crucial role of V-ATPase in calcium homeostasis in fungi and presumably in plants. It is tempting to suggest that a similar Ca^{++}/H^+ exchanger plays a role in calcium homeostasis in mammalian cells by controlling the calcium fluxes across membranes of organelles derived from the vacuolar system.

EXOCYTOSIS

The vacuolar system is very complicated but highly organized machinery. Despite extensive flow of membranes among the intracellular organelles and membranes, the composition of each compartment is strictly preserved. The Golgi apparatus functions in sorting proteins of the vacuolar system and transports secreted proteins into specific delivery vesicles. Specific receptors function in the targeting of luminal and secreted proteins into their respective vesicles (see Figure 1.7). Some but not all vesicles contain V-ATPases, and the trans-Golgi itself is furnished by the enzyme.[15,80,81] The possible involvement of V-ATPase in the sorting of some proteins was recognized by studying the effect of amines on the targeting of vacuolar proteins.[75] It was shown that agents able to neutralize acidic pH in yeast vacuoles bring about the mistargetting of vacuolar enzymes. Mistargetting is also the result of treating yeast cells with specific V-ATPase inhibitors as well as mutant yeast cells in which genes encoding V-ATPase subunits were interrupted.[82] In addition, the processing of the precursors of those vacuolar proteins is inhibited in the V-ATPase disruptant mutants. These mutants accumulate the intracellular precursors within the secretory pathway at some point before delivery to the vacuole and after transit to the Golgi complex.[83] These precursors are accumulated at the trans-Golgi complex or in post-Golgi vesicles. Thus, a picture emerges suggesting that V-ATPase plays a key role in several activities involving the Golgi apparatus including receptor-mediated targeting of luminal proteins as depicted in Figure 1.7. Studies involving antisense mRNA to subunit A of plant V-ATPases suggest the presence of two isozymes,

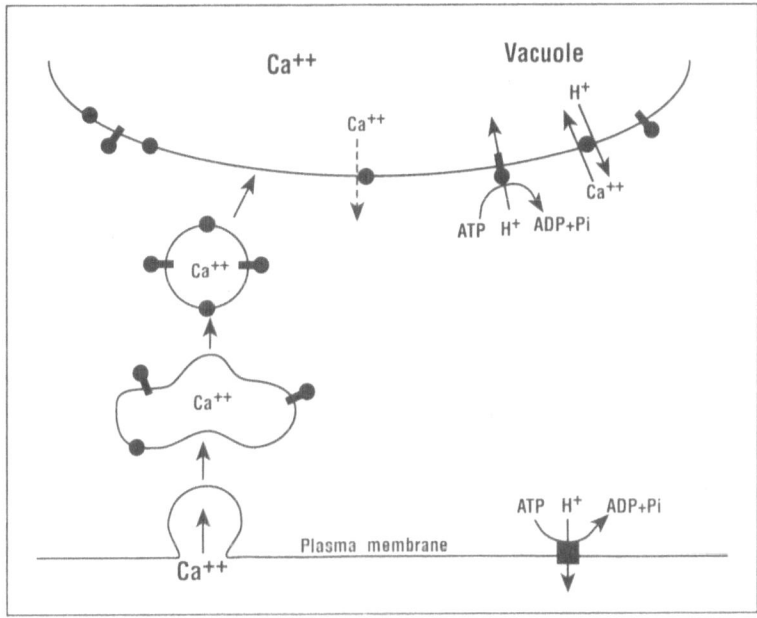

Fig. 1.6. Events in calcium uptake and homeostasis in yeast cells.
A pathway of calcium uptake into the vacuole by endocytosis is proposed. A) Calcium uptake from the external medium into the vacuole where it is concentrated and stored. Calcium leaks and channel activities increase the calcium concentration in the cytoplasm. V-ATPase provides protonmotive force into the vacuole utilized by the calcium/proton exchanger for pumping back excess calcium from the cytoplasm. B) Proposal for the mechanism of calcium uptake in the presence of excess EGTA. The calcium is taken up into the vacuole as a complex with EGTA. In the acidic environment inside the vacuole the calcium is dissociated from the complex and becomes available for the required processes.

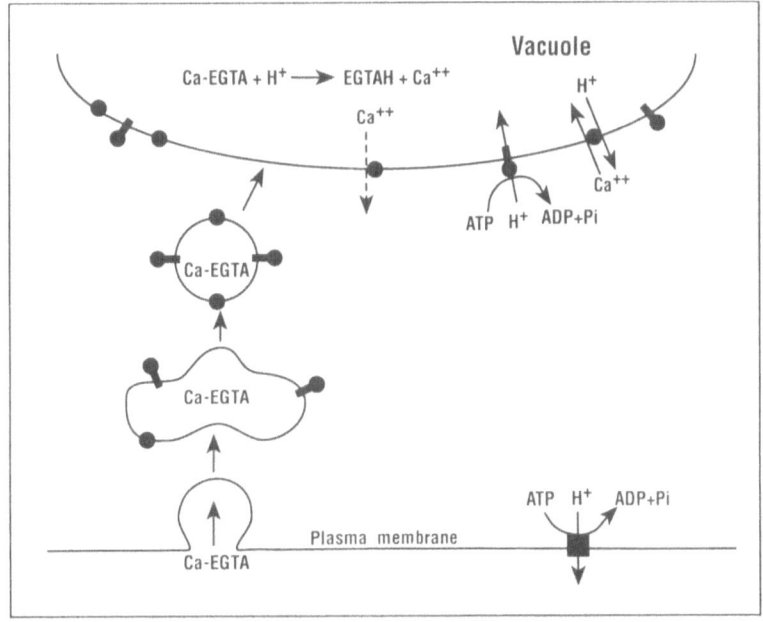

one of which is localized to the Golgi apparatus.[84] This observation poses an important question about the sites and mechanism of bio-genesis and assembly of V-ATPases in eukaryotic cells.

The exocytic pathway starts in the Golgi and is divided into two processes of constitutive and regulated secretions.[85] As shown in Figure 1.7, the regulated secretion contains V-ATPases that provide energy for their uptake systems and may play a role in the sorting of proteins into the system.[75] The constitutive secretory pathway contains no V-ATPase and is used for secretion of several proteins and maintenance of the cell membrane. Synaptic vesicles and chromaffin granules are some of the organelles of the regulated pathway, and the function of the V-ATPase in these organelles has been discussed before. As was

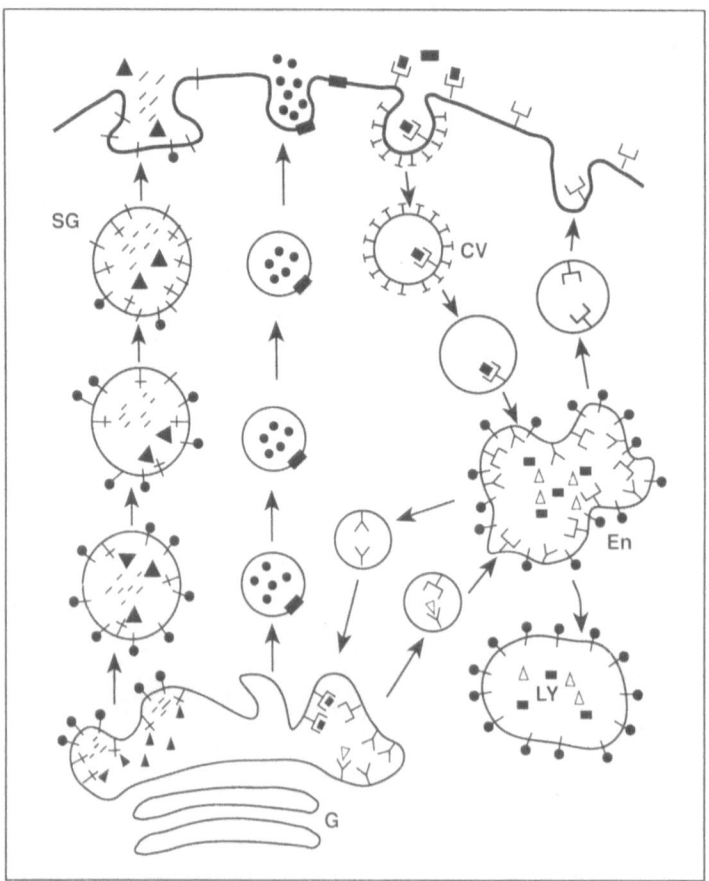

Fig. 1.7. Involvement of V-ATPases in endo and exocytotic processes.
The V-ATPase enzyme (†) is present in the Golgi apparatus (G), Endosome (En), Lysosome (LY) but not in clathrin-coated vesicles (CV). It is present in the Golgi part that gives rise to the regulated secretory pathway but not in the constitutive pathway. Secretory granule (SG) is depicted on its way from the Golgi to the plasma membrane. Young vesicles are loaded with full complement of V-ATPases and during migration to the plasma membrane are losing some of the catalytic sectors of the enzyme. (⊤), Endocytotic receptor with high affinity at neutral pH. (■) A ligand for this receptor. (†) Golgi receptor of the regulated exocytotic pathway with high affinity at low pH. (▲), A ligand for this receptor. (Y), A Golgi receptor for internal traffic into the endosome. (▽), A ligand for this receptor.

discussed above, in receptor-mediated endocytosis a receptor present at the cell surface binds its specific ligand and the binding of the ligand is pH-dependent. At neutral pH it exhibits high affinity to the ligand and at low pH its affinity is markedly lower. Following binding of its ligand, the receptor is internalized by clathrin-coated pits into clathrin-coated vesicles. After the clathrin coat is removed, the uncoated vesicles fuse with endosomes containing V-ATPases. At internal low pH of the endosome the ligand is dissociated from its receptor and the receptor can be either recycled to the plasma membrane or delivered into the lysosomes for distraction.

We propose that an analogous process takes place during exocytosis through the regulated pathway. However, during exocytosis the ligands bind to the receptors at low pH provided by V-ATPases in the trans-Golgi network. This process causes a correct sorting of different proteins and enzymes into their respective organelles. When these organelles fuse with the cytoplasmic membrane their interior pH becomes neutral and the ligands are dissociated from the receptors. Moreover, we foresee a function for the membrane sector of V-ATPase in the fusion process and propose that the presence of the catalytic sector participates in the delay of secretion in the regulated pathway. The biochemical and molecular biology tools developed in recent years make it feasible to prove or discard this hypothesis in the near future.

PERSPECTIVE

Today, the field of vacuolar H^+-ATPase can be compared to a teen-ager. It has the feeling of power without recognition, it foresees a possibility of a bright future but also several uncertainties, and it has the sense of missing ingredients but is not sure of how to go about finding them. Notable among the missing parts in the study of the structure of V-ATPases are missing subunits that may be vital for its function or necessary for its assembly in specific organelles and membranes in the eukaryotic cells. Figure 1.2 depicts a recent schematic proposal of the structure of V-ATPases in light of two newly discovered subunits of the enzyme. One of them is subunit F of the catalytic sector that was identified in the enzymes from insects and shown to be necessary for the activity and/or assembly of V-ATPase in yeast cells. The second is the Ac 48 accessory subunit of the membrane sector that is present in a selective set of organelles in mammalian cells. We anticipate more subunits of this sort to surface in the near future and it will not be a surprise if more than 20 gene products are involved in the structure and proper function or assembly of the enzyme. The yeast system is likely to yield the final account of the subunits necessary for the activity of the enzyme in all eukaryotic cells and several more accessory subunits are expected to participate in specific functions of the enzyme in different organelles in mammalian cells. Therefore, utilizing yeast mutants for the study of the function of specific mammalian subunits will help to shed light on their function.

It was argued in this chapter that every mutation with a deleterious effect is likely to be eliminated during fertilization. However, mutations in accessory polypeptides, that modulate but not eliminate the activity of the enzyme, are likely to be discovered and connected with some diseases. Moreover, modulation of V-ATPase activity in specific tissues is likely to be utilized for controlling osteoporosis, histocompatibility and several other processes in which the enzyme is involved. In addition, the proteolipid subunit of V-ATPase was implicated as the main structural protein in gap junctions and in neurosecretion of acetylcholine.[86,87] Possible involvement of V-ATPase in communication between cells and neurosecretion will have far reaching applications for better understanding physiological processes. Selective inhibition of specialized V-ATPases in specific organelles will enable the manipulation of processes that were not approached previously because of the lack in specificity. Synthetic and natural modulators of proton pumping activity of the enzyme are likely to take center stage in future research. The main lesson to be learned from studies of such a fundamental enzyme as V-ATPase is that nature conserved valuable mechanisms with little changes but may have utilized a conserved system for a wide variety of biochemical reactions by adding specialized subunits that interact with numerous cellular processes.

REFERENCES

1. Pedersen PL, Carafoli E. Ion motive ATPases. II. Energy coupling and work output. Trends Biochem Sci 1987; 12:186-189.
2. Nelson N, Taiz L. The evolution of H^+-ATPases. Trends Biochem Sci 1989; 14:113-116.
3. Sachs G, Besancon M, Shin JM, et al. Structural aspects of the gastric H,K-ATPase. J Bioenerg Biomembr 1992; 301-308.
4. Nelson N. Evolution of organellar proton-ATPases. Biochim Biophys Acta 1992; 1100:109-124.
5. Nelson N. Structural conservation and functional diversity of V-ATPases. Bioenerg Biomembr 1992; 24:407-414.
6. Nelson N. Structure, molecular genetics and evolution of vacuolar H+-ATPases. J Bioenerg Biomembr 1989; 21:553-571.
7. Mellman I, Fuchs R, Helenius A. Acidification of the endocytic and exocytic pathways. Annu Rev Biochem 1986; 55:663-700.
8. Abrahams JP, Lutter R, Todd RJ, et al. Inherent asymmetry of the structure of F1-ATPase from bovine heart mitochondria at 6.5 Å resolution. EMBO J 1993; 12:1775-1780.
9. Nelson N. Organellar proton-ATPases. Curr Opin Cell Biol 1992; 4:654-660.
10. Futai M, Noumi T, Maeda M. ATP synthase (H^+-ATPase): Results by combined biochemical and molecular biological approaches. Annu Rev Biochem 1989; 58:111-136.
11. Bowman BJ, Dschida WJ, Harris T, et al. The vacuolar ATPase of *Neurospora crassa* contains an F_1-like structure. J Biol Chem 1989; 264:15606-15612.
12. Taiz SL, Taiz L. Ultrastructural comparison of the vacuolar and mitochondrial H^+-ATPases of *Daucus carota*. Bot Acta 1991; 104:85-168.
13. Moriyama Y, Nelson N. Cold inactivation of vacuolar H^+-ATPases. J Biol

Chem 1989; 264:3577-3582.

14. Moriyama Y, Nelson N. Lysosomal H⁺-translocating ATPase has a similar subunit structure to chromaffin grauune H⁺-ATPase complex. Biochim Biophys Acta 1989; 980:241-247.

15. Moriyama Y, Nelson N. H⁺-translocating ATPase in Golgi apparatus: Characterization as vacuolar H⁺-ATPase and its subunit structures. J Biol Chem 1989; 264:18445-18450.

16. Schäfer G, Meyering-Vos M. The plasma membrane ATPase of archaebacteria: A chimeric energy converter. In: Scarpa A, Carafoli E, Papa S. eds. Ion-Motive ATPases: Structure, Function, and Regulation (Vol. 671). New York: The New York Academy of Sciences, 1992:293-309.

17. Moriyama Y, Nelson N. Nucleotide binding sites and chemical modification of the chromaffin granule proton ATPase. J Biol Chem 1987; 262:14723-14729.

18. Bowman EJ, Tenney K, Bowman BJ. Isolation of genes encoding the Neurospora vacuolar ATPase. J Biol Chem 1988; 263:13994-14001.

19. Zimniak L, Dittrich P, Gogarten JP, et al. The cDNA sequence of the 69 kDa subunit of the carrot vacuolar H⁺-ATPase. J Biol Chem 1988; 263:9102-9112.

20. Feng Y, Forgac M. Cysteine 254 of the 73-kDa A subunit is responsible for inhibition of the coated vesicle (H⁺)-ATPase upon modification by sulfhydryl reagents. J Biol Chem 1992; 267:5817-5822.

21. Manolson MF, Rea PA, Poole RJ. Identification of 3-O-(4-benzoyl) benzoyladenosine 5'-triphosphate- and N,N'-dicyclohexylcarbodiimide-binding subunits of a higher plant proton-translocating tonoplast. J Biol Chem 1985; 260:12273-12279.

22. Bowman BJ, Allen R, Wechser MA, et al. Isolation of genes encoding the Neurospora vacuolar ATPase. J Biol Chem 1988; 263:14002-14007.

23. Beltrán C, Kopecky J, Pan Y-CE, et al. Cloning and mutational analysis of the gene encoding subunit C of yeast V-ATPase. J Biol Chem 1992; 267:774-779.

24. Hirsch S, Strauss A, Masood K, et al. Isolation and sequence of a cDNA clone encoding the 31-kDa subunit of bovine kidney vacuolar H⁺-ATPase. Proc Natl Acad Sci USA 1988; 85:3004-3008.

25. Gluck SL, Nelson RD, Lee BS, et al. Biochemistry of the renal V-ATPase. J Exp Biol 1992; 172:219-229.

26. Foury F. The 31-kDa polypeptide is an essential subunit of the vacuolar ATPase in Saccharomyces cerevisiae. J Biol Chem 1990; 265:18554-18560.

27. Gräf R, Lepier A, Harvey WR, et al. A novel 14-kDa V-ATPase subunit in the tobacco hornworm midgut. J Biol Chem 1994 (In press).

28. Nelson H, Mandiyan S, Nelson N. The Sacchoromyus ceranisiae VMA7 gene encodes a 14-kDa subunit of the vacuolar H⁺-/ATPase catalytic sector. J Biol Chem 1994 (in press).

29. Fillingame RH. H⁺ transport and coupling by the F_o sector of the ATP synthase: insights into the molecular mechanism of function. J Bioenerg Biomembr 1992; 24:485-491.

30. Girvin ME, Gillingame RH. Helical structure and folding of subunit c of F_1F_o ATP synthase: H NMR resonance assignments and NOE analysis. Biochemistry 1993; 32:12167-12177.

31. Mandel M, Moriyama Y, Hulmes JD, et al. Cloning of cDNA sequence encoding the 16-kDa proteolipid of chromaffin granules implies gene duplication in the evolution of H⁺-ATPases. Proc Natl Acad Sci USA 1988; 85:5521-5524.

32. Arai H, Berne M, Forgac M. Inhibition of the coated vesicle proton pump and labeling of a 17,000-dalton polypeptide by *N,N*'-dicyclohexyl-carbodiimide. J Biol Chem 1987; 262:11006-11011.

33. Sze H, Ward JM, Lai S, et al. Vacuolar-type H⁺-translocating ATPases in plant endomembranes: subunit organization and multigene families. J Exp Eiol 1992; 172:123-135.

34. Nelson H, Nelson N. The progenitor of ATP synthases was closely related to the current vacuolar H⁺-ATPase. FEBS Lett 1989; 247:147-153.

35. Nelson H, Nelson N. Disruption of genes encoding subunits of yeast vacuolar H⁺-ATPase causes conditional lethality. Proc Natl Acad Sci USA 1990; 87:3503-3507.

36. Schneider E, Altendorf K. Bacterial adenosine 5'-triphosphate synthase (F_1-F_o): purification and reconstitution of F_o complexes and biochemical and functional characterization of their subunits. Microbiol Rev 1987; 51:477-497.

37. Perin MS, Fried VA, Stone DK, et al. Structure of the 116-kDa polypeptide of the clathrin-coated vesicle/synaptic vesicle proton pump. J Biol Chem 1991; 266:3877-3881.

38. Wang S-Y, Moriyama Y, Mandel M, et al. Cloning of cDNA encoding a 32-kDa protein: an accessory polypeptide of the H⁺-ATPase from chromaffin granules. J Biol Chem 1989; 263:17638-17642.

39. Bauerle C, Ho MN, Lindorfer MA, et al. The *Saccharomyces cerevisiae* *VMA6* gene encodes the 36-kDa subunit of the vacuolar H⁺-ATPase membrane sector. J Biol Chem 1993; 268:12749-12757.

40. Supek F, Supekova L, Mandiyan S, et al. A novel accessory subunit for vacuolar H⁺-ATPase from chromaffin granules. J Biol Chem 1994 (in press).

41. Manolson MF, Ouellette BFF, Filion M, et al. cDNA sequence and homologies of the "57-kDa" nucleotide-binding subunit of the vacuolar ATPase from *Arabidopsis*. J Biol Chem 1988; 263:17987-17994.

42. Gogarten JP, Kibak H, Dittrich P, et al. Evolution of the vacuolar H⁺-ATPase: implications for the origin of eukaryotes. Proc Natl Acad Sci USA 1989; 86:6661-6665.

43. Denda K, Konishi J, Oshima T, et al. The membrane-associated ATPase from *Sulfolobus acidocaldarius* is distantly related to F_1-ATPase as assessed from the primary structure of its α-subunit. J Biol Chem 1988; 263:6012-6015.

44. Shih C-K, Wagner R, Feinstein S, et al. A dominant trifluoperazine resistance gene from *Saccharomyces cerevisiae* has homology with F_oF_1 ATP synthase and confers calcium-sensitive growth. Mol Cell Biol 1988; 8:3094-3103.

45. Hirata R, Ohsumi Y, Nakano A, et al. Molecular structure of a gene, *VMA1*, encoding the catalytic subunit of H⁺-translocating adenosine triphosphatase from vacuolar membranes of *Saccharomyces cerevisiae*. J Biol Chem 1990; 265:6726-6733.

46. Kane PM, Yamashiro CT, Wolczyk DF, et al. Protein splicing converts

the yeast *TFP1* gene product to the 69-kD subunit of the vacuolar H⁺-adenosine triphosphatase. Science 1990; 250:651-657.

47. Penefsky HS, Cross RL. Structure and mechanism of F_oF_1-type ATP synthases and ATPases. In: Meister A. ed. Advances in Enzymology and Related Areas of Molecular Biology, Vol. 64. New York: John Wiley & Sons, Inc., 1991:173-214.

48. Manolson MF, Rea PA, Poole RJ. Identification of 3-O-(4-benzoyl)benzoyladenosine 5'-triphosphate-and N,N'-dicyclohexylcarbodiimide-binding subunits of a higher plant H⁺-translocating tonoplast ATPase. J Biol Chem 1985; 260:12273-12279.

49. Walker JE, Fearnley IM, Gay NJ, et al. Primary structure and subunit stoichiometry of F_1-ATPase from bovine mitochondria. J Mol Biol 1985; 184:677-701.

50. Puopolo K, Kumamoto C, Adachi I, et al. Differential expression of the "B" subunit of the vacuolar H⁺-ATPase in bovine tissues. J Biol Chem 1992; 267:3696-3706.

51. Nelson H, Mandiyan S, Noumi T, et al. Molecular cloning of cDNA encoding the C subunit of H⁺-ATPase from bovine chromaffin granules. J Biol Chem 1990; 265:20390-20393.

52. Denda K, Konishi J, Hajiro K, et al. Structure of an ATPase operon of an acidothermophilic archaebacterium, *Sulfolobus acidocaldarius.* J Biol Chem 1990; 265:21509-21513.

53. Ho MN, Hill KJ, Lindorfer MA, et al. Isolation of vacuolar membrane H⁺-ATPase-deficient yeast mutants; the *VMA5* and *VMA4* genes are essential for assembly and activity of the vacuolar H⁺-ATPase. J Biol Chem 1993; 268:221-227.

54. Noumi T, Beltrán C, Nelson H, et al. Mutational analysis of yeast vacuolar H⁺-ATPase. Proc Natl Acad Sci USA 1991; 88:1938-1942.

55. Supek F, Supekova L, Beltrán C, et al. Structure, function, and mutational analysis of V-ATPases . Ann NY Acad Sci 1992; 671:284-292.

56. Manolson MF, Proteau D, Preston RA, et al. The *VPH1* gene encodes a 95-kDa integral membrane polypeptide required for *in vivo* assembly and activity of the yeast vacuolar H⁺-ATPase. J Biol Chem 1992; 267:14294-14303.

57. Umemoto N, Ohya Y, Anraku Y. MA*11*, a novel gene that encodes a putative proteolipid, is indispensable for expression of yeast vacuolar membrane H⁺-ATPase activity. J Biol Chem 1991; 266:24526-24532.

58. Stevens TH. The structure and function of the fungal V-ATPase. J Exp Biol 1992; 172:47-55.

59. Gluck SL. The structure and biochemistry of the vacuolar H⁺ ATPase in proximal and distal urinary acidification. J Bioenerg Biomembr 1992; 24:351-359.

60. Cross RL, Taiz L. Gene duplication as a means for altering H⁺/ATP ratios during the evolution of F_oF_1 ATPases and synthases. FEBS Lett 1990; 259:227-229.

61. Beltrán C, Nelson N. The membrane sector of vacuolar H⁺-ATPase by itself is impermeable to protons. Acta Physiol Scand 1992; 146:41-47.

62. Njus D, Knoth J, Zallakian M. Proton-linked transport in chromaffin granules. Curr Topics Biognergetics 1981; 11:107-147.

63. Tabb JS, Kish PE, Van Dyke R, et al. Glutamate transport into synaptic

vesicles. J Biol Chem 1992; 267:15412-15418.

64. Moriyama Y, Nelson N. The purified ATPase from chromaffin granule membranes is an anion-dependent proton pump. J Biol Chem 1987; 262:9175-9180.

65. Nelson N. Structure and pharmacology of the proton-ATPases. Trends Pharmac Sci 1991; 12:71-75.

66. Kanner BI. Ion-coupled neurotransmitter transport. Curr Opin Cell Biol 1989; 1:735-738.

67. Liu Y, Peter D, Roghani A, et al. A cDNA that suppresses MPP⁺ toxicity encodes a vesicular amine transporter. Cell 1992; 70:539-551.

68. Cidon S, Sihra T. Characterization of a H⁺-ATPase in rat brain synaptic vesicles. Coupling to L-glutamate transport. J Biol Chem 1989; 264:8281-8288.

69. Moriyama Y, Maeda M, Futai M. Energy coupling of L-glutamate transport and vacuolar H⁺-ATPase in brain synaptic vesicles. J Biochem 1990; 108:689-693.

70. Liu Q-R, Mandiyan S, Nelson H, et al. A family of genes encoding neurotransmitter transporters. Proc Natl Acad Sci USA 1992; 89:6639-6643.

71. Liu Q-R, López-Corcuera B, Mandiyan S, et al. Molecular characterization of four pharmacologically distinct γ-aminobutyric acid transporters in mouse brain. J Biol Chem 1993; 268:2106-2112.

72. Ohkuma S, Moriyama Y, Takano T. Identification and characterization of a proton pump on lysosomes by fluorescein isothiocyanate dextran fluorescence. Proc Natl Acad Sci USA 1982; 79:2758-2762.

73. Schneider DL. ATP-dependent acidification of intact and disrupted lysosomes. Evidence for an ATP-driven proton pump. J Biol Chem 1981; 256:3858-3864.

74. Reeves JP, Reames T. ATP stimulates amino acid accumulation by lysosomes incubated with amino acid methyl esters. J Biol Chem 1981; 256:6047-6053.

75. Klionsky DJ, Herman PK, Emr SD. The fungal vacuole: composition, function, and biogenesis. Microbiol Rev 1990; 54:266-292.

76. Taiz L. The plant vacuole. J Exp Biol 1992; 172:113-122.

77. Sarafian V, Kim Y, Poole RJ, et al. Molecular cloning and sequence of cDNA encoding the pyrophosphate-energized vacuolar membrane proton pump of *Arabidopsis thaliana*. Proc Natl Acad Sci USA 1992; 89:1775-1779.

78. Anraku Y, Hirata R, Wada Y, et al. Molecular genetics of the yeast vacuolar H⁺-ATPase. J Exp Biol 1992; 172:67-81.

79. Ohya Y, Umemoto N, Tanida I, et al. Calcium-sensitive *cls* mutants of *Saccharmoyces cerevisiae* showing a pet-phenotype are ascribable to defects of vacuolar membrane H⁺-ATPase activity. J Biol Chem 1991; 266: 13971-13977.

80. Chanson A, Taiz L. Evidence for an ATP-dependent proton pump on the Golgi of corn coleoptiles. Plant Physiol 1985; 78:232-240.

81. Young GP-H, Qiao J-Z, Al-Awqati Q. Purification and reconstitution of the proton-translocating ATPase of Golgi-enriched membranes. Proc Natl Acad Sci USA 1988; 85:9590-9594.

82. Klionsky DJ, Nelson H, Nelson N. Compartment acidification is required for efficient sorting of proteins to the vacuole in *Saccharomyces cerevisiae*.

83. Yaver DS, Nelson H, Nelson N, et al. Vacuolar ATPase mutants accumulate precursor proteins in a pre-vacuolar compartment. J Biol Chem 1993; 268:10564-10572.

84. Gogarten JP, Fichmann J, Braun Y, et al. The use of antisense mRNA to inhibit the tonoplast H$^+$-ATPase in carrot. The Plant Cell 1992; 4:851-864.

85. Kelly RB. Pathways of protein secretion in eukaryotes. Science 1985; 230:25-32.

86. Finbow ME. Structure of a 16 kDa integral membrane protein that has identity to the putative proton channel of the vacuolar H$^+$-ATPase. Protein Engineering 1991; 5:7-15.

87. Birman S., Meunier F-M, Lesbats B, et al. A 15 kDa proteolipid found in mediatophore preparations from *Torpedo* electric organ presents high sequence homology with the bovine chromaffin granule protonophore. FEBS Lett 1990; 261:303-306.

88. Hanada H, Hasebe M, Moriyama Y, et al. Molecular cloning of cDNA encoding the 16 kDa subunit of vacuolar H$^+$-ATPase from mouse cerebellum. Biochem. Biophys, Res. Commun. 1991; 176:1062-1076.

89. Meagher L, McLean P, Finbow ME. Sequence of a cDNA from *Drosophila* coding for the 16 kD proteolipid component of the vacuolar H$^+$-ATPase. Nucleic Acids Res. 1990; 18:6712.

90. Lai S, Watson, JC, Hansen JN. Molecular cloning and sequencing of cDNAs encoding the proteolipid subunit of the vacuolar H$^+$-ATPase from a higher plant. J. Biol. Chem. 1991; 266:16078-16084.

91. Nelson H, Mandiyan S, Nelson N. A bovine cDNA and a yeast gene-VMAS encoding subunit D of the vacuolar H$^+$-ATPase 1994 (submitted).

CHAPTER 2

AN OVERVIEW
OF INTRACELLULAR pH
REGULATION: ROLE
OF VACUOLAR H⁺-ATPASES

Gergely Lukacs, Ori D. Rotstein and Sergio Grinstein

INTRODUCTION

The purpose of this chapter is to briefly describe the mechanisms responsible for the regulation of the cellular pH, highlighting the role of vacuolar (V)-type ATPases in this process. The control of the cytosolic pH and that of selected organelles where V-type ATPases play a predominant role will be discussed separately, with emphasis on the molecular mechanisms involved and on the physiological implications of adequate pH homeostasis.

The regulation of intracellular pH is of paramount importance in the maintenance of most cellular functions. Though near neutrality, the cytosolic pH (pH_c) is nevertheless regulated at a level which is significantly different from the extracellular pH. In addition, individual membrane-bound compartments within the cell display pH values that can differ by up to 3 units from the cytosolic pH. The prevalence of unique pH levels within the individual cellular compartments is not fortuitous. Instead, as discussed below in more detail, pH plays an influential role in the functioning of whole cells and of specific organelles. For this reason, even slight deviations from the physiological pH can have catastrophic consequences on cell function and viability. This results primarily from the fact that the activity of most enzymes is highly pH dependent. An example is phosphofructokinase, a cytosolic enzyme which constitutes the rate-limiting step in glycolysis. A marked acceleration of glycolytic activity is observed when cytosolic pH increases slightly. Conversely, increased pH can largely inactivate

Organellar Proton-ATPases, edited by Nathan Nelson; ©1994 R.G. Landes Company.

lysosomal hydrolases that have acidic pH optima. These observations highlight the requirement for accurate and effective regulation of cellular pH and the need for maintenance of differential pH in individual subcellular compartments.

Short term maintenance of intracellular pH can be favored by cellular buffering, which will tend to mitigate rapid departures from the prevailing pH level. The simplest form of intracellular buffering is physicochemical (passive) buffering, i.e. the ability of weak acids and bases to associate with or dissociate from H^+, thereby minimizing shifts in the pH of the surrounding milieu. In most animal tissues, the intracellular buffering power in the physiological pH range has been estimated to vary from about 25 to 100 mM/pH unit (i.e. 25 to 100 mmoles of acid or base equivalents need to be added to one liter of intracellular medium to alter its pH by 1 unit; see ref. 1). It is noteworthy, however that many of the available estimates of buffering power do not include the contribution of CO_2/HCO_3^- to the buffering capacity. Because CO_2/HCO_3^- are always present in animal tissues, their contribution must be estimated and added to the endogenous buffers. In an open system in equilibrium with CO_2, the buffering power will increase by $2.3 \times [HCO_3^-]$[1]. Because at normal pCO_2 the $[HCO_3^-]$ inside cells has been estimated to be in the millimolar range, its contribution to the buffering power is sizable and should not be neglected.

While acute intracellular pH changes can be at least partially counteracted by the cellular buffering capacity, long term regulation of pH requires continuously operating active (i.e. directly or indirectly energy-dependent) transport processes. Such a need arises from the finite permeability of biological membranes to acid equivalents, which, though comparatively small, nevertheless favors the continuous and spontaneous tendency of H^+ (or OH^-) to attain electrochemical equilibrium. Thus, H^+ continuously tend to exit the very acidic intralysosomal compartment down their concentration gradient. Conversely, H^+ tend to enter the cytosol across the plasma membrane, drawn by the internally negative transmembrane electrical potential. An additional challenge to cytosolic pH homeostasis is imposed by a variety of metabolic pathways which yield H^+ or other acid equivalents as their final product.

Transport of H^+ equivalents for the purpose of regulating pH can be "primary active", i.e. directly requiring ATP hydrolysis for the catalytic (transport) event. This is the case for V-type H^+ pumps located on the plasma membrane as well as in endomembranes. These systems are described in detail below. In addition, however, a variety of pH regulating systems exists which can be classified as "secondary active". In these cases the energy for the translocation of H^+ equivalents is provided by coupling their flux to the movement of another solute down its electrochemical gradient. Frequently, the gradient of the "driver" solute (e.g. Na^+) is itself established by a primary active process (e.g. the Na^+/K^+ ATPase). For the sake of completeness, a brief overview of secondary active systems that participate in the regulation of pH is included below.

CYTOSOLIC pH REGULATION: ION EXCHANGERS AND CHANNELS

The absolute value of the cytosolic pH of a variety of animal cell types and the processes underlying its regulation have been extensively investigated. At physiological extracellular pH (i.e. 7.3-7.4), the cytosolic pH has been found in the vast majority of cases to range between 7.0 and 7.3.[1] In principle, accumulation of H^+ in the cytosol and therefore regulation of cytosolic pH can be accomplished either by translocation of acidic or basic equivalents out of the cell across the plasma membrane or by translocation into cytosolic organelles. The importance of translocation into intracellular organelles for the purposes of cytosolic pH regulation is questionable, however, since their capacity to store acid is finite and because the intraorganellar pH itself must be maintained within a narrow, defined range (see below) and cannot be merely subservient to the regulation of pH_c. For these reasons, the present section will emphasize the mechanisms present on the plasma membrane which are involved in the regulation of cytosolic pH. The transfer of acidic and basic equivalents is accomplished by a variety of membrane-spanning proteins (or glycoproteins) which have specific functions, structures and biochemical/pharmacological characteristics. The principal systems involved in cytosolic pH regulation in animal cells are described succinctly below.

Na⁺/H⁺ EXCHANGE

The Na^+/H^+ exchanger (NHE), also termed Na^+/H^+ antiport, is an integral plasma membrane glycoprotein of approximately 100 kDa. The amino-terminal moiety of the protein is largely hydrophobic and is thought to span the membrane 10-12 times. It contains consensus sites for N-glycosylation. The carboxy terminal segment is hydrophilic and immunological evidence suggests that it is located intracellularly. Kinetic and biochemical evidence suggests that the antiport exists in oligomeric (dimeric or higher order) aggregates within the membrane. An excellent review relating the structure and function of the Na^+/H^+ exchanger appeared recently.[2]

The Na^+/H^+ antiport catalyzes the exchange of extracellular Na^+ for intracellular H^+, thereby counteracting the tendency of the cytosol to become acidic. The stoichiometry of the exchange process is one Na^+ for one H^+. Therefore, the transport cycle is electroneutral and insensitive to changes in the membrane potential. This enables excitable cells such as nerve and muscle to undergo electrical potential changes without jeopardizing pH regulation and vice-versa. Transport through the Na^+/H^+ exchanger is driven by the combined concentration gradients of the substrate ions, i.e. Na^+ and H^+. Thus, in theory, net flux through the exchanger should cease when the Na^+ concentration gradient is balanced by an identical H^+ gradient. Under physiological conditions, the net direction of exchange (Na^+ influx and H^+ efflux) is dictated by the prevailing inward Na^+ gradient (extracellular $[Na^+]$ being over ten-fold higher than the cytosolic $[Na^+]$). This gradient is generated and maintained in most animal cells by the Na^+/K^+ pump. Therefore, the continued extrusion of intracellular H^+ is directly fueled by the

Na⁺ gradient and indirectly by the hydrolysis of ATP. In this sense, the Na^+/H^+ antiport can be considered a "secondary" active transport process.

Recent evidence, derived from low stringency hybridization experiments, suggests the existence of at least four mammalian isoforms of the antiport, each with specific pharmacological and biochemical properties, specific cellular/tissue localizations, and specific functions. These isoforms have been termed NHE-1 through NHE-4, reflecting the chronological order of their discovery. NHE-1 is the most studied and best characterized of all the isoforms. The NHE-1 isoform is a rather ubiquitous system, present in the surface membrane of virtually all the mammalian cells studied to date, including the basolateral membrane of epithelial cells. It is characterized by sensitivity to inhibition by amiloride, a potent diuretic, and its analogs and by its ability to respond to growth factors and tumor promoters (e.g. epidermal growth factor, the phosphatase inhibitor okadaic acid, and protein kinase C agonists phorbol esters). The exchange activity of NHE-1 is greatly stimulated when the cytosol is acidified. This peculiar behavior is partly attributable to the increased availability of internal substrate (i.e. H⁺), but is mostly due to protonation of an allosteric H⁺ binding site on the internal surface of the exchanger. When protonated, this allosteric site located within the amino-terminal domain of NHE-1 is thought to activate Na^+/H^+ exchange, thereby counteracting the acidification of the cytosol. For this reason, NHE-1 is central to the regulation of cytosolic pH and is sometimes referred to as the "housekeeping" isoform of the antiporter. The other isoforms of the antiport are far less well characterized and their specific role in cytosolic pH regulation is not yet clear. NHE-2, NHE-3 and NHE-4 have a much more discrete tissue distribution, and can differ from NHE-1 in their susceptibility to amiloride and affinity for the transported cations. In all likelihood, some of these isoforms will prove to be more important for transepithelial transport and less influential in the determination and homeostasis of pH_c.

CL^-/HCO_3^- Exchange

An electroneutral anion exchange process, analogous to the cation antiport, is seemingly also involved in cytosolic pH regulation. This anion exchanger is capable of transporting a variety of halides as well as sulfate and phosphate, yet is essentially independent of the cationic composition of the medium. Under physiological conditions the anion antiporter is believed to transport chiefly Cl^- in exchange for HCO_3^-.[3] The Cl^-/HCO_3^- exchanger translocates anions with a one-to-one stoichiometry and, like the Na^+/H^+ antiport, is therefore electroneutral and insensitive to alterations in transmembrane potential. As before, this enables anion metabolism and pH regulation to proceed without interference to the electrical signalling events of excitable tissues. Cl^-/HCO_3^- exchange is generally inhibited by externally added disulfonic stilbene derivatives. These are impermeant drugs, implying that inhibition involves an exofacial site. Covalently binding disulfonic stilbenes were used initially to label and identify the molecules responsible for anion exchange in red cells.[4] These were called AE-1 (anion exchanger #1). Subsequently, additional isoforms were identified and cloned, and their location defined by in situ hybridization or by immunological means.

As in the case of the cation antiport, anion exchange is driven by the concentration gradients of the substrate ions. In most cell types, the intracellular concentrations of Cl^- and HCO_3^- are lower than those in the external medium, due in part to the electronegativity of the cell interior. However, the inward concentration gradient of Cl^- is generally greater than that of HCO_3^-. This imbalance is predicted to drive the net influx of Cl^- simultaneously with net HCO_3^- efflux. Because HCO_3^- is a base equivalent, this is anticipated to result in cytosolic acidification. Consistent with this prediction, recent studies have indicated that Cl^-/HCO_3^- exchange plays a role in the recovery of pH_c from experimental alkalosis.[3]

In principle, it would appear that continuous operation of a base extrusion (acid loading) mechanism would be detrimental to the maintenance of pH_c, which is already stressed by metabolic acid generation and by electrophoretic accumulation of acid equivalents (see above). However, under physiological conditions, the activity of the Cl^-/HCO_3^- exchanger is curtailed by its steep dependence on the intracellular pH. It is noteworthy that the pH versus activity profile of the anion exchanger is virtually opposite to that of the Na^+/H^+ exchanger. Anion exchange is *inactivated* at acidic pH and greatly stimulated at more alkaline levels. This behavior is consistent with a role in the recovery from alkalosis, since the activated exchange of external Cl^- for cytosolic HCO_3^- would tend to restore the physiological pH. Its pH dependence also dictates that, at or below the normal pH, anion exchange is greatly reduced, precluding Cl^-/HCO_3^- exchanger-mediated acidification, which would further compromise homeostasis of pH_c.

NA⁺-DEPENDENT CL⁻/HCO₃⁻ EXCHANGE

A second type of Cl^-/HCO_3^- exchange mechanism, distinctly different from that described in the preceding section, exists in the plasma membranes of both invertebrates and mammalian cells.[5] It differs from the cation-insensitive anion exchanger described above in that the availability of Na^+ is an essential requirement for anion exchange to proceed. Under physiological conditions, this system exchanges extracellular Na^+ and HCO_3^- for intracellular Cl^-. Electrophysiological determinations found this process to be electroneutral, implying that some other cation exits the cell or that an additional anion is transported inward. Because two acid equivalents have been found to be extruded from the cell per transport cycle, two possible combinations of transported ions have been suggested: extracellular Na^+ could enter the cell accompanied by one CO_3^{2-} or two HCO_3^- ions, rather than a single HCO_3^-. Alternatively, one H^+ could be ejected from the cell along with Cl^-.[5] Regardless of the precise ionic species involved, one complete cycle of the transporter results in the electroneutral ejection of two acid equivalents, with cellular gain of Na^+ and a concomitant loss of Cl^-.

The large inward Na⁺ gradient that prevails in most cells propels the net translocation of HCO_3^- from outside the cell into the cytosol through the Na⁺-dependent Cl^-/HCO_3^- exchanger. Thus, this transporter normally catalyzes the extrusion of acid from the cytosol and is considered to be a functional pH regulatory system, capable of antagonizing the spontaneous tendency of the cytosol to become acidic. Consistent with this view, the dependence of this system on the intracellular pH resembles that of the Na⁺/H⁺ antiport, i.e. the rate of transport is greater at more acidic levels and the system is essentially quiescent at normal cytosolic pH. As is the case for the cation antiport, the regulatory activity of the Na⁺-dependent Cl^-/HCO_3^- exchanger relies strictly on the existence of an inward Na⁺ gradient and is therefore a secondary active transport system, driven indirectly by the Na⁺/K⁺ pump. Despite their similarities, however, Na⁺-dependent Cl^-/HCO_3^- exchange and Na⁺/H⁺ exchange can be readily distinguished by their HCO_3^- dependence as well as pharmacologically. Like the cation-independent Cl^-/HCO_3^- exchanger, the Na⁺-dependent anion exchanger is sensitive to disulfonic stilbene derivatives, but not to amiloride and its analogs. Conversely, the Na⁺/H⁺ exchanger is sensitive to amiloride and its analogs, but not to disulfonic stilbenes. Thus, in all likelihood, the two processes are mediated by distinct molecular entities. At this time, the molecule(s) responsible for Na⁺-dependent Cl^-/HCO_3^- exchange remain to be identified.

Permeation of H⁺, OH⁻ and HCO_3^- Through Channels

Experiments in liposomes and in planar bilayers have unequivocally demonstrated that H⁺ (or OH⁻) can permeate through lipid bilayer membranes, but at comparatively low rates. In addition to permeation through lipids, however, acid equivalents are thought to traverse biological membranes passively through other components, likely proteins or glycoproteins. Indeed, evidence for high-conductance, highly H⁺-selective, voltage dependent channels has recently been obtained for various cell types (e.g. neurons, alveolar cells, oocytes and, of particular relevance to this chapter, phagocytes; see ref. 6 for review). The physiological significance of these channels is still being debated. Under standard, physiological conditions, the existence of such pathways in the plasma membrane would appear to be counter-intuitive. At the normal resting potential of most mammalian cells (more negative than -50 mV), these channels would favor net uptake of acid equivalents into the cell, thereby exacerbating the normal tendency for the cytosol to become acidic. However, as discussed for the ion exchangers, the H⁺ channels are under allosteric control and are not permanently activated. Unique properties of the channels allow for activation only under very defined conditions. Specifically, acidification of the cytosol activates the conductance. Furthermore, the channel is activated only when cells are depolarized. Conversely, acidification of the exterior of the cell reduces the conductance, as does hyperpolarization (i.e. increased internal negativity) of the membrane potential. These properties facilitate unidirectional outward flow of H⁺, while precluding H⁺ influx. Interestingly, these channels have been described primarily in specific cell types in which a depolarizing event is also accompanied by a metabolic burst (e.g. nerves, phagocytes). Because the channels are activated by depolarization, it is possible that efflux of acid occurs during depolarization, facilitating the maintenance of pH. Conversely, the predicted deleterious entry of acid into repolarized cells would be prevented by closure of the channels.[6]

Bicarbonate, a base equivalent, is present at concentrations several orders of magnitude higher than those of free H^+ or OH^-. It has been demonstrated that HCO_3^- can traverse the plasma membrane not only by electroneutral exchange (see above), but also through conductive channels which likely transport also other inorganic anions, primarily Cl^-.[7] At the resting (inside negative) membrane potential, opening of HCO_3^- channels is expected to induce HCO_3^- efflux, with a resulting cytosolic acidification. This phenomenon has indeed been reported to occur, but its physiological significance remains obscure.

REGULATION OF CYTOSOLIC pH BY H⁺-ATPASES

As discussed in greater detail elsewhere in this volume, proton pumps are highly specialized H^+ transporters that utilize cytoplasmic ATP to drive acid equivalents across biological membranes.[8] As such, these transporters exist in the membranes of mitochondria, gastric epithelium, and certain cytoplasmic organelles. In the latter case, the pumps are of the vacuolar or V-type and are responsible for the acidification of compartments such as endosomes, lysosomes, Golgi-derived vesicles and phagosomes (see below for further discussion). In addition to their organellar distribution, recent studies have suggested the presence of V-type H^+ pumps in the plasma membrane of epithelia, osteoclasts and phagocytes. Epithelial plasma membrane pumps are thought to function primarily in the net transcellular displacement of acid equivalents from the blood to urine or vice-versa. In osteoclasts, plasmalemmal pumps are believed to promote the formation of localized acidic lacunae, which are sites of active bone resorption. In contrast, the main role of surface membrane pumps in phagocytes (i.e. neutrophils and macrophages) has been postulated to be in cytoplasmic pH homeostasis.[9] A detailed review of this topic has been published recently.[9]

V-TYPE ATPASES IN pH_c HOMEOSTASIS:
LEUKOCYTES AS A MODEL SYSTEM

In phagocytic leukocytes, cytoplasmic acidification impairs both the ability of the cells to migrate towards an inflammatory focus containing microbial pathogens as well as their capacity to effect killing once the microbe has been ingested.[10] It therefore seems evident that phagocytes must be endowed with effective pH_C regulatory mechanisms to ensure optimal function. This requirement appears particularly crucial for neutrophils and macrophages due to the constant threat of intracellular acid loading within the inflammatory microenvironment. Metabolic activation resulting from the exposure of cells to microbial products and other proinflammatory molecules generates a remarkably large quantity of endogenous acid equivalents. In addition, the acidic extracellular milieu which is characteristic of sites of inflammation tends to impose a gradual cytoplasmic acidification due to the inward leakage of protons, a process which is facilitated by weak organic acids which accumulate as a result of anaerobic bacterial metabolism.

Phagocytes have been shown to possess active Na^+/H^+ exchangers, a cation-independent HCO_3^-/Cl^- exchanger and a Na^+-dependent HCO_3^-/Cl^- exchange mechanism. As mentioned earlier, a voltage and pH-sensitive H^+ conductance has also been detected in these cells. Nevertheless, it appears that active H^+ pumping is required for effective pH_c homeostasis under some of the unique conditions encountered by itinerant cells such as phagocytes.

Using pH-sensitive fluorescent probes, early studies investigating the pH regulatory mechanisms of macrophages observed a significant pH_c recovery from an imposed acid load that persisted even when the Na^+/H^+ antiport and Cl^-/HCO_3^- exchange activities were precluded by ionic substitution or by use of selective inhibitors.[11] This observation suggested that an alternative H+ extrusion process existed in phagocytes, and prompted detailed studies to characterize the underlying mechanism. In peritoneal and pulmonary macrophages, two observations suggested that the observed pH recovery was mediated by a H^+ pump: i) depletion of cellular ATP resulted in a parallel inhibition of the pH_c recovery, and ii) the nonspecific pump inhibitors N-ethylmaleimide and N,N'-dicyclohexylcarbodiimide inhibited the pH recovery.[11] Additional functional evidence supported these preliminary experiments. In macrophages, pH_c recovery from acid loading occurred simultaneously with hyperpolarization of the plasma membrane and both were abolished by pump inhibitor.[12] This observation implied that the H^+ extrusion process mediating the pH_c recovery was translocating net positive charge (likely H^+ ions) across the plasma membrane to the extracellular space. Moreover, depolarization facilitated recovery of pH_c after an acid challenge in both murine macrophages and in human neutrophils, while hyperpolarization was inhibitory. Together, these data indicated that the transport process mediating the recovery from acid loading was electrogenic.

It was also noted that the pH_c recovery can occur against an electrochemical gradient, therefore suggesting an active process, consistent with its dependence on ATP. Thus, in an acidic medium (pH 6.5), acid-loaded human neutrophils recovered their pH to near neutrality ($\approx$pH 6.9), even though the chemical driving force for H^+ was inward. Together, these data provided strong evidence that the acid extrusion observed in phagocytic cells was mediated by an active and electrogenic process, likely an ATP-dependent pump.

Subsequent experiments in both macrophages and neutrophils sought to classify the H^+ pump using pharmacological criteria. These studies revealed that acid extrusion from the cytosol could also be blocked by 7-chloro-4-nitrobenz-2-oxa-1,3-diazole (NBD-Cl) and bafilomycin A_1,[11,13] two compounds that are potent blockers of H^+ pumps of the vacuolar- or V-type. By contrast, inhibitors of the mitochondrial F-ATPases (e.g. oligomycin and azide) and of the gastric E_1E_2-type H^+-ATPases (e.g. vanadate) were without effect in phagocyte pH_c homeostasis, ruling out the involvement of these pumps. The observed pharmacological profile is typical of V-type H⁺-ATPases. Combined with the functional characteristics described above, the pharmacological characterization strongly implicates this class of proton pumps in the cytoplasmic pH regulation of phagocytic cells.

It is noteworthy that, while assumed to be on the plasma membrane, the precise location of the functional H⁺ pumps mediating cytoplasmic pH homeostasis in phagocytes has not been established conclusively. It is conceivable that acid equivalents are translocated into pump-bearing intracellular organelles which then secrete their contents to the extracellular milieu by exocytosis. Alternatively, functional proton pumps may reside in the plasma membrane, translocating H⁺ from the cytosol directly to the extracellular space. The bafilomycin A₁-sensitive appearance of extracellular proton equivalents, with kinetics similar to pH_c recovery, is consistent with either model. However, the sensitivity of the pH_c recovery to the membrane potential (see above) is strong evidence that active pumps reside in the plasma membrane. This does not, however, preclude the possibility that dormant pumps are recruited to the plasma membrane from an intracellular store. It is in fact noteworthy that, as described with regards to urinary epithelia, latent pumps can be mobilized to the plasma membrane from an endomembrane store by a process dependent on an intact microtubule system and requiring a rise in cytosolic Ca^{2+}. However, neither preventing a rise in Ca^{2+} nor disrupting microtubules had an appreciable effect on the H⁺ pump-mediated pH_i recovery of acid loaded macrophages.[12] Thus, the precise cellular location of the H⁺ pumps in phagocytes remains to be established.

Although V-type pumps have been associated with pH_c regulation in both macrophages and neutrophils, the mechanism of activation differs between the two cell types. Under resting conditions, a bafilomycin-sensitive efflux of H⁺ can be measured in macrophages, consistent with constitutive pumping of acid from the cells. When macrophages are acid-loaded under conditions precluding the activity of other mechanisms of H⁺ translocation, pump-mediated recovery of pH_c occurs within minutes and is accompanied by an acceleration in the appearance of extracellular acid equivalents, suggesting direct activation of the pumps. Neutrophils, on the other hand, have little pump activity in the resting state. When acid-loaded under comparable conditions, neutrophil pH_c remains low, and bafilomycin-sensitive H⁺ extrusion is barely measurable. A proton pump-mediated alkalinization (accompanied by bafilomycin-sensitive extracellular acidification) is only observed when neutrophils are activated by agonists of protein kinase C such as 12-O-tetradecanoylphorbol 13-acetate or by chemotactic formyl peptides.[13] Importantly, the V-type proton pump can be activated in neutrophils without prior acid loading, when the cells are stimulated by chemo-attractants or phorbol esters at resting pH_c.

Pharmacological evidence indicates that activation of the neutrophil H⁺ pump by phorbol esters is indeed mediated by protein kinase C. Only β-isomers of phorbol esters are stimulatory, while α-isomers are without effect. Moreover, stimulation can also be observed with other, non-phorbol activators of protein kinase C. Finally, concentrations of staurosporine known to inhibit the kinase prevent the stimulatory effect of phorbol esters.

The implications of the difference in V-type proton pump activities between phagocytic cell types is not known. The elicited macrophage may require the continuous operation of this additional pH regulatory mechanism to cope with a constitutively increased acid production, which results from its elevated basal metabolic activity. The peripheral blood neutrophil, on the other hand, is relatively quiescent with respect to microbicidal activity, and only initiates H^+ pumping when undergoing its respiratory burst: at inflammatory sites in vivo or upon activation by agonists in vitro.

The concurrent presence of multiple pH regulatory mechanisms in phagocytes would superficially appear to be redundant. All the aforementioned pH regulatory mechanisms (exchangers, channels and pumps), with the exception of the cation-independent $HCO3^-/Cl^-$ exchanger, are poised to effect proton extrusion in response to cytoplasmic acidification. The relative roles of each of the pH homeostatic mechanisms under resting, physiological conditions have not been defined. One would expect the Na^+/H^+ antiport, the Na^+-dependent HCO_3^-/Cl^- exchanger and the V-ATPase to play predominant roles, at least in macrophages, while the H^+ conductive pathway would assume lesser importance. In fact, studies performed under nominally bicarbonate-free conditions where the HCO_3^-/Cl^- exchanger is rendered inactive, showed that both the Na^+/H^+ antiport and the V-ATPase contribute to the maintenance of physiological pH_c levels. Further, inhibition of one pathway led to a compensatory acceleration of the other, ensuring continued proton extrusion at the necessary rate.

In acidic extracellular environments, such as those encountered in abscesses and some tumours, local conditions would tend to preclude optimal function of both the $Na+/H^+$ antiport and the Na^+-dependent HCO_3^-/Cl^- exchanger. Low extracellular pH directly inhibits Na^+/H^+ exchange and reduces the local concentration of HCO_3^-, thus impairing HCO_3^-/Cl^- exchange. By contrast, the ability of V-ATPases to translocate H^+ against a concentration gradient suggests their potential importance in pH_c regulation under acidic conditions. Thus, when the relative activity of V-ATPase-mediated proton extrusion and Na^+/H^+ exchange was examined at low external pH using specific inhibitors of each pathway, V-ATPases played a major role in the maintenance of pH_c close to the physiological range, while the Na^+/H^+ antiport was almost inactive. Importantly, under these conditions V-ATPase-mediated pH_c homeostasis was crucial to the preservation of the normal microbicidal function of the cells.[14]

pH REGULATION OF THE VACUOLAR SYSTEM

This section deals with the functional role and regulation of organellar pH along the exocytic and endocytic transport pathways (Fig. 2.1), with emphasis on their specialized function in phagocytic cells. The organelles that constitute the vacuolar system are involved in the synthesis and transport of both soluble and membrane-bound proteins from one intracellular compartment to another and participate in the uptake of nutrients, ligands and various pathogens from the extracellular milieu.

The exocytic pathway consists of smooth and rough endoplasmic reticulum, the Golgi complex, condensing vesicles and secretory granules. Newly synthesized proteins are routed via transport vesicles to the Golgi complex, where post-translational modifications and sorting occurs to their final destination (plasma membrane, secretory vesicles, lysosomes, etc.). The organelles along the endocytic pathway (clathrin-coated vesicles, endocytic vesicles, endosomes, phagosomes and lysosomes) are involved in the transport, sorting and degradation of molecules taken up from the extracellular compartment or deriving from

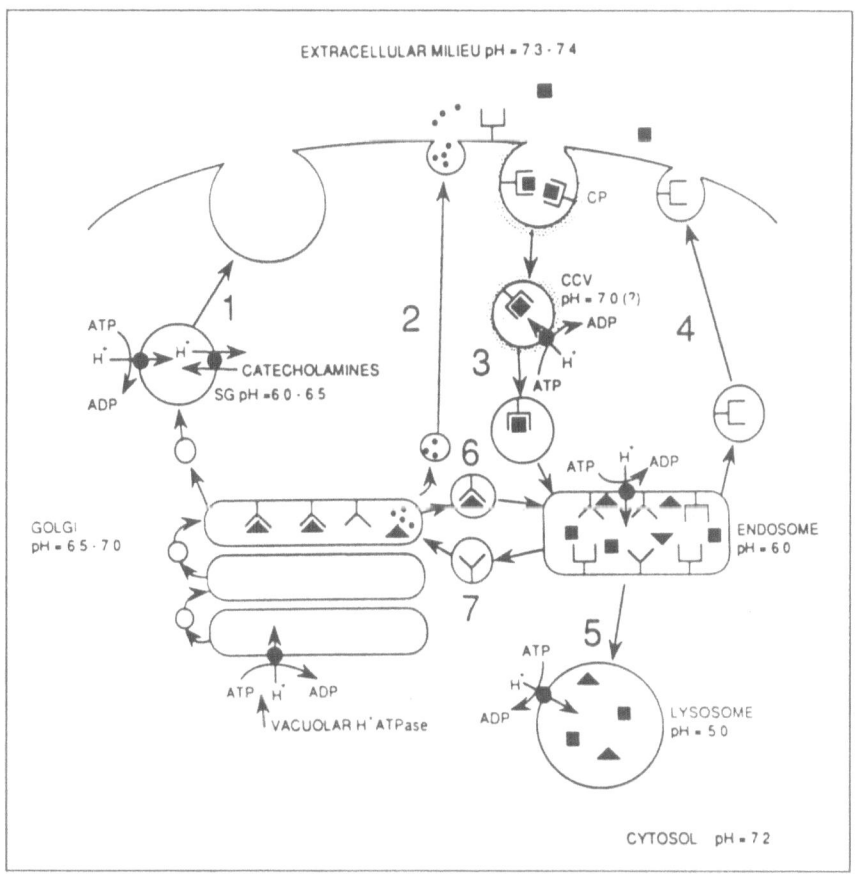

Fig. 2.1. Vacuolar system and intracellular transport pathways of a typical cell. Schematic representation of vacuolar compartments known to have an acidic pH. The luminal pH value for each compartment is shown next to the organelle. Large numbers represent transport pathways between different cellular compartments. Pathway 1, regulated secretion; 2, constitutive secretion; 3, receptor-mediated endocytosis; 4, recycling of certain receptors back to the plasma membrane; 5, delivery of ingested material to lysosomes; 6, transport of lysosomal enzymes from the Golgi to the endosomes; 7, recycling of mannose-6-phosphate receptor to the Golgi. Abbreviations used: SG, secretory granule; CP, coated pits; CCV, clathrin-coated vesicle; solid squares represent a ligand taken up by receptor-mediated endocytosis; solid triangles represent lysosomal enzymes and solid circles represent products secreted constitutively.

the plasma membrane by endocytosis (i.e. the formation of membrane-enclosed transport vesicles by invagination and pinching off of a portion of the plasma membrane). Internalization of particles by phagocytic cells is called phagocytosis and plays an important role in the destruction of pathogenic organisms inside the cells and in antigen processing.

It is becoming clear that differential pH regulation of the vacuolar compartments is a prerequisite for the maintenance of their physiological function. In understanding the mechanisms and regulation of organellar pH, two considerations must be borne in mind. First, that the characteristic morphological, biochemical and functional features of the vacuolar compartments are preserved despite an intensive membrane turnover, mediated by vesicular traffic between organelles. It is also important to realize that the luminal side of these organelles is topologically equivalent to the exterior of the cell.

ORGANELLES OF THE VACUOLAR SYSTEM HAVE AN ACIDIC INTERIOR

The work of Metchnikoff at the turn of the century indicated the ability of vacuolar organelles to generate and maintain an acidic pH. Using crude pH indicators such as neutral red or litmus paper he observed that bacteria ingested by phagocytic cells were transferred to acidic structures in the cytoplasm. He also concluded that the intracellular acidity was maintained by cellular energy and was necessary to digest the microorgansim.

In the past 80 years, in vivo and in vitro studies demonstrated the existence of acidic vacuolar organelles using a variety of methods. Acidification of the luminal space of isolated and purified organelles can be monitored measuring the uptake of radioactive or fluorescent weak bases. In intact cells, the pH of compartments containing high concentration of ATP (such as some secretory granules) can be estimated by ^{31}P-NMR spectroscopy. Acidic organelles have also been detected in intact cells using electron microscopic immunocytochemistry of samples treated with the weak base 3-(2,4-dinitro-anilino-)3'-amino-N-methyldipropilamine (DAMP), which is secondarily detected with an antibody specific for dinitrophenol. Additionally, compartments along the endocytic pathway can be labeled by fluid-phase endocytosis or receptor-mediated endocytosis of fluorescent pH-sensitive probes coupled to macromolecules or ligands (e.g. fluoresceinated derivatives of transferrin, macroglobulin or dextran). The routes and rates of intracellular traffic of these ligands have been well characterized and can be used to selectively label various subcompartments of the endo-lysosomal network. Thus, transferrin conjugates are specific probes for the recycling compartment while dextran conjugates can be used for labelling of early or late endosomes, and/or lysosomes. Highest sensitivity can be obtained during these measurements by choosing the appropriate fluorescein derivative, i.e. that with a pK value closest to the expected physiological pH of the target compartment.

ROLE OF ACIDIFICATION IN THE SECRETORY AND ENDOCYTIC PATHWAYS

Secretory Pathway

Smooth and Rough Endoplasmic Reticulum

An electrogenic proton translocating ATPase was detected in preparations of rat liver rough and smooth microsomes. The authors speculated that the proton gradient could play a role in the translocation of proteins (as occurs in mitochondria) and/or sugars, and that the acidic internal pH might facilitate protein folding. Consistent with this hypothesis, recent metabolic labelling experiments suggest that the assembly of the V-type ATPase is completed in the endoplasmic reticulum.[15] However, the functional competence of the newly synthesized proton pump in the absence of post-translational modification is unknown. It is also noteworthy that the microsomal preparation described above was not tested for possible contamination by Golgi or lysosomal membranes. Therefore, the existence and possible role of V-pumps in the endoplasmic reticulum remains a subject of speculation.

Golgi Apparatus

The Golgi complex is a heterogeneous organelle involved in the post-translational modification and sorting of proteins in the exocytic pathway. It is composed of at least three compartmens (*cis* and *trans* Golgi network and the Golgi stacks) which can be distinguished morphologically and enzymatically. The acidity of the *trans* compartment of the Golgi was suggested initially based on the drastic morphological changes evoked in secretory cells by treatment with the proton ionophore monensin. Accordingly, immunocytochemical electron-microscopy demonstrated the accumulation of DAMP in the *trans*-region of the Golgi complex. The presence of a proton translocating ATPase capable of generating a pH gradient and a transmembrane potential in rat liver Golgi membranes was subsequently documented. The function of the low pH in this compartment is less certain. Conceivably, the acidic pH assists in the condensation of nascent granule contents, in the proteolytic cleavage of membrane proteins and some pro-hormones,[16] or in providing the proper environment for the activity of terminal glycosyltransferases. Indeed, impaired N-linked glycan modification of soluble secretory proteins, seemingly due to impaired sialylation, was detected after inhibition of V-ATPases by the highly specific inihibitor concanamycin B.[17] By contrast, the proton pump inhibitor had no major effect on the posttranslational modifications of secreted proteins that occur in the reticulum and proximal Golgi complex, arguing against significant acidification of these compartments.

Secretory Granules

Proton translocating ATPases have been identified in secretory organelles of a number of different cell types: chromaffin granules of the adrenal medulla, dense granules of platelets, neurosecretory granules of the pituitary gland and cholinergic synaptic vesicles, to mention a few. The substrate specificity and inhibitor susceptibility of the pumps associated with secretory granules are typical of V-ATPases. Indeed, the V-ATPase of chromaffin granules has been purified and succesfully reconstituted in phospholipid membranes. The physiological role of the intragranular acidic pH is well documented. In chromaffin granules, platelet dense granules and synaptic vesicles the electrochemical proton gradient drives the vectorial uptake of epinephrine, serotonin and catecholamines, respectively, from the cytoplasm into the lumen of the organelle. The accumulation of biogenic amines results from a coupled transport process whereby intraluminal H^+ is exchanged for the cytoplasmic amine. For glutamate accumulation, the internally positive membrane potential provides the predominant driving force.

In some endocrine glands the proteolytic conversion of pro-hormones (e.g. pro-insulin) to their active form takes place in acidic compartments. Inhibiton of the V-type proton pump using specific blockers or by ATP depletion often obliterates the conversion of the pro-hormone to its active counterpart.

Endocytic Pathway

V-type ATPases play a wide variety of roles along the endocytic pathway, where acidification is important not only for the normal physiological function of the cell but also for infection by a variety of viral, protozoan and bacterial pathogens.

The Pathway of Receptor-Mediated Endocytosis

Ligand-receptor complexes formed at the plasma membrane are typically internalized in clathrin coated pits and coated vesicles. After their rapid uncoating, the vesicles fuse with early endosomes. The slightly acidic internal pH of early endosomes (pH 5.5-6.5) facilitates the dissociation of ligands from their plasma membrane receptors, allowing the vacated receptor to recycle back to the plasmalemma. The free ligands are transferred to late endosomes where the internal H^+ concentration is higher (pH $\approx$ 5.5) and subsequently to the lysosomes, where they can be degraded (see below). The low-pH mediated discharge of ligands and the subsequent sorting of ligands from their receptors is one of the hallmarks of receptor-mediated endocytosis.

The content of acid hydrolases increases along the endocytic pathway in parallel with the progressive acidification of the vesicular lumen: while early endosomes contain very small amounts of lysosomal-type hydrolases, late endosomes and lysosomes have high concentrations of these enzymes. Hormones (e.g. insulin), growth factors, toxins and foreign antigens are cleaved by endosomal proteolysis. Interestingly, proteolysis can occur at neutral or acidic pH, depending on the nature of the substrate. Cleavage of mannose-albumin in endosomes is inhibited by weak bases, by protonophores or in the absence of ATP, indicating the activity of proteases with an acidic pH optimum. Similarly, cells defective in endosomal acidification have a reduced capacity to process foreign antigens, an event thought to require endosomal proteolysis. In contrast, the endosomal cleavage of ricin A and insulin persists at neutral luminal pH. These differences may reflect subcompartmentation of the endosomal complex and/or the existence of multiple proteases with different pH sensitivity.

Acidification of the lumen of the endosomal compartment is also a requisite for the migration of certain viruses and toxins into the cytoplasmic space. Several membrane-enclosed viruses, such as influenza, Semliki forest and vesicular stomatitis virus bind to receptors on the surface membrane and are taken up by receptor-mediated endocytosis. The acidic environment of endosomes favors a conformational change of viral spike glycoproteins, which is thought to expose a domain that facilitates the fusion of the viral envelope with the endosomal membrane with subsequent exposure of the nucleocapsid to the cytosol, thus initiating cellular infection.

Finally, the role of acidic endosomal pH in lysosomal enzyme delivery must also be mentioned. Newly synthesized lysosomal enzymes bearing the mannose-6-phosphate residue associate with mannose-6-phosphate-receptors in the *trans* Golgi network. The receptor-ligand complex is delivered by vesicular transport to late endosomes. In the latter compartment, the acidic environment facilitates ligand-receptor dissociation, allowing the return of mannose-6-phosphate-receptor back to the Golgi and the delivery of the enzyme to lysosomes.

Lysosomes

The lysosomal compartment is the most acidic vacuolar compartment (pH< 5). It is also the site of degradation of materials arriving via a variety of internalization processes (receptor-mediated endocytosis, phagocytosis and pinocytosis). The lysosomal compartment contains a variety of hydrolytic enzymes (e.g. proteases, lipases, ribonucleases and deoxyribonucleases) that exhibit optimal activity at acidic pH. The acidic pH optima of these enzymes provides a safety device to prevent degradation of cellular constituents elsewhere. Lysosomes also participate in the degradation of certain cytoplasmic proteins that are specifically targeted for lysosomal breakdown by virtue of a unique pentapeptide signal. In addition to supporting the activity of lysosomal hydrolases, the low intralysosomal pH may function in the coupled transport of breakdown products (e.g. amino acids, sugars, nucleotides) from the lysosomes into the cytoplasmic compartment, where they can be reutilized by the biosynthetic machinery of the cell.

Phagosomes

Phagocytosis is the process of engulfment of relatively large extracellular particles. The resulting intracellular vacuole, the phagosome, is subsequently transformed into an acidic, oxidizing, hydrolase-rich phagolysosome that is hostile towards invading microorganisms. The elevated proton concentration in phagosomes is directly lethal for certain microorganisms, promotes the spontaneous dismutation of superoxide to hydrogen peroxide, provides optimal conditions for the activity of acid hydrolases and appears to facilitate phagosome-lysosome fusion. Indeed, certain microorganisms such as *Legionella pneumophilia* and *Toxoplasma gondii* appear capable of preventing phagosomal acidification, thereby evading killing by phagocytic cells of the host.

We have examined the mechanism responsible for phagosomal acidification. To measure phagosomal pH in situ, macrophages (or human neutrophils) were allowed to phagocytose fluorescein-labeled, opsonized *Staphylococcus aureus* and the intraphagosomal pH was monitored fluorimetrically. Two lines of evidence indicated that V-type ATPases are largely, if not entirely responsible for phagosomal acidification.[19] After selective electropermeabilization of the plasma membrane, the phagosomal pH remained near neutrality due to washout of cytoplasmic ATP. Reintroduction of both ATP and Mg^{2+} was necessary to initiate phagosomal acidification. That acidification required ATP hydrolysis was indicated by the failure of non-hydrolyzable ATP analogues (AMP-PNP or ATP-γ-S) to support proton accumulation. The second finding indicating involvement of V-ATPases was mainly pharmacological. Phagosomal acidification was insensitive to oligomycin, azide and vanadate ruling out the participation of F-or P-type ATPases. In contrast, a variety of well established V-ATPase inhibitors (N-ethylmaleimide, N,N'-dicyclohexylcarbodiimide and nitrate) diminished or prevented phagosomal acidification. The most convincing evidence was obtained with bafilomycin A_1, which completely abrogated phagosomal acidification at submicromolar concentrations both in intact and permeabilized cells. Consistent with these functional measurements, appearance of the 31 kDa subunit of the V-ATPase was demonstrated by immunoblot analysis during the process of phagosomal maturation.[18]

REGULATION OF THE INTRALUMINAL pH
OF THE VACUOLAR COMPARTMENTS

In general terms, the luminal pH of the vacuolar compartment is more acidic than that of the cytoplasm. As should be apparent from the preceding sections, however, the steady state pH of individual components of the vacuolar complex differs substantially (Fig. 2.1). This is readily exemplified by comparison of the very acidic lysosomes (pH< 5) with the mildly acidic endosomes (pH 6-7). The mechanism underlying the differential pH regulation in the vacuolar compartments is not fully understood. In principle, several factors could control the pH of subcellular organelles. The number and activity of proton pumps in an individual organelle would represent the first level of control. Direct modulation of the isolated pump activity by anions, boundary lipids, intramolecular disulfide-bond formation and cytosolic protein inhibitors has been reported. Counterions can also influence the rate of pumping. Because the proton pump mechanism is electrogenic, anion uptake or cation efflux must accompany H^+ accumulation, in order to prevent the development of an opposing membrane potential. Therefore the passive ionic conductance of the membrane and the activity of other electrogenic ion transporters (e.g. the $Na^+/K+$ ATPase in early endosomes) could also modulate the activity of the pump. Finally, the luminal acidity can also be controlled by the rate of passive proton efflux, which depends on the endogenous membrane proton conductance and/or the activity of ion-coupled proton transporters (e.g. Na^+/H^+ antiporters).

Since preliminary experiments indicated that subcellular fractionation perturbs the physiological pH regulation of certain vacuolar compartments, we studied the determinants of phagosomal pH in situ in intact macrophages.[20] In the following paragraph we summarize studies of the relative contribution of the passive proton leak, counterion-conductance and pump activity in the pH regulation of phagosomes.

Passive H⁺ efflux (leak) was evaluated by measuring the rate of dissipation of the phagosomal acidification upon complete inhibition of the proton pump with bafilomycin A_1. The proton permeability was found to be relatively low. Dissipation of the phagosomal pH gradient was not limited by the counterion permeability, since addition of exogenous protonophores induced a rapid H⁺ efflux. Thus phagosomes appear to exhibit relative tightness to protons, at least in intact cells. Interestingly, when phagosomes were isolated by nitrogen cavitation or studied in permeabilized cells, they demonstrated markedly increased proton efflux, suggesting that soluble cytosolic factors contribute to the regulation of phagosomal proton permeability. Similar phenomena were observed in endosomes of Chinese hamster ovary cells and in macrophage lysosomes.

The role of counterion conductance was studied next. We assumed that if the steady-state rate of pumping was limited by counterion flux, an increase in counterion permeability would promote acidification. However, increasing the cation permeability with ionophores such as valinomycin, nonactin or gramicidin had no effect on the rate of acidification or the steady-state pH attained. This finding is indicative of a low phagosomal membrane potential and suggests that the intrinsic conductance for ions other than protons is comparatively high. The observation that addition of protonophores to bafilomycin-treated cells induced a rapid phagosomal alkalinization provided further support for this concept. Both cations and anions appear to contribute to the intrinsic phagosomal ionic permeability, as shown by the ability of quinine (a K⁺ channel blocker) and 5-nitro-2-(3-phenylpropyl-amino) benzoic acid (a Cl⁻ channel blocker) to reduce the rapid H⁺ efflux evoked by the protonophore. Thus, the passive proton permeability of phagosomes is markedly smaller than the counterion permeability. Similar findings were made in endosomes from macrophages and Chinese hamster ovary cells.

Finally, since the counterion conductance was shown to be high, the protonmotive force, specifically the pH gradient across the phagosomal membrane, was predicted to be the primary determinant of the rate of proton pumping. This hypothesis was tested by measuring the rate of proton pumping as a function of phagosomal pH at a constant cytosolic pH. As predicted, pump activity in the steady state was low, just matching the minimal backward H⁺ leak. Small increases in phagosomal pH markedly increased the rate of pumping, indicating that the pump virtually "turns off" when the desired pH set point is attained. Hence, the luminal pH sensitivity of the pump appears to be the primary determinant of the phagosomal pH.

In summary, the membrane of phagosomes in murine macrophages is comparatively impermeable to H⁺, yet quite permeable to counterions. As a result, the transmembrane potential is minimal and the pH gradient or intraphagosomal pH is the primary determinant of the rate of pumping, possibly due to an allosteric effect on the H$^+$-ATPase. Inactivation of the pump by intraphagosomal acid could result from increased slippage, altered stoichiometry or direct inhibition of the H$^+$-ATPase activity.

CONCLUDING REMARKS

Precise control over intracellular pH, both cytosolic and intraorganellar, is critical to the proper physiology of the cell. Traffic of intracellular membranes, endocytosis and receptor recycling, and cytosolic enzyme function are but a few of the myriad processes which are pH dependent. In compiling this review, it was our intent to summarize recent advances in our understanding of some of the major mechanisms responsible for this precise regulation. Clearly, much remains to be learned regarding the existence of multiple isoforms of the different subunits of the H$^+$-ATPase, their cellular and compartmental distribution and their differential role in the regulation of cytosolic and intraorganellar pH.

REFERENCES

1. Roos, A and Boron,WF. Intracellular pH. Physiol Rev 1981; 61:296-434.
2. Wakabayashi, S, Sardet, C, Fafournoux, P et al. Structure and function of the growth factor activatable Na/H exchanger NHE-1. Rev Physiol Biochem Pharmacol 1992; 119: 157-186.
3. Madshus, IH. Regulation of intracellular pH in eukaryotic cells. Biochem J 1988; 250:1-8.
4. Cabantchik, ZI, Knauf, PA and Rothstein, A. The anion transport system of the red blood cell. Biochim Biophys Acta 1978: 515: 239-302.
5. Boron, WF. Transport of H$^+$ and of ionic weak acids and bases. J Membrane Biol 1983; 72: 1-14.
6. Lukacs, G, Kapus, A, Nanda, A, et al. Proton conductance of the plasma membrane: properties, regulation and functional role. Am J Physiol 1993; 265: C3-C14.
7. Bretag, AH. Muscle chloride channels. Physiol Rev 1987; 67: 618-645.
8. Forgac, M. Structure and function of a vacuolar class of ATP-driven proton pumps. Physiol Rev 1989; 69:765-769.
9. Grinstein, S, Nanda,A, Lukacs,G, et al. J Exp Biol 1992; 172: 179-192.
10. Rotstein OD, Fiegel VD, Simmons RL, Knighton DR. The deleterious effect of reduced pH and hypoxia on neutrophil migration in vitro. J Surg Res 1988; 45:298-303.
11. Swallow CJ, Grinstein S, Rotstein OD. Cytoplasmic pH regulation in macrophages by an ATP-dependent and N,N'-dicyclohexylcarbodiimide-sensitive mechanism. Possible involvement of a plasma membrane proton pump. J Biol Chem 1988; 263:19558-19563.
12. Swallow CJ, Grinstein S, Rotstein OD. A vacuolar type H$^+$-ATPase regulates cytoplasmic pH in murine macrophages. J Biol Chem 1990; 265:7645-7654.
13. Nanda, A, Gukovskaya, A, Tseng, J, et al. Activation of vacuolar type proton pumps by protein kinase C. Role in pH regulation. J Biol Chem

1992; 267: 27740-27747.

14. Swallow CJ, Grinstein S, Sudsbury RA et al. Relative roles of Na⁺/H⁺ exchange and vacuolar-type H⁺ ATPases in regulating cytoplasmic pH and function in murine peritoneal macrophages. J Cell Physiol 1993; 157:453-460.

15. Myers M and Forgac M Assembly of the peripheral domain of the bovine vacuolar H⁺-ATPase. J Cell Physiol 1993; 156:35-42

16. Blum JS, Diaz R, Mayorga LS et al. Reconstitution of endosomal transport and proteolysis. In: Bergerac JJM and Harris JR eds. Subcellular Biochemistry, Vol.19, 1993:69-93

17. Yilla M, Tan A, Ito K et al. Involvement of the vacuolar H⁺-ATPases in the secretory pathway of HepG2 cells. J Biol Chem 1993; 268: 19092-19100.

18. Pitt A, Mayorga LS, Stahl PD et al. Alterations in the protein composition of murine phagosomes. J Clin Invest 1992; 90:1978-1983.

19. Lukacs GL, Rotstein, OD, and Grinstein, S. Phagosomal acidification is mediated by a vacuolar-type H⁺-ATPase in murine macrophages. J Biol Chem 1990; 265: 21099-21107.

20. Lukacs, GL, Rotstein, OD, and Grinstein, S. Determinants of the phagosomal pH in macrophages. In situ assessment of H-ATPase activity, counterion conductance and H⁺ "leak". J Biol Chem 1991; 266: 24540-24548.

ACIDIFICATION AND BONE RESORPTION: THE ROLE AND CHARACTERISTICS OF V-ATPASES IN THE OSTEOCLAST

Roland Baron, Marcjanna Bartkiewicz, Pe'er David
and Natividad Hernando-Sobrino

BONE RESORPTION: NORMAL FUNCTION AND PATHOPHYSIOLOGY

The purpose of this chapter is to briefly describe the mechanisms involved in the resorption of bone, highlighting the role of vacuolar ATPases and pH, both intracellular and extracellular, in this process. The cell responsible for the resorption of bone, i.e. for the degradation of the mineralized extracellular matrix that forms bone tissue, is the osteoclast.

Bone resorption is a cell-mediated event that is an integral part of the normal growth and remodeling of the skeleton. Bone remodeling is responsible for the adaptability of the skeleton and for the quantitative and qualitative maintenance of bone mass throughout life.

Under normal conditions, bone resorption by the osteoclast is tightly coupled to bone formation by the osteoblast, the amount of bone resorbed in a given period of time and at a given site being roughly equal to the amount of bone that is formed. It is the balance between these two cellular activities that determines skeletal mass and shape at any point in time.

Besides its role in bone growth and remodeling, bone resorption is also involved in calcium homeostasis, mobilizing the calcium stored in the skeleton for use by the whole organism when necessary. This occurs when changes in Ca intake, adsorption in the gut, or readsorption in the kidney, whether primary or secondary to changes in levels of calciotropic hormones (primarily calcitonin, parathyroid hormone and vitamin D3), lead to a negative systemic Ca balance.

Several diseases result from an abnormal regulation of bone resorption, either by alterations in the number of cells that are formed from their hematopoietic precursors in the bone marrow or by alterations in the activity of the mature osteoclasts, or a combination of both. The changes in osteoclast-mediated bone resorption occur either in absolute terms (increased bone resorption) or relative to the amount of new bone formed (decreased bone formation relative to bone resorption). Among the most widespread diseases which involve such an imbalance between bone resorption and bone formation, osteoporosis is the most prominent and has attracted a lot of attention in recent years due to its extreme epidemiology and cost to governments. Other diseases with alterations in bone resorption include Paget's disease, hyperparathyroidism, humoral hypercalcemia of malignancy and various cancers where resorption is increased as well as various forms of osteopetrosis, where bone resorption is genetically impaired.

BASIC MECHANISM: ROLE OF ACIDIFICATION

Although this chapter will focus on the process of acidification by the osteoclast, we will first summarize our current undestanding of the cellular and molecular biology of this cell and the molecular mechanisms involved in the process of bone resorption. The osteoclast is a multinucleated cell (although mononuclear osteoclasts are also encountered) which is formed by the asynchronous fusion of mononuclear precursors derived from the hematopoietic bone marrow and which differentiate within the granulocyte-macrophage lineage. The mature osteoclast, and probably its precursor, is a highly motile adherent cell, like the closely related cells of the monocyte-macrophage series. The particularity of the osteoclast is that it attaches to, and migrates along, the surface of bone. This occurs mostly along the interface between bone and bone marrow (endosteum) but also at interfaces with fibrous connective tissues such as the periosteum at the periphery of bones and the periodontal ligament, which serves to attach the teeth to their bony support. Conceptually, the osteoclast can therefore be considered as a specialized tissue macrophage which is resident in bone. It therefore shares several features and functional characteristics with other macrophages but also differs markedly from these cells in several respects.

The osteoclast attaches to the mineralized bone matrix that it resorbs by forming a tight ring-like zone of adhesion, the sealing zone. This attachment involves the specific interaction between adhesion molecules of the integrin family in the cell's membrane and proteins found in the bone matrix or at the surface of bone and which contain amino acid motifs such as RGD. The space contained inside this ring of attachment and between the osteoclast and the bone matrix constitutes the bone resorbing compartment. The osteoclast then synthesizes several proteolytic enzymes which are vectorially transported and secreted into this extracellular bone resorbing compartment. Simultaneously, the osteoclast lowers the pH of this compartment by extruding protons across its apical membrane (facing the bone matrix). The concerted action of the enzymes and the low pH in the bone resorbing compartment leads to the extracellular digestion of the mineral and organic phases of the bone matrix. After resorbing to a certain depth, determined by mechanisms that remain to be elucidated, the osteoclast detaches and moves along the bone surface before reattaching and forming another resorption lacuna, usually in close proximity to the first one. In the process, a certain volume of bone matrix has been removed, only to be replaced, under normal circumstances, by newly formed matrix a few days later. The calcium, phosphate and other components of the matrix, most of which have been completely digested but some of which may have been only mobilized during this process, will be either eliminated, serve locally as messengers, reutilized at sites where bone formation and mineralization occur or else used to maintain the proper ionic concentrations in the extracellular fluids. In the following paragraphs, the various elements involved in the process of bone resorption will be discussed in greater detail.

THE OSTEOCLAST FORMS A SEALED-OFF EXTRACELLULAR COMPARTMENT BY ATTACHING TO THE BONE MATRIX

The osteoclast is characterized by its size (50-100 μm in average), its multinucleation (usually two to ten nuclei), and its presence within a resorptive (Howship's) lacuna along the edge of the calcified matrix-bone marrow interface. The apical area of the cell, closest to the matrix, is characterized by an organelle-free zone of attachment (sealing zone) and a highly vacuolated area corresponding to the ruffled-border and its associated transport vesicles (Fig. 3.1). In addition, the osteoclast is characteristically rich in mitochondria.

The peripheral zone of the apical domain of the plasma membrane of the osteoclast is closely apposed to the extracellular matrix. This so-called "sealing zone"[1] is characterized by a very narrow space (0.2 to 0.5 nm) between the plasma membrane of the cell and the calcified matrix and by the presence of an organelle-free area in the adjacent cytoplasm, the "clear-zone" (Fig. 3.1), which is characteristically enriched in contractile proteins.[2,3] Toward the center of the apical domain, the plasma membrane of the osteoclast develops progressively deeper infoldings, which reach a high degree of geometric complexity. The cytoplasmic side of the membrane forming the ruffled-border is lined by small, regularly spaced studs.[4] In contrast, the basolateral domain of the plasma membrane is relatively smooth and is not lined on its cytoplasmic face by any defined structure.

The structural features of the osteoclast reflect its specific function. The clear-zone and the sealing-zone are responsible for the attachment of the osteoclast to the bone matrix, the ruffled-border corresponds to the area of proton transport and enzyme secretion, the basolateral membrane is a major site for both ion transport and regulatory molecules and receptors. Mitochondria provide the ATP required for the various energy-dependent systems within the cell, especially the several ion-transporting ATPases, and are a source of CO_2, which is used by carbonic anhydrase to produce protons for the acidification of the bone resorbing compartment.

By analogy with epithelia, the osteoclast is therefore a cell that is morphologically and functionally polarized (Fig. 3.1), with a pole facing the bone matrix, where attachment occurs and towards which most of the secretion is targeted (the apical pole, with the ruffled-border and the resorbing compartment), and a pole facing the soft tissues in the local microenvironment (bone marrow or periosteum) and which provides mostly, but not exclusively, regulatory and support transport functions (the basolateral pole). As discussed later, this analogy is mostly functional but the osteoclast is not an epithelial cell and thereby differs markedly from cells of the kidney tubule in other aspects.

THE CLEAR-ZONE

The most striking and unique feature of the osteoclast actin cytoskeleton is at the site of cell contact with the substratum. In osteoclasts observed under a variety of conditions, there is a prominent peripheral band of F-actin which contains actin filaments oriented parallel to the plane of the underlying substrate and running around the cell

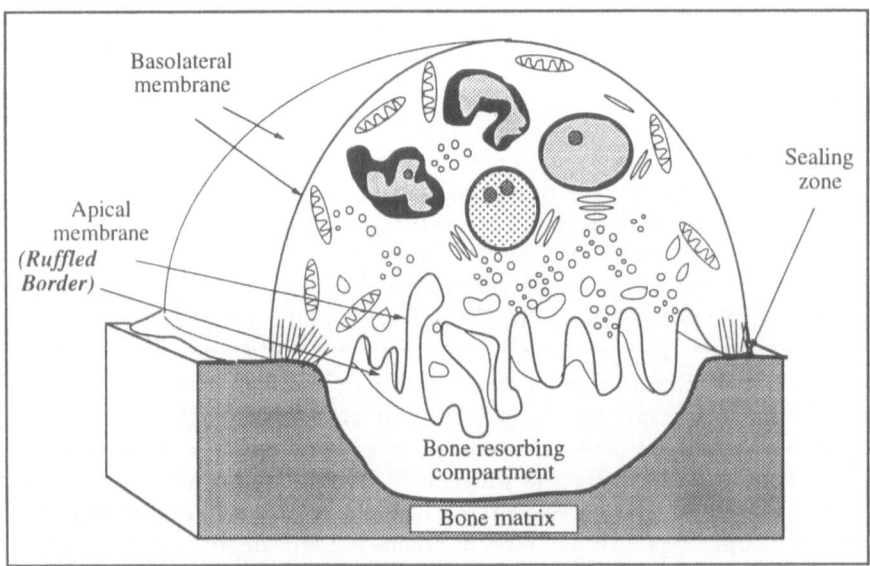

Fig. 3.1. Morphological characteristics and functional organization of the osteoclast: The osteoclast is a multinucleated cell rich in mitochondriae and closely apposed to the mineralized bone surface; the plasma membrane facing the bone surface (apical) is highly folded (ruffled-border); the membrane at the opposite side, facing the bone marrow, is smoother (basolateral); the membrane at the attachment area is closely apposed to the matrix (sealing zone), with actin filaments on the cytoplasmic side; the sealing zone delimits an extracellular compartment where bone resorption occurs (bone resorbing compartment), forming lacunae along the bone surface (Howship's lacunae).

periphery as well as numerous punctate structures where the actin filaments are organized in bundles perpendicular to the plane of the substratum. When osteoclasts are cultured on bone, the band circumscribes the area of active bone resorption[5-7] and corresponds to the clear zone, where the high density of cytoskeletal elements excludes organelles from the region of the cytoplasm immediately adjacent to the plasma membrane. In electron micrographs, bundles of actin filaments can be observed oriented perpendicular to the bone surface and extending into short cell processes that enter irregularities of the bone surface.

THE SEALING-ZONE

The punctate actin structures in the peripheral band, termed "podosomes",[6] apparently occur only in cells of monocytic origin (osteoclasts and monocytes) and in cells which have been transformed by the *src*, *fps* and *abl* oncogenes.[8] In addition to the bundles of actin filaments, the podosomes contain a number of other proteins which have been reported to occur at sites of cell-substratum or cell-cell interaction (see ref. 6 for review). These include fimbrin, α–actinin, cortactin and gelsolin, which are closely associated with the actin filaments in the core of the podosome, as well as vinculin and talin, which appear to form rosette structures surrounding the podosome cores.

The osteoclast attachment goes through a cyclic reorganization, alternating from a motile pattern and a resorbing pattern. In highly motile osteoclasts, few podosomes are observed and seem to be confined to the irregularly shaped leading edge of the cell, or lamellipodium. Upon arrest and attachment, much more numerous podosomes are formed and are organized in a peripheral ring, forming the sealing zone.

THE ROLE OF INTEGRINS

The cytoskeletal complexes described above are cytoplasmic and stabilize the interaction of the osteoclast with the bone surface, but they are not directly responsible for that interaction This role is filled by integral membrane proteins whose cytoplasmic domains interact with the cytoskeleton while their extracellular domains bind to bone matrix proteins. The molecular recognition and interaction of the osteoclast with bone is mediated by transmembrane proteins of the integrin family of adhesion molecules, which mediate cell-substratum interactions.[9-11] Integrins are heterodimeric molecular assemblies of an α subunit and a β subunit with specific, receptor-like, extracellular binding sites which recognize specific sequences in matrix proteins such as the Arg-Gly-Asp (RGD) sequence. The amino acid sequence surrounding the RGD motif determines the specificity and the affinity with which a specific integrin will recognize and bind a specific matrix protein.

Osteoclasts express at least two α subunits, α_1 and α_v, and at least two β subunits, β_1 and β_3, implying that multiple integrins may be involved in osteoclast adhesion to the bone matrix. The α_v and β_3 proteins form a dimer which is closely related, if not identical, to the vitronectin receptor (VNR), and is expressed at high levels in osteoclast membranes.[12] Although the detailed receptor-ligand interactions at the attachment site are only beginning to be elucidated, several RGD-containing matrix proteins have been identified.[6] Of these, collagen-type I, osteopontin and bone sialoprotein II are the most likely candidates to fill the role of integrin-binding proteins in bone.[6] Most interestingly, data is now accumulating that suggests that the osteoclast synthesizes and secretes both osteopontin and BSPII, raising the possibility that the osteoclast itself deposits along the bone surface the adhesion molecules required for its own attachment and for the establishment of the sealed-off bone resorbing compartment, into which protons will ultimately be transported and segregated.

OSTEOCLAST ATTACHMENT IS CALCIUM AND pH-DEPENDENT

When osteoclasts are activated or inhibited, rapid and dramatic changes occur in their cytoskeleton and attachment structures, further demonstrating the functional importance of these structures in bone resorption.[6] Several of these cytoskeletal changes might be associated with the regulation of the osteoclast's intracellular calcium levels and/or pH.[6] Inhibition of bone resorption is associated with an elevation of cytoplasmic Ca^{2+} levels after calcitonin treatment[13] and similar cytoskeletal changes are seen when cytoplasmic Ca^{2+} is increased by membrane depolarization, elevating extracellular Ca^{2+} levels or treating osteoclasts with a calcium channel agonist.[14] Reducing intracellular calcium, in contrast, promotes the formation of podosomes[15] as does the decrease in cytosolic pH, possibly also via decreased cytoplasmic Ca^{2+}.[15] Furthermore, the binding of RGD-containing matrix proteins to their integrin receptors induces an increase in intracellular Ca^{2+} and may differentially regulate bone resorption.[16]

Hence, the cytoskeleton and integral membrane receptors of the integrin family play essential roles in osteoclast attachment to the bone surface, establishing the seal which will allow the segregation of protons and enzymes within the extracellular bone resorbing compartment.

THE OSTEOCLAST SYNTHESIZES ENZYMES WHICH ARE SECRETED INTO THE BONE RESORBING COMPARTMENT

ACID PHOSPHATASE AND OTHER LYSOSOMAL ENZYMES ARE FOUND IN THE SECRETORY PATHWAY

One of the main cytochemical characteristics of the osteoclast is its enrichment in lysosomal enzymes. Ultrastructural studies have shown, however, that this high concentration of lysosomal enzymes in the osteoclast is not due to the presence of phagocytic structures such as secondary lysosomes. Instead, these enzymes are found, for the most part, in elements of the secretory pathway. Thus, localization of arylsulfatase, beta-glycerophosphatase and acid phosphatase by enzyme cytochemistry and localization of beta-glucuronidase, cathepsin C and tartrate-resistant acid phosphatase by immunocytochemistry[17-20] have shown the abundant concentration of these enzymes in the lumen of the ER cisternae, in the cisternae of the Golgi complexes and in very numerous small (50 to 75 nm), coated vesicles.[17,18] Consequently, this enrichment in lysosomal enzymes does not reflect a high phagocytic activity, but rather a high biosynthetic activity.

The newly synthesized lysosomal enzymes are vectorially secreted into the bone resorbing compartment (Fig. 3.4). Antibodies to TRAP and cathepsin C have allowed the localization of these enzymes in the extracellular bone resorbing compartment, hereby demonstrating that the osteoclast is actively engaged in the synthesis and secretion of lysosomal enzymes. These enzymes proceed through the Golgi and are vectorially transported from the trans-Golgi region to the ruffled-border apical membrane in coated transport vesicles which fuse exclusively with the ruffled border's plasma membrane and release their content into the bone-resorbing compartment.

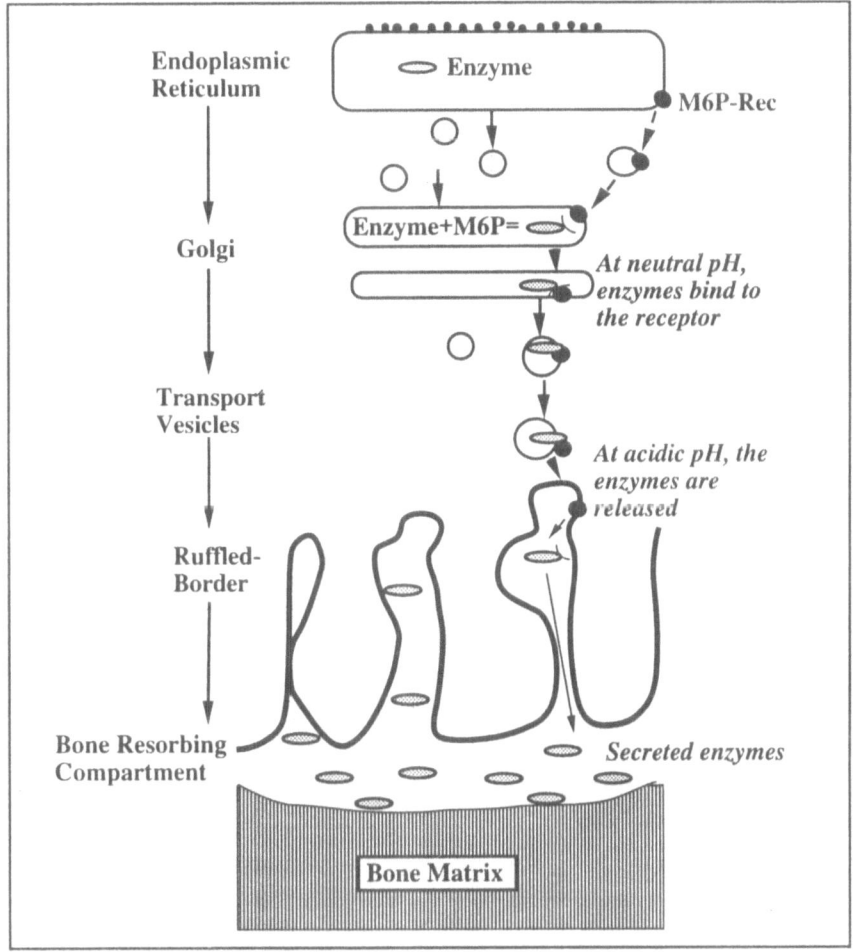

Fig. 3.2. Synthesis and secretion of lysosomal enzymes by the osteoclast is polarized and mannose 6-phosphate- and pH-dependent. The osteoclast synthesizes large amounts of lysosomal enzymes which are transported from the ER, through the Golgi and in very numerous trans-Golgi transport vesicles (primary lysosomes) towards the ruffled border (bottom) and the bone resorbing compartment. In the Golgi, the newly synthesized enzymes bind, at neutral pH, to Mannose-6-phosphate receptors; they are vectorially transported bound to the receptors and are released due to the low pH in the bone resorbing compartment.

THE DELIVERY OF ENZYMES TO THE BONE RESORBING COMPARTMENT IS pH-DEPENDENT

This specific release of the enzymes after targeting of secretory transport vesicles to the apical domain of the cell's membrane is accomplished by their association with the mannose-6 phosphate receptor[18] (Fig. 3.4). In most cells, lysosomal enzymes are sorted out of the main flow of secretory and membrane proteins at the level of the Golgi apparatus and specifically targeted to late endosomes along the endocytic pathway.[21] The newly synthesized enzymes are recognized by a specific binding protein, the mannose-6-phosphate receptor, which is highly concentrated in membranes of the Golgi compartment and binds mannose-6-phosphate, a specific marker added to the lysosomal enzymes in the *cis*-Golgi elements. This mechanism ensures that the enzymes are sorted out of the flow of other proteins synthesized by the cells. The enzymes remain bound to the mannose-6-phosphate receptor until they reach an acidified compartment where the low pH induces

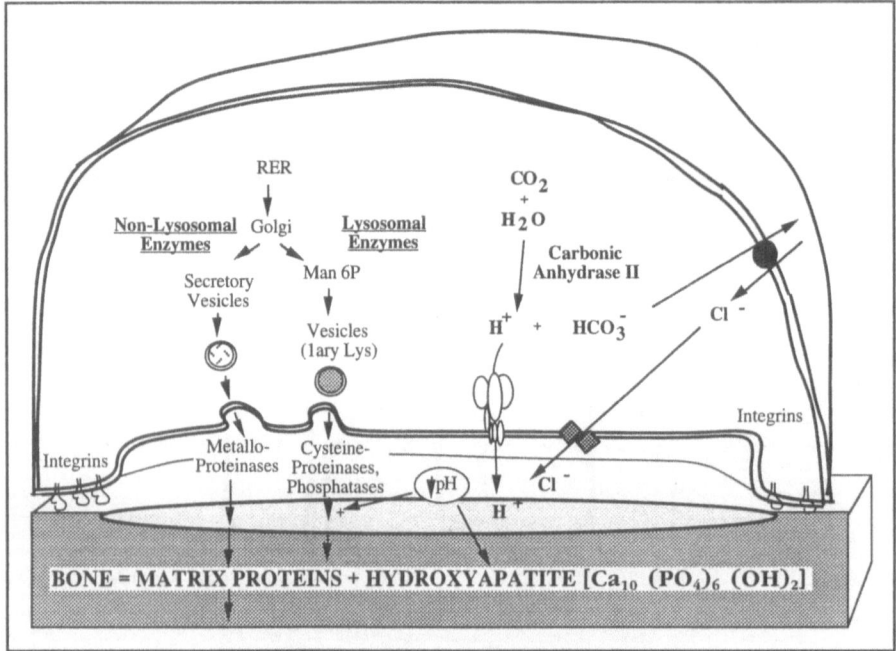

Fig. 3.3. Schematic representation of the relationship of enzyme secretion, attachment and acidification in the mechanisms of bone resorption. Sealing at the periphery of the resorbing compartment is assured by binding of integrin receptors to RGD-containing extracellular proteins; lysosomal enzymes, including cysteine-proteinases, and metalloproteases are secreted into this sealed-off compartment, possibly transported in two separate secretory vesicles; acidification is ensured by a vacuolar-like proton ATPase in the ruffled-border membrane; the protons are generated intracellularly by carbonic anhydrase and the by-product of the reaction, bicarbonate, is secreted basolaterally in exchange for chloride by a band 3-like anion exchanger; the chloride is then transported into the resorbing compartment by chloride channels, thereby avoiding hyperpolarization of the osteoclast's membrane that would result from proton transport. The low pH favors the action of lysosomal enzymes and dissolves the hydroxyapatite, and the presence of metalloproteases broadens the pH range at which resorption can occur.

their dissociation from the receptor. In this manner, enzymes are released in the proper compartment, which, in all cells, is the late endosome.

Despite the fact that the situation is quite different in the osteoclast, which secretes the newly synthesized enzymes instead of targeting them to an intracellular organelle, the cation-independent mannose-6-phosphate receptor co-localizes with lysosomal enzymes along the exocytic pathway.[18] Both the enzymes and the mannose 6-phosphate receptors are present in the transport vesicles and, being in a post-Golgi compartment and at neutral pH, the enzymes must be bound to their receptor during transport towards the ruffled-border. These transport vesicles then fuse with the ruffled border membrane, probably via specific interactions, and the enzymes dissociate from their receptors upon exposure to the low pH in the bone-resorbing compartment. If this process is similar in the osteoclast and in other cells, the receptors are most likely recycled to the Golgi for other rounds of transport after delivery.

This could represent the mechanism by which the osteoclast ensures that the lysosomal enzymes are not secreted towards the surrounding tissues at the basolateral pole of the cell : a transport vesicle that would reach the basolateral membrane would not release the enzymes in the microenvironment since the ligands do not dissociate from the receptors at neutral pH. It is also possible that mannose-6 phosphate receptors present at the basolateral domain of the cell play a role in recapturing enzymes that would leak from the resorbing compartment, particularly when the cell detaches and enters in a motile phase.[18,22]

Hence, besides its role in the extracellular dissolution of the mineral phase, acidification of the bone resorbing compartment plays a direct role in the targeted secretion of the enzymes which will degrade the organic phase of bone matrix.

THE SECRETED ENZYMES COVER A BROAD SPECTRUM OF pH OPTIMA

Although the best characterized osteoclast enzyme is the tartrate-resistant acid phosphatase type 5,[23,24] other acid phosphatases as well as aryl-sulfatase, β-glucuronidase and β-glycerophosphatase have been localized in the osteoclast's biosynthetic pathway.[7] The osteoclast also synthesizes and secretes several cysteine-proteinases, among which the cathepsins B, L and C. The importance of these enzymes is that they are capable of degrading collagen in an acidic environment, such as the one encountered in the bone resorbing compartment. As such, the presence of these cysteine-proteinases resolved the apparent contradiction between the low pH at the resorbing site and the inability of neutral collagenase to efficiently degrade collagen at such a pH in vitro. Furthermore, inhibition of these enzymes led to a decrease in bone resorption and to the accumulation of partially digested collagen in the resorbing lacunae.[25]

Although the cathepsins can digest collagen at acidic pH, collagenase still plays an important role in bone resorption.[26] First, and de-

spite the fact that the exact source of the collagenase is not yet firmly
established, collagenase has been detected in the osteoclast and along
the bone matrix underlying resorbing osteoclasts. As for cysteine proteinases, inhibition of matrix metalloproteinases leads to an accumulation of partially digested collagen in the bone resorbing compartment.[25]
Furthermore, tissue-plasminogen activator and stromelysin have also
been found in osteoclasts and they as well as lysosomal cysteine proteinases can activate latent collagenase.

Hence, the osteoclast synthesizes and secretes several classes of
enzymes: phosphatases, sulfatases, cysteine-proteinases and matrix
metalloproteinases. Although the pH optimum of the cysteine- and
metalloproteinases are very different (4.0 to 5.0 and 7.5 respectively),
collagenase still has an important activity around pH 6 and stromelysin
can still degrade proteoglycans at pH 5. The actions of the lysosomal
acid cysteine proteinases and of collagenase may therefore, as discussed
below, be exerted sequentially.[26]

During active osteoclastic bone resorption, the buffering capacity
exerted by the solubilized bone salts is likely to cause a gradient of
pH extending from the most acid zone, in the immediate vicinity of
the ruffled border and its proton pumps, to a more neutral zone, deeper
in the resorbing lacuna. The higher pH in these regions would both
favor the activation of procollagenase by the lysosomal cysteine-proteinases, as this process is more efficient around pH 6 than at lower
pH, and render the collagenolytic action of collagenase predominant,
since cysteine-proteinases are quite inefficient near neutral pH. As discussed before, the osteoclast moves along the bone surface. The de-
tachment of the cell at the end of a resorptive phase causes the sud-
den neutralization of the pH of the resorbing compartment, resulting
from its opening to outside extracellular fluids. This would immediately prevent further actions of the cysteine proteinases, leaving a fringe
of already demineralized but as yet undegraded collagen, as seen at
the bottom of resorption pits eroded by isolated osteoclasts. The role
of collagenase could then be envisioned as being the digestion of that
collagen fringe at neutral pH after displacement of the osteoclast, so
as to allow the completion of the resorbing process despite the absence of a sealed-off subosteoclastic acidic microenvironment. Thus, it
is the cooperative action of acidic- and neutral-pH optima enzymes
that leads to the complete degradation of the extracellular matrix at
the resorbing site.

The motile osteoclast thus attaches to the bone surface, seals off
an extracellular compartment and secretes into this compartment several enzymes necessary for bone resorption. As discussed below, acidification of the bone resorbing compartment is necessary for bone resorption to occur.

THE OSTEOCLAST ACTIVELY ACIDIFIES THE BONE
RESORBING COMPARTMENT: PROPERTIES
AND STRUCTURE OF THE OSTEOCLAST V-ATPASE

Acidification of the extracellular bone resorbing compartment has
emerged over the last few years as one of the most important features

of the biology of the osteoclast.[17,27-32] Osteoclast-mediated acidification is required for the dissolution of the mineral phase, for the delivery of enzymes to the bone surface and for the enzymatic degradation of the organic phase of the extracellular matrix. The organization of the osteoclast membrane domains is consequently predominantly dictated by the mechanisms of acid secretion. Briefly, proton pumps are present at the ruffled-border plasma membrane and the cell is enriched in carbonic anhydrase II, which generates protons from carbon dioxide and water. The apical membrane of the osteoclast, or ruffled-border, is directly involved in the molecular mechanisms of bone resorption. It is the target for the specific delivery of the newly synthetized secretory enzymes and it is the site of proton extrusion for acidification of the bone resorbing compartment. The extensive folding of this domain of the plasma membrane is probably due both to the intense vesicular traffic associated with secretion and to the need for increasing the number of proton-pumps through amplification of the membrane surface.

THE OSTEOCLAST PLASMA MEMBRANE PROTON PUMP IS A V-ATPASE

As discussed extensively in other chapters of this monograph, the H^+ ATPases responsible for acidification by cells specialized in acid secretion and in various organelles are distinguished from each other on the basis of their structure and/or their sensitivity to specific inhibitors. The osteoclast proton transport system was therefore characterized accordingly.

In our original report showing that osteoclasts actively acidified the bone surface,[17] based upon the cross-reactivity of an antibody to a lysosomal membrane protein with the H^+/K^+ ATPase, the possibility that the osteoclast proton pump could be of the P-type was suggested. Further studies with antibodies to the gastric proton pump failed to demonstrate its presence in osteoclast membranes and the pharmacological characteristics of proton transport in osteoclast-derived vesicles clearly established that the enzyme was not a P-type ATPase.[33] More recently, based upon pharmacological and immunochemical data, the H^+-ATPase present in osteoclast membranes has been shown to be of the V-type, closely resembling the pumps present in kidney membranes.[30,33-35] Thus, proton transport in inside-out vesicles prepared from osteoclast-enriched cell fractions is sensitive to NEM, bafylomicin A1 and other V-ATPase inhibitors but not to vanadate at low μM concentration, azide or oligomycin. These preparations have been shown to contain several of the subunits that constitute V-ATPases[33] and several of these subunits have been immunolocalized at the apical ruffled-border membrane of the osteoclast.[7,30,35] Furthermore, high magnification electron microscopy on negatively stained preparations of osteoclast-derived microsomes showed high densities of ball-and-stalk structures,[33] similar to those observed in kidney tubule apical membranes which are enriched in V-ATPases,[36] in 30-40% of the vesicles. Similar "studs" are routinely observed on the cytoplasmic side of the ruffled-border membrane in situ.[4]

As discussed throughout this monograph, there is however evidence in the literature which shows subtle differences in the properties and structure of mammalian V-ATPases.[37-39] The vacuolar proton pumps contained in coated vesicles, endosomes, chromaffin granules, and kidney tubule plasma membranes may somewhat differ in their subunit composition.[37,39] These observations have led to the hypothesis that variations in isoforms or, possibly, differences in post-transcriptional processing of the multisubunit vacuolar proton pump(s) may constitute the basis for the differential targeting, properties and regulation of the H⁺-ATPases present in different organelles in the same cell, in different cells or in different organs. Recently, this concept has been further established by the fact that one of the subunits forming the catalytic domain (subunit B) is encoded by two distinct genes, one apparently exclusively expressed in kidney and the other probably ubiquitous.[40,41] For these reasons, further characterization of the osteoclast proton pump was necessary.

Proton Transport in Membranes Derived From Osteoclasts Reveals an Increased Sensitivity to Vanadate

For this purpose, chicken osteoclasts (the only source of cells that is abundant enough to perform these studies) were highly purified and the microsomal fraction was further fractionated, achieving a 10-fold enrichment in plasma membrane markers relative to the initial microsomal preparation. Most of the proton transport system(s) present in this microsomal fraction were derived from osteoclasts. ATP was found to be the only substrate for transport in the vesicles, confirming the negligible level of contamination with endocytic vesicles. These microsomal preparations were then used to study the pharmacological properties of proton transport and for immunochemical characterization of the enzyme subunits.

The proton transport assay revealed a sharp and narrow pH dependence, with a pH optimum of 7.4. Analysis of H⁺ transport as a function of ATP concentration demonstrated the presence of a single Km for ATP with a Hill coefficient of 0.9, suggesting that only one type of H⁺-ATPase was present and consistent with the reported Km for other proton ATPases.[33] Pharmacologically, inhibitors of V-ATPases inhibited 100% of the acidification by osteoclast membrane vesicles. The k1/2 for NEM, DCCD and Bafilomycin A1 were 0.1 μM, 35 μM and 6 nM respectively. The mitochondrial proton ATPase inhibitors, oligomycin, azide and fluoride had no effect. This data confirmed that, as reported by others,[30,34,35] the osteoclast proton pump exhibits properties of the vacuolar type, a finding in agreement with the high density of ball and stalk structures observed in electron microscopy on the osteoclast-derived microsomal vesicles used in this assay.

However, and unlike any other reported V-ATPases, the osteoclast -ATPase could also be inhibited by vanadate, which blocked 100% of the acidification at a concentration of 1mM, with an k1/2 of 100 μM[33] and by nitrate at similar concentrations,[42] suggesting a unique pharmacological profile.

The same osteoclast membranes were then used for purification of NEM/vanadate sensitive H+-ATPase by immunoaffinity column[42] as well as by glycerol gradient as described by Moriyama and Nelson[43] for the purification of NEM-sensitive H+-ATPase from chromaffin granule membranes. By affinity purification, the osteoclast membrane H+-ATPase specific activity was enriched 85-90 fold compared to that in the total cell homogenate. The H+-ATPase activity was then assayed according to Xie X-S et al[44] and demonstrated that the increased sensitivity to vanadate was not only observed in the proton transport assay but also in the ATPase assay. Since these results suggested that the V-ATPase present in osteoclast membranes might differ from other V-ATPases, we then studied the structure of the enzyme, particularly the structure of its catalytic domain.

THE OSTEOCLAST V-ATPASE
MAY DIFFER IN ITS CATALYTIC DOMAIN

Two-dimensional polyacrylamide gel electrophoresis of the purified osteoclast H+-ATPase reveals six major polypeptides of molecular weights 100kD, 63kD, 60kD, 42kD, 31kD and 16kD.[45] To analyse the relationship of the OC-H+-ATPase to other V-ATPases, antibodies against several vacuolar proton pump subunits were used in immunoblots of osteoclast membrane proteins obtained at different stages of purification. Antibodies against the 115, 60 and 39 kD of the chromaffin granule ATPase and the 16kD DCCD-binding proteolipid of a plant ATPase detected proteins of the expected molecular weights which co-purified with the osteoclasts, further confirming the vacuolar-like nature of the OC- H+-ATPase. In contrast to the 60kD subunit mentioned above, another 60kD subunit was recognized by antibodies against the *N crassa* ATPase in unpurified fractions and decreased with osteoclast purity. These results suggested the existence of two distinct isoforms of subunit B in our microsomal preparations : A 60kD subunit associated with the OC- H+-ATPase and another 60kD subunit present in contaminating cells.

Similarly, and unlike other subunits, the 70 kD protein detected by antibodies against the 70 kD catalytic subunits from chromaffin granules, coated vesicles or *N crassa* was present in unpurified fractions but, surprisingly, its amount decreased with osteoclast purity, suggesting that this isoform of subunit A was mostly present in contaminating cells. Most interestingly, the antibodies against *N crassa* V-ATPase 70kD subunit also detected a second protein, Mr ~ 63kD, which was undetectable in unpurified fractions and increased during osteoclast purification. The immunological similarity of this 63kD polypeptide to the *N. crassa* 70kD subunit was confirmed by immunoblotting with antibodies raised against the *N crassa* protein expressed in *E. coli* as a recombinant fusion protein. The 63kD protein was not detectable in microsomes from kidney, bone marrow macrophages or peripheral blood monocytes. Hence, the osteoclast V-ATPase may differ in its catalytic domain.

Molecular Cloning of the Catalytic Subunits
of the Osteoclast Enzyme

Current efforts are therefore aimed at determining by molecular cloning whether isoforms of the catalytic A and B subunits exist and which ones are specifically found in the osteoclast enzyme.[46] Screening a chicken osteoclast cDNA library, we have isolated a clone encoding the B subunit. Analysis of the deduced amino acid sequence showed >90% identity with other B subunit sequences in the central region of the molecule but both the N- and C- terminal ends were markedly divergent from the *N. crassa*, bovine or human kidney isoform. In contrast, the osteoclast-derived B subunit is closely related to the recently reported human and bovine brain isoform of subunit B.[40,41] Regarding the A subunit, and despite the immunochemical differences that we previously reported,[33] both PCR and screening of chicken and human osteoclast libraries so far generated only multiple clones which were identical to the previously reported mammalian A subunit.[46] Using a heterogeneous small sample of a human osteoclastoma as a source of RNA to construct a cDNA library, another group[47] recently reported the isolation of a clone encoding a different A subunit. By Northern blots, RNA encoded by this sequence was found only in the osteoclastoma tissue, thereby suggesting that it may constitute an osteoclast-specific catalytic subunit. Although it is quite tempting to link the presence of this putative isoform to the differences in pharmacology we have reported, several inconsistencies remain to be elucidated: (1) The novel A subunit sequence has been found only in the original tumor from which the library was made: extensive screening, PCR with primers derived from the new sequence, Northern and Southern blots with specific oligonucleotides have failed, in our hands, to show this sequence in any other human osteoclastomas, even when using highly purified human osteoclast-like cells as our source of RNA;[46] (2) Southern blot analysis of bovine, human and chicken genomic DNA has shown so far the presence of only one gene encoding subunit A.[46,48] Hence, and although the pharmacological and immunochemical data suggests differences in the catalytic properties of the osteoclast V-ATPase, it has not yet been possible to determine the structural basis for such differences. The existence of another gene encoding subunit A, as suggested by van Hille et al[47] yet uncovered posttranslational modifications of the same gene product or the existence of an associated regulatory protein may indeed constitute the basis for an increased vanadate sensitivity of the osteoclast enzyme.

The osteoclast is therefore a polarized acid-secreting cell whose plasma membrane V-type ATPase may differ in its properties and, possibly, structure, from the V-ATPases found in other acidifying cells and/or organelles. The next section of this chapter is aimed at comparing the osteoclast and these other cells and organelles.

FUNCTIONAL SIMILARITIES BETWEEN THE OSTEOCLAST AND EPITHELIAL ACID-SECRETING CELLS: POLARIZED ION-TRANSPORT AND THE BASOLATERAL MEMBRANE

The functional organization of the osteoclast, being so strongly dominated by the polarized extrusion of large quantities of protons, is quite similar to that of other acid-secreting cells in terms of ion transport.

The apparently simple process of pumping protons across the ruffled border membrane to acidify the resorption lacuna imposes, as it does

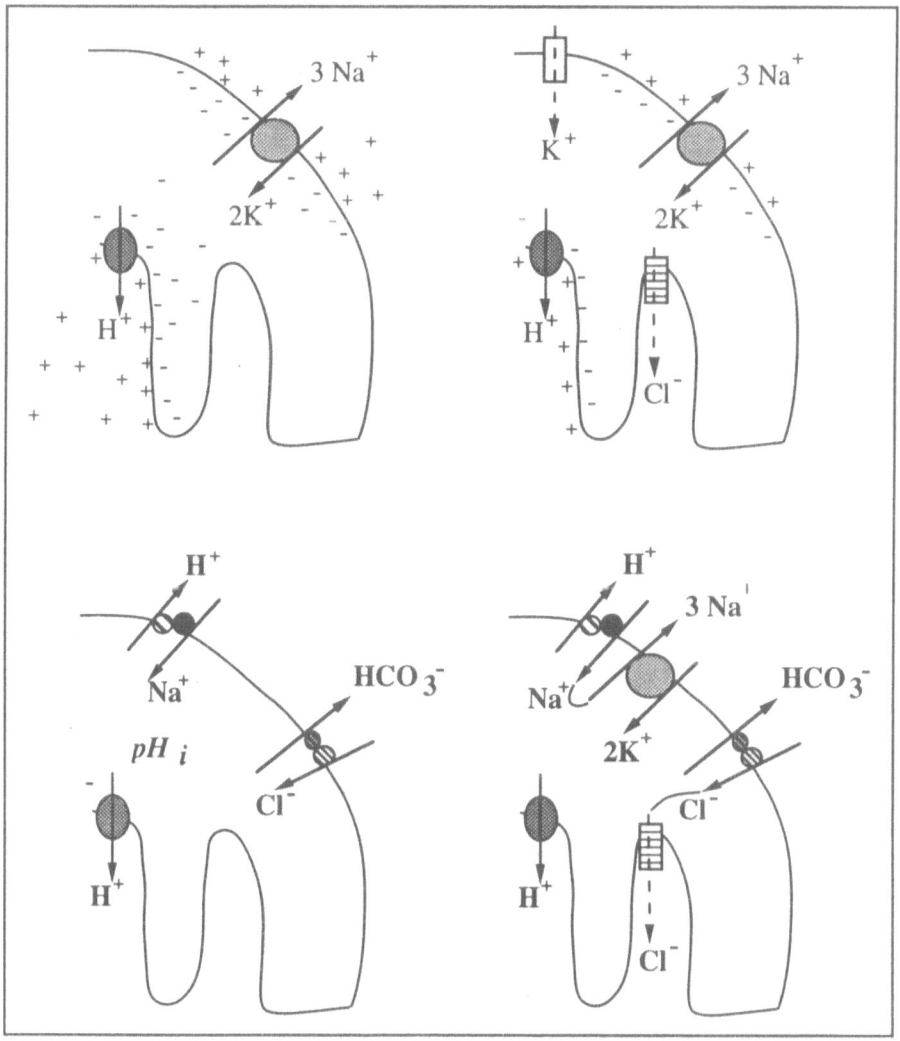

Fig. 3.4. Role of ion channels and acid-base exchangers in the regulation of membrane potential and intracellular pH in the osteoclast. Top: The activities of the apical V-ATPase and of the basolateral Na,K-ATPase lead to membrane hyperpolarization; this is prevented by basolateral inward-rectifying K channels and apical Cl channels. Bottom: The dissociation of carbonic acid into bicarbonate and protons would lead to increase HCO₃ and pH in the cell; this is prevented by a basolateral HCO₃/Cl exchanger; intracellular pH homeostasis and regulation of cell volume are also ensured by the actions of a Na/H exchanger and the Na,K-ATPase.

in cells of the stomach or the kidney tubule, strict ionic requirements in order to maintain the electrochemical balance of the osteoclast during bone resorption. Electrogenic ion pumps, which include the V-ATPase and the Na,K-ATPase ion channels and electroneutral ion exchangers are all present in the plasma membrane of the osteoclast in order to maintain the cytoplasmic pH, the transmembrane electrical potential and the volume of the cell within narrow physiological ranges.[7]

The two main components involved in acidification, the apical electrogenic proton pump itself and carbonic anhydrase II in the cytosol, form the main axis around which ion transport is organized in the osteoclast. Briefly, the protons that the H^+-ATPase transports across the apical membrane are generated in the cytoplasm by the carbonic anhydrase II-dependent reversible hydration of CO_2 to produce carbonic acid, which ionizes to form H^+ and HCO_3^- (Fig. 3.5). The generation of bicarbonate and the transfer of the protons out of the cell by theV-ATPase alter both the cytoplasmic pH, which becomes more alkaline, and the membrane potential, which becomes hyperpolarized as progressively more positive charges are transported out. Cytoplasmic pH and membrane potential therefore have to be tightly controlled in an acid-secreting cell such as the osteoclast. Cytoplasmic pH is maintained near neutrality by the action of electroneutral ion exchangers present in the basolateral membrane (Fig. 3.5), probably at a higher number of copies in the osteoclast than in other cell types due to its very active proton transport. We have recently identified and characterized both an acid extruder (the Na^+/H^+ antiporter) and an acid-loader (i.e. base-extruder, the HCO_3^-/Cl^- exchanger) in the osteoclast.[49] When the proton pump is active, the increasing alkalinity in the cytosol activates the HCO_3^-/Cl^- exchanger, leading to the extrusion of the excess HCO_3^- in a one-to-one exchange for extracellular Cl^-, preventing the intracellular pH from rising excessively. If, on the other hand, intracellular pH should fall, for example when the proton pumps are acutely internalized upon binding of the peptide hormone calcitonin,[50] the Na^+/H^+ antiporter is activated and extrudes the excess protons across the basolateral membrane in exchange for extracellular Na^+.

It is, in fact, the gradient of Na^+ across the membrane, established by the activity of the sodium pumps found in high numbers in the basolateral membrane of the osteoclast,[28] which drives the antiporter. The activity of the Na^+/K^+ ATPase will allow the sodium which enters the cell to be transported out by exchanging $3Na^+$ for $2K^+$, thereby resulting, like the proton pump, in hyperpolarizing the membrane. Furthermore, the regulation of intracellular Na^+ by the balance between the Na^+/H^+ antiporter and the Na,K-ATPase is probably necessary in order to prevent massive losses of water from the osteoclast cytosol at the apical membrane: the ionic strength of the bone resorbing extracellular compartment is probably very high due to the dissolution of hydroxyapatite crystals in the low pH environment, which would lead to passive loss of water across the ruffled-border membrane. It is most likely then that the two Na transporters are involved in maintaining the osteoclast cellular volume despite the demands imposed by the process of bone resorption.

The hyperpolarization resulting from the activities of the H⁺ pump and the Na⁺/K⁺ ATPase, to which the activity of the Ca²⁺ ATPase also contributes,[51] would rapidly reach a level making it nearly impossible to transport any more positively charged ions out of the cell.[7] This would eventually hamper proton generation and electrogenic proton transport, if they were not alleviated by the activity of voltage-sensitive ion channels, which transport positively charged ions into the cell and/or negatively charged ions out of the cell and thus discharge the electrical potential.

Two such ionic conductances have indeed been described in freshly isolated osteoclasts. First, whole-cell patch-clamp studies have shown that osteoclasts possess inwardly rectifying K⁺ channels.[52-55] These channels are activated (opened) by hyperpolarization of the membrane. In addition to inwardly rectifying K⁺ channels, osteoclasts express two other types of K⁺ channels which are both activated by depolarizing membranes.[56] Second, chloride conductances have been found in osteoclast membranes. Using on-cell and cell-free patch-clamp techniques high-conductance anion channels and a voltage sensitive chloride channel with a very high conductance have been identified.[55,57,57] Furthermore, chloride conductances have also been identified in the osteoclast-derived inside out vesicles used for studies of the V-ATPase,[33,58] where the presence of Cl⁻ in the acidification buffers is necessary for H⁺ transport.[30,33-35,59] It is not known whether the implied chloride conductance in vesicles corresponds to the conductances that were electrophysiologically characterized.

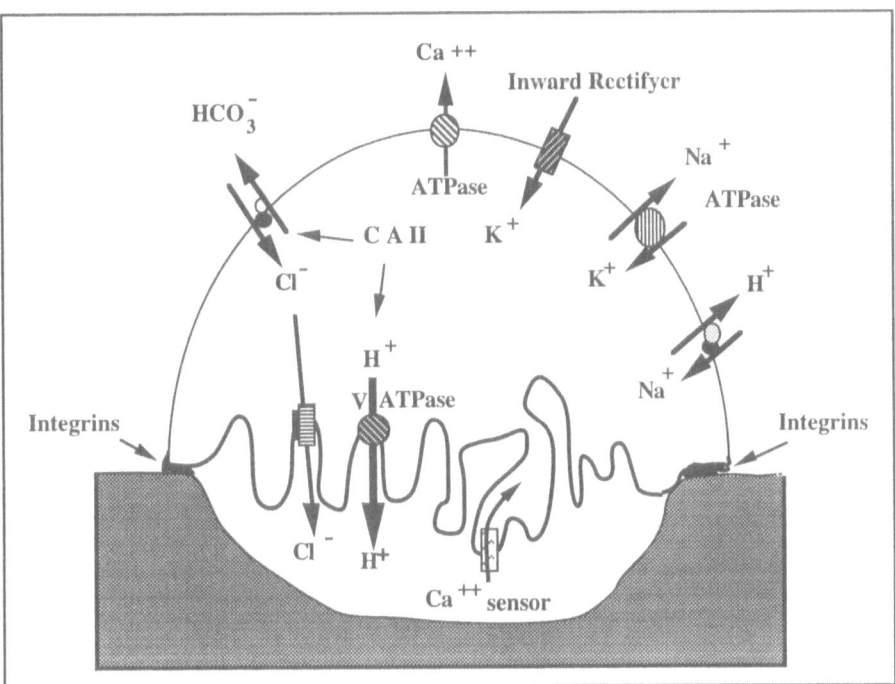

Fig. 3.5. Summary of the major ion transport systems that have been identified in the osteoclast. Single circles represent ion-motive ATPases, double circles represent ion exchangers and rectangles represent ion channels (see text for detailed explanations).

Whether chloride and potassium channels are present in the apical and/or basolateral domains of the plasma membrane of the osteoclast is also not yet known. It is however most logical to hypothesize that at least some of the chloride conductance will be directly associated with proton transport at the apical surface, preventing hyperpolarization and leading to the formation of HCl in the resorbing compartment. The inwardly rectifying K^+ channels, on the other hand, would be functionally most appropriate in the basolateral domain of the cell.

Thus, H^+ transport at the apical surface of the osteoclast is tightly linked to the regulation of intracellular pH and membrane potential, mostly accomplished by exchangers, pumps and channels present in the basolateral membrane of the cell. These functional features are indeed very similar to those observed in the acid-secreting cells of the epithelia lining the kidney tubule and the stomach. As discussed below, this analogy is however only functional and the osteoclast differs markedly from these epithelial cells in several important aspects.[60]

THE OSTEOCLAST IS NOT AN EPITHELIAL CELL

The first, and probably most important, difference between the osteoclast and all other cells specialized in the acidification of an extracellular milieu is the fact that the osteoclast is not part of an epithelium or, for that matter, not of epithelial origin. It is now widely accepted that osteoclasts are of hematopoietic origin and probably of the CFU-M-derived monocyte-macrophage family.[61] Thus, transplants of spleen or bone marrow hematopoietic cells can restore bone resorption in irradiated osteopetrotic mutants; direct cellular filiation has been demonstrated with quail-chick chimerae and unselected bone marrow cells, colonies of the CFU-GM or CFU-M lineage or, even, mature macrophages can lead to the formation of osteoclasts under adequate culture conditions. Cells of the monocyte-macrophage lineage can therefore probably form osteoclasts at most stages of their differentiation. On the one hand, the hematopoietic origin of the osteoclast constitutes a major difference with cells of the kidney tubule. On the other, the close relationship of this cell type with macrophages raises the possibility that the plasma membrane V-ATPase discussed by Lukacs et al. in this monograph is similar to that of the osteoclast, only expressed at much lower levels.

THE OSTEOCLAST DOES NOT HAVE TIGHT JUNCTIONS AND THE RUFFLED-BORDER IS NOT BIOCHEMICALLY EQUIVALENT TO AN EPITHELIAL APICAL DOMAIN

A very important consequence of the fact that the osteoclast does not belong to an epithelium, but rather functions as an individual cell attached to a substratum, is that it cannot form tight junctions with neighboring cells. In epithelial cells, it is the junctional apparatus which determines the polarity of the cell and allows the differential targeting and segregation of membrane proteins to their respective functional domains. However, as the junctional complex forms a band around epithelial cells (zonula adherens), the sealing-zone forms a band around the osteoclast-bone matrix interface (sealing zone) and, thereby, fulfills the same functional role of isolating an "apical" compartment from a "basolateral" compartment. The parallel between the osteoclast and the epithelial acidifying cells of the kidney and the stomach is therefore valid, as it applies to the functional aspects of the cell's organization, but not as it applies to the molecular organization of the cell. Indeed, as the sealing-zone comprises molecules of a different nature than the ones present in the junctional complex of epithelial cells, the composition of the apical membrane of the osteoclast differs from that of epithelial cells. In our hands for instance, infection of osteoclasts with influenza viruses failed to lead to the targeting and expression of this apical marker in the ruffled-border membrane.[60] In contrast, as discussed below, this membrane expresses several markers of membranes limiting organelles of the endocytic pathway, which are not found in epithelial apical domains.

THE OSTEOCLAST RUFFLED-BORDER MEMBRANE EXPRESSES SEVERAL ENDOSOMAL MARKERS

Indeed, there are several reasons why the apical ruffled-border membrane of the osteoclast could be expected to resemble the limiting membrane of vesicular elements of the endocytic pathway in other cells rather than an epithelial apical domain. First, this membrane limits a compartment that is the extracellular functional equivalent of a lysosome, with high concentrations of enzymes and a low pH,[62] and which is very different from the luminal environment of the kidney tubule. Second, the ruffled-border membrane is the target for delivery of newly synthesized and mannose-6 phosphate receptor-bound lysosomal enzymes by post-Golgi transport vesicles, as are late endosomes in other cells.[63] Finally, lysosomal and endosomal membranes are also responsible for establishing and maintaining the pH gradient necessary for the degradation of internalized material via vacuolar ATPases. Hence, it is not surprising that immunocytochemical localization of lysosomal and/or endosomal integral membrane proteins has revealed that the ruffled-border expresses several lysosomal-endosomal membrane proteins,[17,18,64] which are restricted to the apical domain of the osteoclast's plasma membrane.

Thus, the osteoclast ruffled-border membrane is probably quite different from the kidney tubule cells in its molecular organization but could be very similar and even identical to endosomal membranes. This observation therefore raises two very important, and not mutually exclusive, possibilities. The first, is that the V-ATPase of the osteoclast plasma membrane might be an overexpressed form of the endosomal proton pump. The second is that it might be an overexpressed form of the macrophage plasma membrane ATPase (discussed in detail in chapter 2), especially since these two cell types are closely related. Indeed these two possibilities are not exclusive and the ATPase found in the plasma membrane of the macrophage and the osteoclast could very well be the same as the one found in endosome and lysosome membranes. Not enough detailed information on the respective structures and properties of the V-ATPases found in these various membranes is yet available to get a clear answer to this question.

Hence, the parallel between osteoclasts and kidney acidifying cells is probably only functional and the two cell types differ extensively in their molecular phenotype. Obviously, the V-ATPase or some of its subunits (as already shown for subunit B) might constitute one of these molecular differences.

CONCLUSION AND PERSPECTIVES: THE OSTEOCLAST V-ATPᴀsᴇ AS A TARGET FOR THERAPEUTIC INTERVENTION

Due to the fact that, on the one hand, increased bone resorption is involved in several systemic and local diseases of the skeleton, including the joints, and that, on the other hand, acidification is critical for the osteoclast to perform its function, inhibition of the osteoclast proton pump has become a major target for therapeutic intervention.

It has now been well established that inhibiting acidification by the osteoclast V-ATPase blocks bone resorption, at least in vitro. This has been accomplished by inhibiting carbonic anhydrase with acetazolamide,[65] inhibiting the V-ATPase directly by bafilomycin A1[66] or by inhibiting proton transport indirectly by blocking the activity of the basolateral transporters.[67-69] It is however of major concern that these various inhibitors are not acting specifically on the osteoclast and would be expected to affect other functions in vivo, such as the kidney tubule and the endocytic and secretory pathways of most other cell types. It therefore becomes of major importance to determine whether differences exist between these V-ATPases that could serve as a basis for the design of specific means of inhibiting the osteoclast proton pump.

In this context, the potential structural and pharmacological differences that we have observed may indeed permit such specificity to be attained. It is intriguing that several studies have reported inhibition of bone resorption by omeprazole, a drug which supposedly inhibits only the gastric P-type ATPase, although the observed effects were modest.[66,70,71] If indeed the osteoclast V-ATPase is more sensitive to vanadate than other V-ATPases in mammalian cells, this would suggest a difference in the catalytic cycle of the enzyme and, consequently, open the possibility of more specific inhibition of the osteoclast function.

The second possibility, already suggested for another class of compounds which inhibit the osteoclast, is that specificity may come from the very characteristic properties of the bone resorbing compartment. The low pH, combined with the high ionic strength and the fact that bone resorption occurs at the surface of bone, may indeed allow the use of bone-seeking compounds such as bisphosphonates and tetracyclines to target drugs aimed at inhibiting bone resorption to the osteoclast. Thus, these compounds, by their ability to accumulate at the bone surface, would target the drug to an area where only the osteoclast would be capable of mobilizing it, via the very process of acidification and dissolution of the bone matrix mineral. The attachment apparatus would, by sealing-off the bone resorbing compartment, ensure that the concentration of the inhibitor seen by the osteoclast is several fold higher than what other cells would be exposed to.

The acidification process could therefore be at the same time a potential target for and a means of targeting specific inhibitors of bone resorption, thereby allowing prevention and/or treatment of widespread diseases such as osteoporosis, Paget's disease, osteoarthritis and various manifestations of cancer in the skeleton.

In conclusion, the maintenance of a normal skeletal shape and mass, as well as bone growth and repair, are closely dependent upon a normal osteoclastic activity, which depends entirely on the acidification process, allowing bone resorption. This puts the osteoclast vacuolar ATPase at the center stage of a critical biological event and may lead to the development of novel approaches for the treatment of skeletal and joint diseases.

Acknowledgements: The work discussed in this chapter has been mostly supported by grants from the National Institutes of Health to R. Baron (DE-04724 and AR-41339).

REFERENCES

1. Schenk RK, Spiro D and Wiener J. Cartilage resorption in tibial epiphyseal plate of growing rats. J Cell Biol 1967;34:275-291.
2. King GJ and Holtrop ME. Actin-like filaments in bone cells of cultured mouse calvaria as demonstrated by binding to heavy meromyosin. J Cell Biol 1975;66:445-451.
3. Marchisio PC, Cirillo D, Naldini L, Primavera MV, Teti A and Zambonin-Zallone A. Cell-substratum interaction of cultured avian osteoclasts is mediated by specific adhesion structures. J Cell Biol 1984;99:1696-1705.
4. Kallio DM, Garant PR and Minkin C. Evidence of coated membranes in the ruffled border of the osteoclast. J Ultrastruct Res 1971;37:169-177.
5. Kanehisa J, Yamanaka T, Doi S, et al. A band of F-actin containing podosomes is involved in bone resorption by osteoclasts. Bone 1990; 11:287-293.
6. Teti A, Marchisio PC and Zambonin Zallone A. Clear zone in osteoclast function: role of podosomes in regulation of bone-resorbing activity. Am J Physiol 1991;261:C1-C7.
7. Baron R, Ravesloot J-H, Neff L, et al. Cell and molecular biology of the osteoclast. In: Noda, M. Cellular and Molecular Biology of the Bone, Orlando, Florida: Academic Press, Inc., 1993, p. 445-495.
8. Marchisio PC, Cirillo D, Teti A, Zambonin-Zallone A and Tarone G. Rous sarcoma virus-transformed fibroblasts and cells of monocytic origin display a peculiar dot-like organization of cytoskeletal proteins involved in microfilament-membrane interactions. Exp Cell Res 1987;169:202-214.
9. Hynes RO. Integrins: A Family of Cell Surface Receptors. Cell 1987;48:549-554.
10. Hynes RO. Integrins: Versatility, modulation, and signaling in cell adhesion. Cell 1992;69:11-25.
11. Ruoslahti E, Noble N, Kagami S and Border W. Integrins. Kidney Intl 1994;45:s17-s22.
12. Davies J, Warwick J, Totty N, Philp R, Helfrich M and Horton M. The osteoclast functional antigen, implicated in the regulation of bone resorption, is biochemically related to the vitronectin receptor. J Cell Biol 1989;109:1817-1826.

13. Malgaroli A, Meldolesi J, Zambonin Zallone A and Teti A. Control of Cytosolic Free Calcium in Rat and Chicken Osteoclasts:the role of extra-cellular calcium and calcitonin. J Biol Chem 1989;264:14342-14347.

14. Miyauchi A, Hruska KA, Greenfield EM, et al. Osteoclast cytosolic cal-cium, regulated by voltage-gated calcium channels and extracellular cal-cium, controls podosome assembly and bone resorption. J Cell Biol 1990;111:2543-2552.

15. Teti A, Blair HC, Schlesinger P, et al. Extracellular protons acidify osteo-clasts, reduce cytosolic calcium, and promote expression of cell-matrix attachment structures. J Clin Invest 1989;84:773-780.

16. Miyauchi A, Alvarez J, Greenfield E, et al. Matrix protein binding to the osteoclast adhesion integrin (a_vb_3) mediates a reduction in $[Ca^{2+}]_i$. J Bone Miner Res 1991;6:S96.(Abstract)

17. Baron R, Neff L, Louvard D and Courtoy PJ. Cell-mediated extracellular acidification and bone resorption: Evidence for a low pH in resorbing lacunae and localization of a 100 kD lysosomal membrane protein at the osteoclast ruffled border. J Cell Biol 1985;101:2210-2222.

18. Baron R, Neff L, Brown W, Courtoy PJ, Louvard D and Farquhar MG. Polarized secretion of lysosomal enzymes: Co-distribution of cation-inde-pendent mannose-6-phosphate receptors and lysosomal enzymes along the osteoclast exocytic pathway. J Cell Biol 1988;106:1863-1872.

19. Andersson G, Ek-Rylander B, Hammarstrom LE, Lindskog S and Toverud SU. Immunocytochemical localization of a tartrate-resistant and vanadate-sensitive acid nucleotide tri- and diphosphatase. J Histochem Cytochem 1986;34:293-298.

20. Reinholt FP, Mengarelli Wildhom S, Ek-Rylander B and Andersson G. Ultrastructural localization of a tartrate-resistant acid ATPase in bone. J Bone Miner Res 1990;5:1055-1061.

21. Kornfeld S. Trafficking of lysosomal enzymes in normal and disease states. J Clin Invest 1986;77:1-6.

22. Blair HC, Teitelbaum SL, Schimke PA, Konsek JD, Koziol CM and Schlesinger PH. Receptor-mediated uptake of a mannose-6-phosphate bear-ing glycoprotein by isolated chicken osteoclasts. J Cell Physiol 1988; 137:476-482.

23. Hammarstrom LE, Hanker JS and Toverud SU. Cellular differences in acid phosphatase isoenzymes in bone and teeth. Clin Orthop 1971; 78:151-167.

24. Ek-Rylander B, Bill P, Norgard M, Nilsson S and Andersson G. Cloning, sequence, and developmental expression of a type 5, tartrate-resistant, acid phosphatase of rat bone. J Biol Chem 1991;266:24684-24689.

25. Everts V, Delaissé JM, Korper W, Niehof A, Vaes G and Beertsen W. Degradation of collagen in the bone-resorbing compartment underlying the osteoclast involves both cysteine-proteinases and matrix metallo-proteinases. J Cell Physiol 1992;150:221-231.

26. Delaisse J, Eeckhout Y, Neff L, et al. (Pro) collagenase (matrix metalloproteinase-1) is present in rodent osteoclasts and in the underly-ing bone-resorbing compartment. J Cell Sci 1993;106:1071-1082.

27. Vaes G. On the mechanisms of bone resorption: the action of parathy-roid hormone on the excretion and synthesis of lysosomal enzymes and on the extracellular release of acid by bone cells. J Cell Biol 1968;

39:676-697.

28. Baron R, Neff L, Roy C, Boisvert A and Caplan M. Evidence for a high and specific concentration of (Na$^+$,K$^+$)ATPase in the plasma membrane of the osteoclast. Cell 1986;46:311-320.

29. Vaes G. Cellular Biology and Biochemical Mechanism of Bone Resorption. Clin Orthop 1988;231:239-271.

30. Blair HC, Teitelbaum SL, Ghiselli R and Gluck S. Osteoclastic Bone Resorption by a Polarized Vacuolar Proton Pump. Science 1989; 245:855-857.

31. Baron R. Molecular mechanisms of bone resorption by the osteoclast. Anat Rec 1989;224:317-324.

32. Arnett TR and Dempster DW. The effect of pH on bone resorption by rat osteoclasts in vitro. Endocrinology 1986;119:119-124.

33. Chatterjee D, Chakraborty M, Leit M, et al. Sensitivity to vanadate and isoforms of subunits A and B distinguish the osteoclast proton pump from other vacuolar H$^+$-ATPases. Proc Natl Acad Sci USA 1992;89:6257-6261.

34. Bekker PJ and Gay CV. Biochemical characterization of an electrogenic vacuolar proton pump in purified chicken osteoclast plasma membrane vesicles. J Bone Miner Res 1990;5:569-579.

35. Vaananen HK, Karhukorpi EK, Sundquist K, et al. Evidence for the presence of a proton pump of the vacuolar H$^+$-ATPase type in the ruffled border of osteoclasts. J Cell Biol 1990;111:1305-1311.

36. Brown D, Gluck S and Hartwig J. Structure of the Novel Membrane-coating Material in Proton-secreting Epithelial Cells and Identification as an H$^+$-ATPase. J Cell Biol 1987; 105:1637-1648.

37. Forgac M. Structure and function of vacuolar class of ATP-driven proton pumps. Physiol Rev 1989;69:765-796.

38. Nelson N. Structure and pharmacology of the proton-ATPases. Trends in Pharm Sci 1991;12:71-75.

39. Wang Z-Q and Gluck S. Isolation and properties of bovine kidney brush border vacuolar H$^+$-ATPase. A proton pump with enzymatic and structural differences from kidney microsomal H$^+$-ATPase. J Biol Chem 1990;265:21957-21965.

40. Puopolo K, Kumamoto C, Adachi I, Magner R and Forgac M. Differential Expression of the "B" Subunit of the Vacuolar H$^+$-ATPase in Bovine Tissues. Journal of Biological Chemistry 1992;267:3696-3706.

41. Nelson RD, Guo X-L, Masood K, Brown D, Kalkbrenner M and Gluck S. Selectively amplified expression of an isoform of the vacuolar H$^+$-ATPase 56-kilodalton subunit in renal intercalated cells. Proc Natl Acad Sci USA 1992;89:3541-3545.

42. Chatterjee D, Neff L, Chakraborty M, Fabricant C and Baron R. Sensitivity to nitrate and other oxyanions further distinguishes the vanadate-sensitive osteoclast proton pump from other vacuolar H$^+$-ATPases. Biochemistry 1993;32:2808-2812.

43. Moriyama Y and Nelson N. The purified ATPase from chromaffin granule membranes is an anion-dependent proton pump. J Biol Chem 1987;262:9175-9180.

44. Xie XS and Stone DK. Isolation and reconstitution of the clathrin-coated vesicle proton translocating complex. J Biol Chem 1986;261:2492-2495.

45. Chatterjee, D., Chakraborty, M. and Baron, R. , 1994. (UnPub)

46. Hernando-Sobrino N, Bartkiewicz M, Fabricant C, et al. The osteoclast H⁺-ATPase: isolation of the genes encoding the catalytic subunits and studies of their expression in chicken tissues. J Bone Miner Res 1993;8 (Suppl. 1):S128. (Abstract)

47. van Hille B, Richener H, Evans DB, Green JR and Bilbe G. Identification of Two Subunit A Isoforms of the Vacuolar H⁺-ATPase in Human Osteoclastoma. J Biol Chem 1993;268:7075-7080.

48. Puopolo K, Kumamoto C, Adachi I and Forgac M. A Single Gene Encodes the Catalytic "A" Subunit of the Bovine Vacuolar H⁺-ATPase. J Biol Chem 1991;266:24564-24572.

49. Ravesloot J-H, Eisen T, Baron R and Boron WF. Role of Na-H exchangers and vacuolar H⁺ pumps in intracellular pH regulation in neonatal rat osteoclasts. J Gen Physiol 1994;(In Press)

50. Baron R, Neff L, Brown W, Louvard D and Courtoy PJ. Selective internalization of the apical plasma membrane and rapid redistribution of lysosomal enzymes and mannose 6-phosphate receptors during osteoclast inactivation by calcitonin. J Cell Sci 1990;97:439-447.

51. Bekker PJ and Gay CV. Characterization of a calcium ATPase in osteoclast plasma membrane. J Bone Miner Res 1990;5:557-567.

52. Ravesloot JH, Ypey DL, Vrijheid-Lammers T and Nijweide PJ. Voltage-activated K⁺ conductances in freshly isolated embryonic chicken osteoclasts. Proc Natl Acad Sci USA 1989;86:6821-6825.

53. Ravesloot JH, Ypey DL, Nijweide PJ, Buisman HP and Vrijheid-Lammers T. Three voltage-activated K⁺ conductances and an ATP-activated conductance in freshly isolated embryonic chick osteoclasts. Pfluegers Arch 1989;414:S166-S167.

54. Sims SM and Dixon SJ. Inwardly rectifying K⁺ current in osteoclasts. Am J Physiol 1989;256:C1277-C1282.

55. Sims SM, Kelly MEM and Dixon SJ. K⁺ and Cl⁻ currents in freshly isolated rat osteoclasts. Pflugers Arch 1991;419:358-370.

56. Sims SM, Kelly MEM, Arkett SA and Dixon SJ. Electrophysiology of osteoclasts. In: edited by Rifkin, B.R. and Gay, C.V. Biology and Physiology of the Osteoclast, Boca Raton, Fla.: CRC Press, 1991.

57. Schoppa NE, Su Y, Baron R and Boulpaep EL. Identification of single ion channels in neonatal rat osteoclasts. J Bone Miner Res 1990;5:S204.

58. Blair HC, Teitelbaum SL, Tan HL, Koziol CM and Schlesinger PH. Passive chloride permeability charge coupled to H⁺-ATPase of avian osteoclast ruffled membrane. Am J Physiol 1991;260:C1315-C1324.

59. Gillespie J, Ozanne S, Percy J, Warren M, Haywood J and Apps D. The vacuolar H⁺-translocating ATPase of renal tubules contains a 115-kDa glycosylated subunit. FEBS Lett 1991;282:69-72.

60. Baron R, Neff L and Vukasin A. Basolateral targeting of influenza virus and endolysosomal nature of the apical ruffled-border membrane distinguish osteoclasts from polarized epithelial cells. J Bone Min Res 1993;8:S377.

61. Suda T, Takahashi N and Martin TJ. Modulation of osteoclast differentiation. Endocrine Reviews 1992;13:66-80.

62. Mellman I, Fuchs R and Helenius A. Acidification of the endocytic and exocytic pathway. Annu Rev Biochem 1986;55:663-700.

63. Brown WJ, Goodhouse J and Farquhar MG. Mannose-6-phosphate re-

ceptors for lysosomal enzymes cycle between the Golgi complex and endosomes. J Cell Biol 1986;103:1235-1247.

64. Baron R, Neff L, Lippincott-Schwartz J, et al. Distribution of lysosomal membrane proteins in the osteoclast and their relationship to acidic compartments. J Cell Biol 1985;101:53a.

65. Waite LC, Volkert WA and Kenny AD. Inhibition of bone resorption by acetazolamide in the rat. Endocrinology 1970;87:1129-1139.

66. Mattsson JP, Vaananen HK, Wallmark B and Lorentzon P. Omeprazole and bafilomycin, two proton pump inhibitors: differentiation of their effects on gastric, kidney and bone H⁺-translocating ATPases. Biochim Biophys Acta 1991;1065:261-268.

67. Prallet B, Beresford J, Neff L and Baron R. Ouabain inhibits bone resorption in organ culture and in isolated rat osteoclasts. Calcif Tissue Int 1988;Abs. Davos Meeting:(Abstract)

68. Hall TJ and Chambers TJ. Na⁺/H⁺ antiporter is the primary proton transport system used by osteoclasts during bone resorption. J Cell Physiol 1990;142:420-424.

69. Hall TJ and Chambers TJ. Optimal bone resorption by isolated rat osteoclasts requires chloride/bicarbonate exchange. Calcif Tissue Int 1989;45:378-380.

70. Tuukkanen J and Vaananen HK. Omeprazole, a specific inhibitor of H⁺-K⁺-ATPase, inhibits bone resorption in vitro. Calcif Tissue Int 1986;38:123-125.

71. Sarges R, Gallagher A, Chambers T and Yeh L. Inhibition of bone resorption by H⁺/K⁺-ATPase inhibitors. Journal of Med Chem 1993;36:2828-2830.

CHAPTER 4

V-ATPASES IN INSECTS

Julian A.T. Dow

HISTORY

WHY INSECTS?

Insects are one of the most successful classes of organism in the world. Their small size, impermeable exocuticle and short generation times have allowed them to adapt to exploit a huge range of ecological niches, many of which place them in conflict with humans. It has been estimated from our rate of discovery of new species that as many as 30 M species of insect exist, comfortably more than all other living species put together. So a study of how insects exploit V-ATPases is a study of how most organisms exploit them.

EVOLUTIONARY DIVERGENCE

The vertebrate and insect lineages are thought to have separated around 400 M years ago. Accordingly, it might be expected that the ways in which insects do things would by now not resemble the analogous processes in humans very closely. This turns out not to be the case, and an insect/human comparison is often particularly informative, because protein domains which are conserved over so great an evolutionary distance must be vital to the functioning of the whole molecule. Although there has been ample opportunity for molecular divergence, the remarkably high conservation between insects, plants, yeasts and higher animals reinforces, for example, the impression that the constraints on how a V-ATPase can be assembled are very tight indeed.

AGROCHEMICAL & BIOMEDICAL SIGNIFICANCE

Insects impinge on humans in two spheres. The first is agricultural, where monocultures (i.e. fields planted with a single crop species) can be devastated when discovered by an appropriate insect pest species. It has been estimated that as much as a third of the world's food is lost either in the fields, or in storage, to insect pests.

Organellar Proton-ATPases, edited by Nathan Nelson; ©1994 R.G. Landes Company.

The second area is biomedical. Insects are vectors for many of the world's most serious diseases of humans and animals. There is thus not just a pure scientific interest in how insects use V-ATPases, but an applied interest that any property unique to insect V-ATPases might provide a target for a novel control strategy.

Having set the stage for studies on insects, we will trace the discovery of the insect V-ATPase, and its characterization as a major transport process in insect epithelia.

DISCOVERY OF THE "K⁺-PUMP"

In 1963, Harvey, Zerahn and Nedergaard, working in the world's leading frogskin laboratory, performed a serendipitous experiment on a nearby insect, which led to the discovery, 25 years later, of the insect epithelial V-ATPase. The classical frogskin technique is to dissect out the loosely attached abdominal skin, and mount it in an aperture between two perfusion chambers of an "Ussing chamber". The Na⁺, K⁺ ATPase-driven uptake of Na⁺ ions from the "pond" side to the inside can be monitored either as a potential difference, as a short-circuit current, or as a net flux of radio labeled sodium ions. In principle, such techniques can be applied to any epithelium, provided only that it is large enough to handle. Most insect epithelia obviously do not fall into this class; however, the unfortunate insect that day in 1963 was a larva of the large silk-moth, *Hyalophora cecropia*, quite as heavy

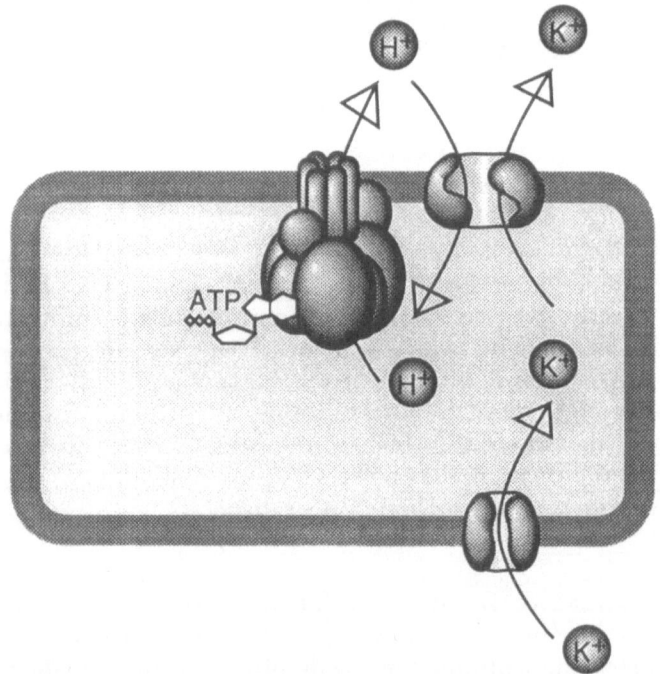

Fig. 4.1. Generalized model for insect epithelia. An apical plasma-membrane V-ATPase pumps protons out of the cell. These are exchanged for alkali metal cations (Na⁺ or K⁺) to produce a net ATP-dependent flux. Entry through the basal plasma membrane is not defined in the basic model, but is thought to be via channels, cotransports or ATPases in various insect tissues.

as a small rodent. It was the work of only a few minutes to remove its gut, mount it on the chamber, and measure the electrical signature of what we now know to be the epithelial V-ATPase. These signatures are pretty impressive; a larval Lepidopteran midgut can deliver 150 mV of potential difference, or 1 mA of current for each square centimeter of area. In the unlikely event that they would cooperate in the experiment, an array of ten such caterpillars with appropriately-placed silver wires could deliver ample power to drive a wristwatch, a calculator or even a small radio.

Initially, the caterpillar midgut was characterized as a "K^+ pump" or "common cation pump", in that precisely 100% of the short-circuit current could be reconciled with the net flux of radio labeled K^+ ions.[1,2] To an electrophysiologist, the significance of this parity is obvious; it implies that the only active transport with a net electrical signature must be for potassium. That, as we now know, is not the case. The precise mechanism by which a vacuolar H^+ pump is able to completely disguise its presence at the epithelial level will be discussed more fully below; in passing, it should be noted that this precise correspondence between short-circuit current and K^+ flux was misleading. Although other active transport processes were subsequently discovered in the midgut, it made the necessary mental jump to implicate a primary proton pump in active transport of K^+ that much harder.

THE WIECZOREK MODEL

The Wieczorek model for the K^+ pump in insect midgut[3-5] is now generally accepted for all insect epithelia which appear to have an apical, electrogenic pump for sodium or potassium. Essentially, it is that an apical plasma membrane V-ATPase energizes an exchanger more or less similar to the vertebrate Na^+/H^+ exchanger, and that this coupling is normally so tight that on a macroscopic scale, the ion pumped appears to be the metal ion, rather than the proton (Fig. 4.1). Unlike the vertebrate use of the pump in kidney epithelium and plasma membrane, the V-ATPase does not appear to be used directly to acidify the extracellular space; rather, it is used as a driving force, employed to move a different ion. In many ways, this is reminiscent of the position in prokaryotes, which use a proton, rather than a sodium, gradient to energize secondary active transport processes; however, it should be recalled that this application to vectorial transport across epithelia is necessarily unique to multicellular eukaryotes.

DISTRIBUTION

Major Epithelial Roles

It is appropriate to review briefly the physiology of insect homeostasis. The impermeability of cuticle -at least of terrestrial insects- means that all the significant fluxes of nutrients, waste, ions and water take place along the length of the alimentary canal (Fig. 4.2).[6] The very small size of insects implies a very high surface-to-volume ratio, which in turn places uniquely heavy osmoregulatory demands on their transporting epithelia. The unique feature of insects—and the justification for this chapter—is that nearly every transporting epithelium, at least in some orders of insects, is energized by an apical plasma-membrane V-ATPase. This thus represents a unique adaptation of an endomembrane pump, and one which is now finding correlates in vertebrates: the intercalated cells of kidney collecting duct,[7] the interstitial cells of frogskin,[8] and the plasma membrane of active osteoclasts.[9]

The functioning of the alimentary canal has been reviewed elsewhere.[6] Structurally, it can be divided into three regions: an ectodermally derived foregut, lined with cuticle and very impermeable, with paired salivary glands feeding into the buccal cavity; an endodermally-derived, selectively permeable midgut, across which most nutrient and water uptake takes place; and an ectodermally-derived hindgut with intermediate permeability properties, comprising Malpighian tubules, an anterior hindgut or ileum, and a posterior hindgut or rectum (Fig. 4.2).

Salivary Glands

Insect salivary glands are usually tubular or acinar simple epithelia (Fig. 4.3). They can be induced to secrete fluid, usually in response to catecholamines, such as 5-HT or dopamine.[10] Although not as robust as other classical model insect epithelia, they have been important in studies of intracellular signaling.[11,12] V-ATPase activity has been documented immunocytochemically to the apical surface of the epithelium,[13,14] which is decorated with the 10 nm "portasomes" characteristic of V-ATPase-packed membranes,[15] although there are no physiological or biochemical demonstrations of its functional role (Table 4.1).

Midgut (of Lepidoptera)

The Lepidopteran larval midgut is the classic model for the K⁺ pump (Fig. 4.4). Originally discovered in the silkworm *Hyalophora*

(A)

(B)

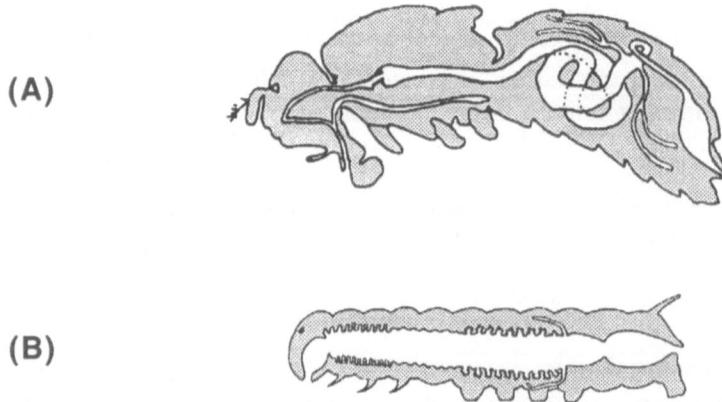

Fig. 4.2. Insect alimentary canal. (A) Typical insect structure, showing a pharynx, crop (diverticulate), foregut, midgut, Malpighian tubules, hindgut and rectum. (B) Adaptation of the body plan for Lepidopteran larvae. The foregut is massively reduced and the midgut (clearly distinguishable into three zones) dominates the body cavity. For clarity, the full ramifications of the Malpighian tubules are not shown. Reproduced, with permission, from Knowles and Dow, 1993. Reprinted with permission from: B.H. Knowles and J.A.T. Dow, BioEssays 1993;15: © by BioEssays.

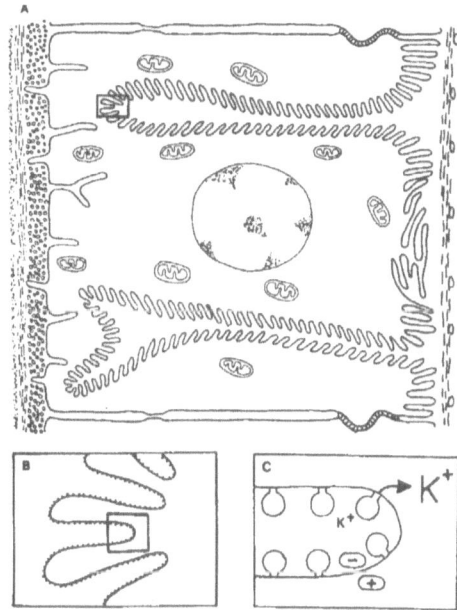

Fig. 4.3. Cross-section of salivary gland of Calliphora. Basal (hemolymph) side is to the left, the apical salivary gland lumen is to the right. (B) and (C) V-ATPases decorate the apical microvillar membranes. Reproduced, with permission, from Harvey et al., 1981. Reprinted with permission from: W.R Harvey, M. Cioffi and M.G. Wolfersberger, Amer Zool 1981;21: © by American Zoologist.

Table 4.1. Insect tissues in which plasma-membrane V-ATPases have been implicated

Order	Species	Tissue	Evidence	Source (ref. no.)
Lepidoptera	Manduca sexta	Midgut	Biochemical	19
			Pharmacological	4
			Physiological	30
		Malpighian tubule	Immuno-cytochemical	14
	Anthaeraea pernyi	Antennal sensillum	Immuno-cytochemical	14
Diptera	Drosophila hydei	Malpighian tubule	Pharmacological	39
	Drosophila melanogaster	Malpighian tubule	Pharmacological	40
Hemiptera	Rhodnius prolixus	Malpighian tubule	Pharmacological	31
Hymenoptera	Formica polyctena	Malpighian tubule	Pharmacological	42
Orthoptera	Schistocerca gregaria	Rectum	Immuno-cytochemical	14
	Periplaneta americana	Salivary glands	Immuno-cytochemical	13

cecropia, the transport properties appear fundamentally similar in all Lepidopteran species so far studied, namely: *Spodoptera littoralis*, an agriculturally important pest;[16,17] and in the tobacco hornworm moth larva, *Manduca sexta*, the major contemporary model.[18]

The V-ATPase is concentrated in a very unusual cell type, the goblet cell.[19] Unlike its vertebrate counterparts, the goblet is not an intermediate in mucus secretion, but forms as an intracellular vacuole which fuses with the apical membrane of the developing cell.[20] Goblet cells are distributed among columnar cells in an overdisperse pattern,[21] suggesting that, by analogy with the patterning of insect cuticular bristles,[22] the commitment of a stem cell to form either goblet or columnar cells is based on the distribution of some cell marker or morphogen.[23] A conspicuous feature is the apical "valve" separating the goblet cell from the apical gut lumen[24] which appears to pose a formidable permeability barrier, even to the K^+ ions which must pass through.[25]

Basal K^+ entry is thought to be via a heterogeneous population of K^+ channels[26] and possibly via an active pathway; although K^+ is near equilibrium across the basal membrane under physiological conditions, there can be a significant electrochemical gradient under some conditions in vitro.[27,28]

The larval Lepidopteran midgut is also distinguished by an extremely high luminal pH which can exceed pH 12.[29] Inconveniently,

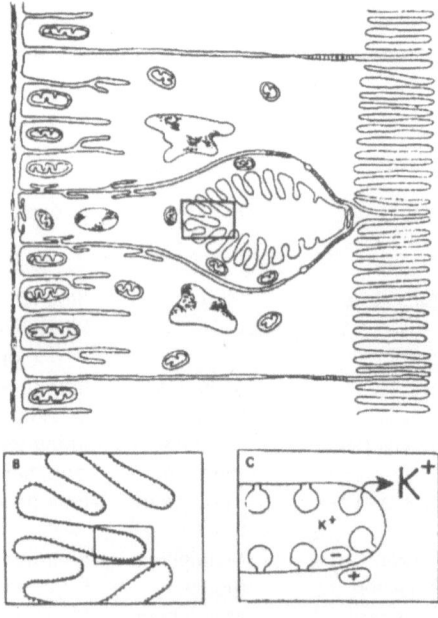

Fig. 4.4. (A) Generalized structure of the larval Lepidopteran midgut. Basal (hemolymph) side is to the left, apical gut lumen to the right. A K^+-pumping goblet cell (with morphology characteristic of the posterior midgut) is flanked by two reabsorptive columnar cells. (B) and (C) V-ATPases decorate the goblet cavity microvillar membranes. Reprinted with permission from: W.R Harvey, M. Cioffi and M.G. Wolfersberger, Amer Zool 1981;21: © by American Zoologist.

this gradient is in direct opposition to the direction in which the V-ATPase pumps across the midgut! This thus poses a fascinating series of questions: (i) why does the insect pump protons across its midgut, (ii) why does it maintain an alkaline midgut, and (iii) how does it reconcile the two? There are no firm answers, but the development of the argument has been summarized elsewhere.[30]

Malpighian Tubules

The insect renal (Malpighian) tubule is responsible for clearance of waste products from the hemolymph (Fig. 4.5). It achieves this with the aid of an apical V-ATPase and one or more amiloride-sensitive alkali-metal/proton exchangers[31] and an apical chloride channel[32] with properties that identify it as a homologue of the maxi-chloride channel family of vertebrates.[33,34] On the basal membrane, things are more varied. In at least one insect, a bumetanide-sensitive $Na^+:K^+:2Cl^-$ cotransporter is important in permitting ion entry,[31] whereas in other species ions are thought to enter[35] through a set of channels.

Most insects secrete a fluid rich in K^+, the major exception being those insects that feed on animals, where the urine is Na^+ rich. There is evidence that the basal ion permeability may determine the ionic ratio of the secreted fluid, as permeabilization of the basal membrane with the non-specific ionophore gramicidin reduces the K^+ content of the fluid secreted by either *Rhodnius* tubules or *Calliphora* salivary glands.[36]

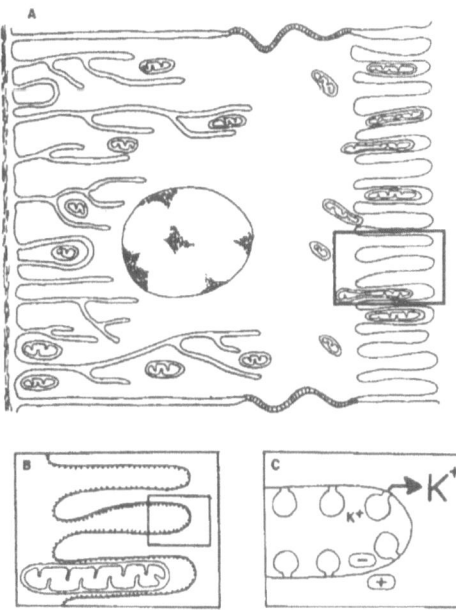

Fig. 4.5. (A) Cross section of a Malpighian tubule. Basal (hemolymph) side is to the left: apical the tubule lumen is to the right. (B) and (C) V-ATPases decorate the tubule apical microvillar membranes. Reprinted with permission from: W.R Harvey, M. Cioffi and M.G. Wolfersberger, Amer Zool 1981;21. © by American Zoologist.

A common theme is that, while they may contain very high levels of Na[+], K[+]-ATPase,[37] fluid secretion by tubules of many insect species is relatively insensitive to ouabain.[38-42] This may reflect the possibilities that (i) some insect Na[+], K[+]-ATPases are known to be resistant to ouabain;[43] (ii) some insect tubules are specialized rapidly to transport alkaloid toxins;[44] or (iii) that Na[+], K[+]-ATPases are simply not important in short term homeostasis or fluid transport in tubules.

Hindgut

During the passage of dietary wastes and tubule secretion through the hindgut, the insect has a final opportunity to rescue desirable solutes and maintain osmoregulatory equilibrium. There are many remarkable specializations of the hindgut (Fig. 4.6), for example an insect (*Thermobia domestica*) capable of absorbing water vapor directly from the atmosphere, to a vapor pressure of 45%, by a process of reverse flatulence. However, the industrial standard rectal preparation is that of the locust, *Schistocerca gregaria*.[45,46] Although the precise nature of the active transport processes across locust rectum has remained controversial, involving an apparent electrogenic active transport of chloride,[35] it now seems clear from immunocytochemical evidence that an apical plasma-membrane V-ATPase must be numbered among them.[47] Recently, this has been confirmed by studies showing locust rectum shares a characteristic sensitivity to amiloride with other insect V-ATPase driven epithelia.[48]

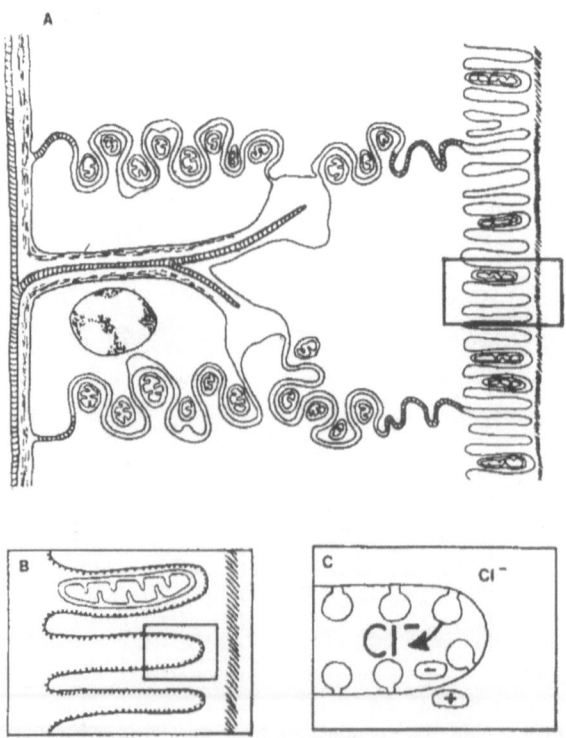

Fig. 4.6. (A) Cross-section of cockroach rectum. Basal (hemolymph) side is to the left: apical rectal lumen is to the right. (B) and (C) V-ATPases decorate the apical microvillar membranes. Reprinted with permission from: W.R Harvey, M. Cioffi and M.G. Wolfersberger, Amer Zool 1981;21: © by American Zoologist.

Tormogen Cells of all Trichoid Sensilla

The insect sensory system is built around a relatively limited set of designs.[49] In *Drosophila*, chemosensory, humidity-sensitive and mechanosensitive hairs of the peripheral nervous system are all based on clones of four cells derived from single neuroepidermoblasts[49] and are known as *Trichoid sensilla* (Fig. 4.7). Specificity is provided by appropriate receptors on the surface of the sensory neuron, which is surrounded by a glial cell. The hair is produced by a trichoid cell, while the sensitive surface of the neuron is bathed in a receptor lymph cavity, lined by the tormogen cell.[50] The receptor lymph cavity is known to be rich in potassium and highly polarized electrically, exactly as if the "K+-pump" were located on the apical surface of the tormogen cells. This was born out by biochemical studies on the labella of adult flies[51,52] which are packed with chemosensory hairs. Recently, it has been shown immunocytochemically that tormogen cells of one of the more elaborate manifestations of trichoid sensilla, the antennae of a male moth, are indeed loaded with V-ATPase.[66]

The function of the V-ATPase in this tissue is thought to be to hyperpolarize the sensory neuron receptive membrane. In this way, a given change in membrane resistance ΔR, effected by the action of a ligand on a receptor or by mechanical deformation of the membrane, is transduced into the largest possible receptor current ΔI. This simple application of Ohm's law, $\Delta I \propto V/\Delta R$ has the effect of increasing the

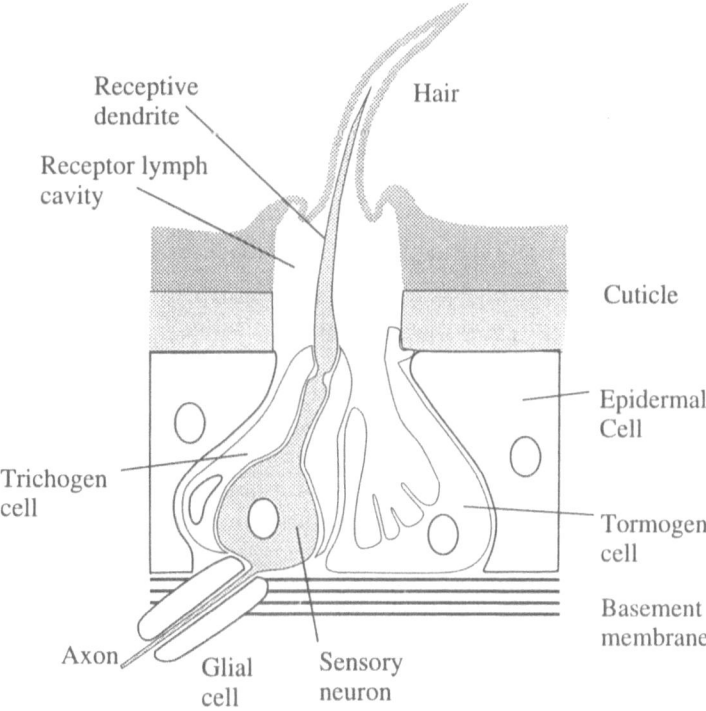

Fig. 4.7. General plan for the insect Trichoid sensillum. The modality sensed by the receptor is dependent on the nature of the receptors in the sensory neuron plasma membrane. V-ATPases decorate the apical infoldings of the tormogen cells.

sensitivity of the receptor. The effects of this optimization can be rather significant. For example, male moths use their extraordinarily elaborate antennae to home in on the pheromone released by receptive females. Their mating search can be initiated by reception of few or even single molecules of pheromone at a range of several miles from the female.

THE PRESENT

PROPERTIES OF ATPASE

Morphology & Energization

Insect epithelia in which V-ATPases are known to operate share a diagnostic ultrastructure: their apical plasma membranes are packed with 10 nm particles –"portasomes"– on the apical face. These are close-packed to a density of 12-13,000 μm^3, densities comparable with those observed in, for example, vertebrate vesicles. Usually, the apical surface is microvillate, to maximize the available area for V-ATPase insertion. This would suggest that the epithelia would be extremely active, and the mitochondrial placement confirms this. It is generally known that the number and subcellular distribution of mitochondria reflects the distribution of ATP consuming processes in cells. In several insect epithelia, such as the Lepidopteran goblet cavity and Malpighian tubules, mitochondria are located not just close to the apical surface, but are actually inserted into every microvillus. In fact, mitochondria have been shown to migrate into microvilli of *Rhodnius prolixus* lower tubule in response to hormonal stimulation.[53] The implication from these ultrastructural data is that, where a V-ATPase is active in an insect epithelium, it is a major consumer of cellular ATP, and we can predict that the epithelium will be capable of impressive work, in terms of fluxes.

Physiology and Pharmacology

The impression of high activity from ultrastructural analysis is amply borne out by the physiological signature of the V-ATPase in insects. In lepidopteran midgut, as discussed earlier, net fluxes of K^+ driven by goblet cavity V-ATPase activity exceed 1 mA cm^{-2}, and can generate PDs in excess of 150 mV in vitro. Significantly, this flux is attributable to a cell type which makes up -at most- a third of the population. In addition, it has been argued that a large fraction of the pumped K^+ circulates directly between goblet and columnar cells, recycling via gap junctions, without ever appearing as a net transepithelial flux.[30,54] This circuit plays dual roles: it is thought to contribute to the alkalinization of the gut lumen to pH>12 (in some species), the highest seen in biology[30,55,56] and a process calculated to use 10% of the insect's total ATP budget[57] and it drives K^+:amino acid symports to energize nutrient uptake.[58] By all accounts then, the lepidopteran midgut is an impressive exemplar of V-ATPase action.

The other well characterized V-ATPase role is in generation of urine by the Malpighian tubules. In *Rhodnius*, the electrical signature is usually suppressed by a high chloride conductance (presumably via channels), but this then allows the V-ATPase to manifest as a bulk flow of solutes and thus fluid. Again, this makes the *Rhodnius* tubule cell the fastest known secreting cell, able to pump its own volume of fluid every 15 s.[59] This has a clear adaptive role in the lifestyle of *Rhodnius*, as it allows the insect to void its enormous load of fluid after a blood meal, and so retreat quickly to the safety of a crack or recess to molt. More surprisingly, recent work on adults of *Drosophila* has shown them to be equally competent to handle fluid, and estimates suggest they may even be able to pump their cell volumes of fluid faster than *Rhodnius*.[40] This is remarkable in an insect which is not famed for having a particular problem with excess dietary water: indeed, *Drosophila* in the laboratory are rather prone to desiccation. Our explanation is that the peak rates of secretion observed in this tissue in vitro require the concerted action of the cAMP, NO/cGMP and Ca^{2+} signaling pathways[40,60,61] and are probably only replicated in vivo once in the insect's life cycle, at pupal/adult ecdysis. This means that, under normal conditions, diuresis is tightly controlled to a level appropriate to the insect's osmoregulatory status.

The Wieczorek model has also received support from pharmacological studies on insect transport. *Manduca* midgut is sensitive to bafilomycin and to less specific inhibitors of V-ATPase activity, such as N-ethyl maleimide (NEM) and dicyclocarbodiimide (DCCD). Although it is effective in isolated vesicle preparations,[4] demonstration of amiloride sensitivity of the exchanger has not been possible in vivo, presumably because of problems of access to the goblet apical membrane. In tubules of *Rhodnius prolixus*, *Drosophila hydei* and *Drosophila melanogaster*, fluid transport is abolished after application of bafilomycin.[31,39,40] Similarly, amiloride is a potent inhibitor of fluid secretion,[31,39,40] although a small but measurable residual fluid flux remains even after application of high concentrations of amiloride in *D. melanogaster*.[40]

The model would also predict that, if alkali metal fluxes could be inhibited, then the apical exchanger would run down and the presence of an apical V-ATPase would be revealed as an acidification of the apical lumen.[74] This can be shown in *Manduca sexta* midgut; goblet cavities normally exclude the weak base acridine orange, but accumulate it when bathed in K^+-free saline.[30] In *Formica polyctena* tubules, barium (an inhibitor of certain classes of K^+ channel) reduces fluid secretion,[42] again suggesting the presence of both an apical V-ATPase and exchanger.

Midgut V-ATPase Biochemical Characterization

Generally, insects are smaller than the preferred size for biochemical investigations. The larval Lepidopteran midgut, however, is an exception to this rule: its area can exceed 3 cm², and its weight 100 mg. Nonetheless, the purification of membrane fractions enriched in V-ATPase activity is a daunting task. The purification is only possible because of a morphological quirk: in the posterior midgut, the goblet cell mitochondria lie just below the microvilli containing the V-ATPase, rather than within them, as is the case in the rest of the midgut. This means that this is the only gut region from which V-ATPase can be purified without contamination by mitochondria. The purification process is painstaking, involving firstly the fine dissection of the posterior midgut to unfold it; then the removal of columnar microvilli by sonication; then the removal of goblet cavities by trituration through a Pasteur pipette. The cavities are then further purified by sucrose density centrifugation. Although tedious, and producing relatively little protein, this procedure can separate columnar apical membrane, goblet cavities, lateral membranes and basal membranes to a purity rarely encountered in animal biochemistry (Fig. 4.8).[80]

The membranes can be used for several purposes. As vesicles, they can be employed for classical proton transport experiments, employing fluorescence quench of the weak base acridine orange dye.[5,62] Alternatively, the ATPase activity can be assayed biochemically to investigate enzyme substrate requirements. The protein can also be analyzed on native and denaturing gels, to study subunit composition, and can also be probed with antibodies from other organisms to investigate the cross-reactivity, and thus the homology, of the subunits between species. For example, that vertebrate V-ATPase subunits cross-react with antibodies against plant tonoplast homologues. Similarly, plant antibodies recognize insect V-ATPase subunits on Western blots.[63]

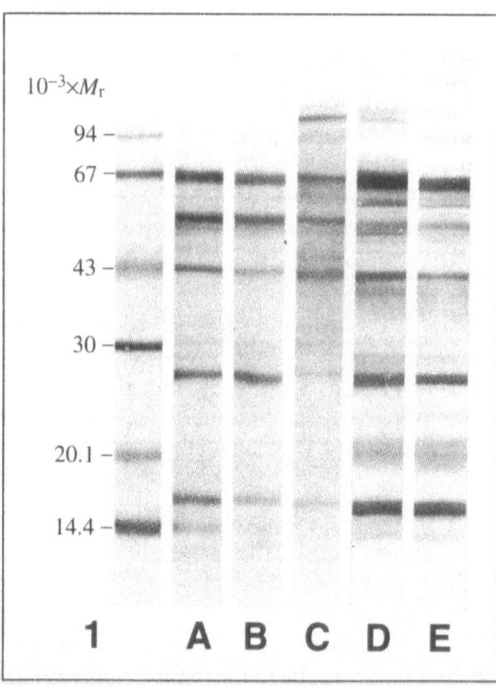

Fig. 4.8. SDS-PAGE analysis of purified goblet cavity membrane from Manduca sexta posterior midgut. Amido Black stained (A-C), or Western blotted with polyclonal antiserum against cavity membrane proteins (D-E). Reprinted with permission from: B.H. Knowles and J.A.T. Dow, BioEssays 1993; 15: © by BioEssays.

Subunit Composition

Manduca sexta goblet cell plasma membrane V-ATPase is the only insect representative yet purified biochemically. It is similar to that of plants, fungi and animals. Native gels suggest a holoenzyme size of 600-900 kDa, and denaturing SDS-PAGE analysis reveals subunits of 67, 57, 43, 28, 20, 17, 16 and 14 kDa (Fig. 4.8). In addition, there appear to be glycoproteins of 40 and 20 kDa. Polyclonal antisera against a highly purified membrane fraction also reveal additional bands on Western blots, although this cannot prove that they are functionally associated with V-ATPase activity.

The lepidopteran midgut V-ATPase subunits have now been sequenced more completely than any other system except yeast. Given that this is a candidate for a plasma membrane V-ATPase, is there any significant difference from endomembrane isoforms which could suggest a specialization for the role? Interestingly, one of the isoforms of the A-subunit of osteoclasts,[99] another cell type in which V-ATPase has a plasma membrane role, is more similar to the *Manducta sexta* A-subunit[55] than any other sequence.

Unique Insect Subunits

The spectrum of insect subunits is broadly comparable with those seen in other systems, although the existence of the 100 kDa subunit seen in some other systems[64] has yet to be demonstrated in insects. However, a novel 14 kDa subunit has been identified in *Manduca* midgut and a cDNA characterized, using antibodies against the purified V-ATPase.[65] A *Drosophila* homologue has also been cloned using the *Manduca* cDNA as a probe, and shown to be almost identical at the deduced amino-acid level (Y. Guo, J.A.T. Dow, H. Wieczorek, and K. Kaiser, in preparation; EMBL accession number Z26918). This subunit may not be insect-specific; however, initial studies have failed to identify vertebrate homologues by Western blotting,[65] although antibodies do cross-react with a third insect preparation, the locust rectum.

Is the 14 kDa protein a genuine V-ATPase subunit, or does it coincidentally copurify with the goblet cavity membrane? At this stage, it is hard to exclude the possibility that it may be an accessory protein, perhaps part of the cytoskeleton. However, monospecific antibodies purified by adsorption to 14-kDa fusion proteins were shown to inhibit purified V-ATPase from goblet cavities,[65] suggesting that the 14 kDa protein must either be a functional subunit, or so closely associated that antibody binding can disrupt function of the overall enzyme. Significantly, antibody inhibition required the enzyme to be functioning, suggesting that the epitope on the 14 kDa subunit recognized by the antibody is physically close to the active site.[65]

Close Conservation of Known Sequence

Not only are most of the V-ATPase subunits similar sizes to those seen in plants and animals, but their sequences once deduced from cDNA as very similar to those from other species. Within insects, the genes encoding the 57 kDa subunit have been characterized in the lepidopteran larvae *Manduca sexta, Helicoverpa virescens*[67] and the Dipteran *Drosophila melanogaster.*[68] Similarly the genes encoding the 17 kDa proteolipid subunit have also been characterized in the lepidopteran larvae *Manduca sexta,*[69] *Helicoverpa virescens*[87] and the Dipteran *Drosophila melanogaster.*[70] At the deduced amino acid level, these insect subunits are more than 90% identical, and are more than 80% identical to the analogous human subunit, despite a 400 M year evolutionary separation.

A Dual Role for the V-ATPase Pore in Gap Junctions?

Gap junctional coupling is widespread in epithelia, where it is thought to allow cells to synchronize their transporting activities, perhaps by averaging out second messenger levels, or recycling ions or small solutes between neighboring cells. The lepidopteran midgut is a good example; it is clear from the complexity of radioactive pool kinetics that potassium is recruited to the goblet cells not just through the basal membranes, but also via columnar cells[2] and it has been argued that much of this K[+] may derive from the apical side in a "local circuit" of K[+] flux.[30]

Surprisingly, it has proved difficult to isolate connexins, the building blocks of vertebrate gap junctions,[71] from insect tissues.[72] Instead, when gap junction-rich membranes are purified from a range of tissues like Lepidopteran midgut or lobster hepatopancreas,[73] a small proteolipid is found. Direct microsequencing of the proteins showed virtual identity with vertebrate proteolipid subunits of the V-ATPase, and when the *Manduca* proteolipid gene was sequenced, the deduced amino acid sequence[69] was identical to the microsequence previously obtained from *Manduca* gap junctional preparations (M.E. Finbow, Pers. comm.). The heretical suggestion therefore is that, at least in invertebrates, the actual pore in gap junctions might be formed of V-ATPase 16 kDa proteolipid rather than the connexins favored in vertebrates, and the proteolipid subunit has accordingly been renamed "ductin".[75] Small 16 kDa proteins colocalizing with gap junctions have been reported in bovine brain tissue,[76] in *Drosophila* CNS[77,78] and in the *Torpedo marmorata* mediatophore preparation which represents the fusion event between cholinergic vesicles and the plasma membrane.[79] The orthodox view, by contrast, is that the invertebrate proteolipid is a purification artifact from a procedure in which the true gap junctional protein was lost.[72]

What other evidence is available? Cell-cell coupling in *Drosophila* embryos can be blocked by microinjection of antibodies against lobster hepatopancreas ductin, although anomalously the antibody immunoblots to a 29 kDa, rather than a 16 kDa, *Drosophila* protein.[81] A small proteolipid has been implicated in gap junctional coupling in *Drosophila* brain, although its identity with ductin is not yet established.[77,78] Gap junctional coupling in a range of vertebrate tissues has been shown to be sensitive to N, N'-dicyclohexyl carbodiimide (DCCD), an inhibitor of the V-ATPase proteolipid,[82] but is unaffected by bafilomycin, confirming that V-ATPase inhibition alone does not cause uncoupling.

It is now clear that V-ATPase can be found on plasma membranes, as well as in endomembranes, of insect cells. Gap junctions are known to form hexameric structures (connexons) in each of the coupled plasma membranes, and the same can be said of the V-ATPase proteolipid. Indeed, detailed modeling of proteolipid structure shows that the pore through the middle of the hexamer is far larger than would be needed for proton flux, and would be capable of carrying solutes to the size limit ($\approx$ 1 kDa) associated with gap junctional coupling.[83] It is known that, unlike the analogous F_o membrane domain of F-ATPases, the V_o proteolipid domain does not pass ions when the V_1 head group is removed. This of course is a prerequisite for gap junctional coupling: half connexons must seal, and must only conduct when associated with a partner structure in the other plasma membrane.

Perhaps the biggest remaining problem is that ductin is required to sit in opposite orientations in membranes for its gap junctional and ion transporting roles. The termini and loop between transmembrane domains 2 & 3 are poorly conserved, while the loops between domains 1 and 2 and between 3 and 4 are highly conserved, and are known to be essential for V-ATPase assembly. The termini and central loop are thus believed to face away from the cytoplasm for the V-ATPase role (Fig. 4.9). However, antibodies against the N-terminal can immunoprecipitate gap-junction-containing membranes, implying

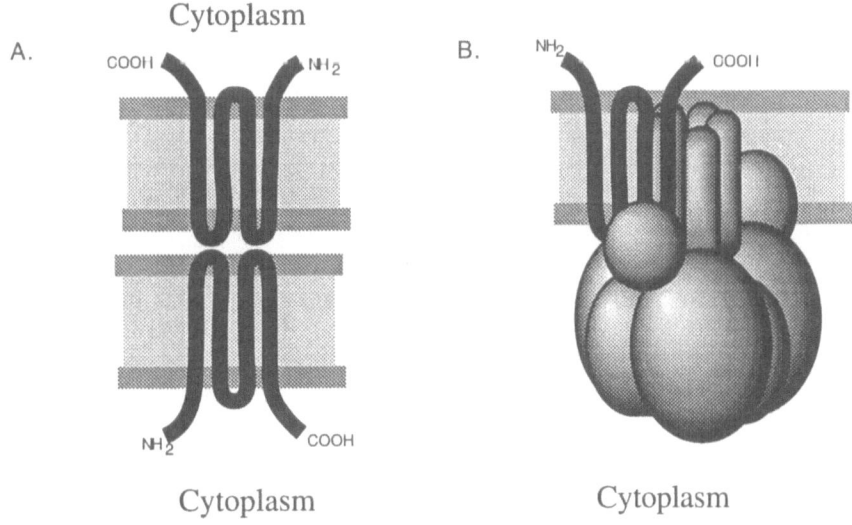

Fig. 4.9. If the 16 kDa V-ATPase proteolipid fulfills a second role as the pore in gap junctions, then it is likely to be inserted into the plasma membrane in both possible orientations. (A) In the proposed gap junctional role, the conserved loops are assumed to mediate gap junctional connection, while in isolated junctional complexes the N terminus is accessible to either antibodies or proteases, implying that it must be cytoplasmic. (B) In the V-ATPase role, the conserved loops mediate binding to the cytoplasmic V_o complex. In this figure, only one of the six proteolipid subunits has been drawn in detail.

that for this role the termini face the cytoplasm.[82] As "flipping" of mature integral membrane proteins is likely to be a rare event, it is necessary to argue that ductin is inerted into membranes in two distinct orientations, perhaps with the aid of chaperoning or accessory proteins, and that the two pools of ductin do not normally interconvert.

Ductin is thus argued to represent an ancient gene family, based on gene duplication of the 8 kDa F_o subunit of F-ATPases, which has been adapted to serve multiple roles in transmembrane pores.[84] Its identity as the transmembrane channel of V-ATPases may not thus be its ancestral role; it may also serve as a gap junctional pore, or in other fusion events such as the mediatophore of acetylcholine vesicle fusion.[79]

It might be expected that, to serve related but distinct roles, the ductin gene might have specialized following gene duplication. However, the ductin gene is known to be single copy both in *Drosophila melanogaster* (L. Meagher and M.E. Finbow, Pers. comm.) and in *Manduca sexta*.[69] However, the *Manduca sexta* gene is expressed as two messages of 1.4 and 1.9 kb, reflecting alternative choice of polyadenylation signals.[69] It will be interesting to see whether there is any tissue specificity in patterns of expression of the two transcripts.

The topic is likely to remain controversial for some time, but until convincing connexins are isolated from insects, shown to copurify with gap junctions, and shown to be sufficient for coupling in a heterologous system, the possibility that the V-ATPase proteolipid serves multiple roles must remain.[75]

CONTROL OF V-ATPASE

It is not yet clear how the V-ATPase is controlled, although insect systems may be able to provide useful prototypes. Presumably, it is not allowed to run unchecked at full speed at all times in every cell type, as the metabolic cost would be prohibitive. Exquisite control mechanisms have been discovered for other transport proteins, and so it is reasonable that this principle will extend to V-ATPases. There are several potential mechanisms by which such multi-subunit ATPases can be controlled:

(1) Transcriptional or translational control. The nearest approach to transcriptional sophistication yet has been for the 16 kDa proteolipid subunit, which is known to be a multi-gene family in oat.[86] In *Manduca sexta*, where the corresponding gene is known to be single copy, alternative polyadenylation sites are used to produce messages of 1.4 or 1.9 kb in *Manduca* midgut.[69] Similar variations of size have been observed in another Lepidopteran insect.[87] However, the transcripts differ only in the length of their 3' untranslated regions, and our understanding of the function of such sequences is not yet good enough to suggest a role in (for example) transactivation.

(2) Phosphorylation. Multiple phosphorylation consensus sites can be found in most published sequences for V-ATPase subunits. However, there is not yet evidence for phosphorylation as a controlling mechanism.

(3) Recruitment of pumps to the plasma membrane. So far, this appears to be a major mechanism for regulation in the kidney interca-

lated cells.[88,89] V-ATPases are held in intracellular vesicles, and recruited into apical or basal surfaces in response to acid or alkali loading.[90] Although there is not yet direct evidence for such a mechanism in insect plasma membranes, an increase in apical membrane surface area of Malpighian tubules by recruitment from intracellular vesicles after fluid stimulation has been noted.[91] It will be interesting to see, using immunocytochemistry, whether such recruitment brings more V-ATPase to the apical membrane surface.

(4) Post-translational modifications by alteration of subunit composition. In this class, we consider peptide activators or inhibitors of V-ATPases, or the re-arrangement of subunit composition to alter the activity of the ATPase. The existence of activator or inhibitor proteins has been suggested in kidney intercalated cells, in addition to the membrane recruitment model discussed above.[7] In addition, an inhibitor has been suggested to act by binding to the 16 kDa proteolipid subunit from the abcytoplasmic face.[92] Further possibilities for subunit shifts during inactivation of the insect V-ATPase will be discussed below.

Insect Models for Hormonal Control of V-ATPase

As insects rely so heavily on V-ATPases to energize their epithelial transport, and as their epithelia allow them to exploit some rather extreme ecological niches, it is not surprising that they provide some useful models for ion transport and its control. For example, the blood-sucking bug *Rhodnius prolixus* can take in a blood meal corresponding to ten times its bodyweight, then void the excess water in two or three hours to allow it to molt safely in cracks in the floorboards. Its peak rate of urine production is 10^3 times higher than the basal rate[59] and must be attributed in large part to the upregulation of tubule V-ATPase activity.[31]

Fluid secretion by *Drosophila melanogaster* tubules is known to be stimulated by cAMP, or by cGMP which in turn is elevated by nitric oxide.[60,61] These two second messengers are not additive and act in parallel.[40] This would not be surprising if the messengers acted through their respective kinases, which are both serine/threonine kinases with overlapping consensus sequences.[93] Our present understanding is that both of these messengers act to increase V-ATPase activity, whereas intracellular calcium selectively opens apical chloride channels[41] which have recently been characterized in *Drosophila*.[32]

Unfortunately insect tubules as mentioned above are not promising candidates for biochemical studies. However, the larval lepidopteran midgut is relatively massive. Surprisingly, despite the known high metabolic cost of running the pump,[57] there appears to be no diurnal or feeding-related control of the pump until its activity is lost at the larval/pupal molt. Although there is evidence for control both by a blood-borne factor and by the cAMP pathway, these effects are extremely modest.[94,95] A previous study of insect tubules had shown that they were also downregulated during the larval/pupal molt by a process ascribable to the action of the molting hormone, 20-hydroxyecdysterone.[96,97] The molts from larva to pupa and then to adult are associated with extensive tissue remodeling, and in the case of Lepidopteran midgut, the V-ATPase containing goblet cells are lost for ever.[98] We thought, therefore, that the larval/larval molts were more likely candidates for a "clean" study of reversible V-ATPase control. This appears to be the case; around 30 h before the fourth/fifth instar molt, the electrical gradient across *Manduca* midgut collapses suddenly. This is associated with loss of the pH gradient across the midgut, which allows the transport status of the V-ATPase to be monitored non-invasively by the simple expedient of adding pH indicator to the larval diet.[30] Starting at the moment of ecdysis, these parameters recover in time for the start of appetitive behavior. Fortunately, the fourth instar larvae are just large enough to allow biochemical purification of the goblet cavity membranes, and so we have been able to show that these physiological changes reflect a total loss of V-ATPase activity, measured either by vesicle acidification studies or by ATPase assays (J.-P. Sumner, J.A.T. Dow, H. Wieczorek and F. Earley, submitted). Accordingly, this presents a unique opportunity to study the long-term regulation of V-ATPase activity, with possible general applicability.

FUTURE DIRECTIONS

THE PROBLEM

One of the most informative analytical techniques to which a gene can be exposed is to mutagenise it and express it in a cellular or organismal context, to study any phenotypic effects of the changes. This allows a functional mapping of protein domains, and complements the other powerful techniques which can be brought to bear.

There are a number of criteria which must be met in such conditions. It must be possible to distinguish the transgene product from any endogenous gene product. For most purposes, this means that the expression system is heterologous, that is, it is not the natural system for the gene. For example, it is relatively safe to assume that acetylcholine receptors on the surface of a microinjected *Xenopus* oocyte have been expressed from microinjected human mRNA, as the expression of the analogous *Xenopus* gene in the oocyte is negligible. Another requirement is that the heterologous cell is competent to express the gene in its physiological context. For an ion channel, this is relatively straightforward; it need only be targeted to the plasma membrane. However, for multi-subunit assemblies, which may normally be targeted specifically to one surface of an epithelium, things rapidly become very unsatisfactory. It can be seen rapidly that the standard heterologous expression systems, namely *Xenopus* oocytes and vertebrate fibroblasts, fail on both counts when the target for expression is a V-ATPase subunit. The high level of conservation and present lack of pharmacological difference in specificity between plasma-membrane and endomembrane V-ATPases compel us to seek a system in which null mutants are readily available, and yet not necessarily lethal. For many purposes, the pH-sensitive conditional mutants of yeast are ideal[43,100,101] and the elegance of the screen by which V-ATPase mutants can be detected has proved enormously useful in the advancement of the field. However, yeast mutants are of endomembrane function, and my interest was in an epithelial context. How could it be possible to provide a range of null mutants for heterologous expression of V-ATPase subunits, in an epithelial context, and thus in a higher organism?

In principle, the availability of genetic knockouts by homologous recombination might provide a useful model to study the roles of V-ATPases in a higher organism. However, this is an expensive technology which can produce sick animals which often do not survive long enough to perform an experiment, and so is likely to prove a troublesome technology when addressing questions involving novel roles of a housekeeping gene.

A SOLUTION: *Drosophila* MOLECULAR GENETICS

Drosophila melanogaster is a valuable model system for the analysis of genes.[85] At present, it has perhaps the best physical mapping of any animal genome,[102] based both on classical and molecular genetic techniques. Additionally, the chromosomal localization of a newly-characterized gene is particularly straightforward, as the salivary gland polytene chromosomes are large and easy to map. This means that a newly discovered gene can be reconciled rapidly with the sum of existing knowledge of the *Drosophila* genome: in such a way, suitable existing mutants at the locus of the gene under study may be discovered rapidly.

P-Elements

Transposons are ubiquitous, but in *Drosophila* they are particularly well characterized. In particular, the P-element has been harnessed as a molecular genetic tool. Normally, P-elements contain the gene for the enzyme (transposase) which catalyses their own excision and reinsertion within the genome. The splicing which produces active transposase occurs only in the germ cells, and so is stable within somatic cells of each individual.

Drosophila lines carrying marked, mutant P-elements have been produced which either lack functional transposase, or which lack the terminal repeat sequences which are essential for mobilization. Both of these lines are stable, as no transposition can occur. However, when crossed, the progeny produce a series of pseudo-randomly transposed offspring. This can be harnessed in several ways.

(1) P-elements can be harnessed to produce mutations in genes of choice by the novel technique of "site-selected" mutagenesis.[103]

(2) If DNA is injected at the syncytial blastoderm stage of the early *Drosophila* embryo, it can become stably incorporated into the genome of cells as they form. If this occurs in the germ cell precursors, then the DNA can be stably inherited, and transgenic flies will have been produced by germ-line transformation. Rubin and Spradling discovered that the efficiency of such incorporation could be hugely increased if the DNA were inserted into a P-element transposon cassette.[104] This now means that the production of transgenic flies is a tedious, but essentially routine, process.

(3) Given that gene constructs can be stably incorporated into *Drosophila* with relative ease, which constructs would be relevant to our

study? We would like either to suppress the expression of wild type genes, either by mopping up their mRNAs with a huge excess of antisense RNA, as has recently been accomplished in carrot for the V-ATPase A-subunit,[105] or by employing "hammerhead" ribozymes[106] which cleave other RNA molecules (such as mRNAs) in a site-specific fashion. Alternatively, we would like to express heterologous genes ectopically, either in the whole fly, or in tissues of our choice. Fortunately, one technique to accomplish this is well established in *Drosophila*. Genes placed downstream of the heat-shock promoter are not expressed to any great extent under normal conditions, but are massively overexpressed if the flies are "heat shocked" to 37°C for an hour.[107] This technique has proved invaluable in dissecting the interactions between segment patterning genes in the developing *Drosophila* embryo.[108]

(4) Enhancer traps. If a P-element, containing a reporter gene such as β-galactosidase coupled to a very weak ("permissive") promoter, is mobilized then it will on occasion land close enough to a gene to be influenced by the same control elements (promoters and enhancers), and will take on the same pattern of expression. The neighboring gene can then be identified and sequenced after (with luck) a short chromosome walk. This has been an important general tool for the identification of interesting genes, based on their pattern of expression, not just in *Drosophila*[109,110] but now in mouse. Furthermore, if the P-element construct reporter gene encodes the yeast transcription factor GAL4, then the histochemical stain only becomes visible after crossing the flies to lines containing β-gal downstream of the yeast UAS_G binding site for GAL4. Why the extra sophistication? The advantage is that a given enhancer trap line can be used to direct expression of any genetic construct (for example antisense RNA or mutated V-ATPase subunit genes) in a highly tissue specific fashion. This "binary" technique[111] thus provides the enormously powerful tool to ask questions about what a particular transcript does in a particular tissue of a higher organism, as it obviates the need for cloned and fully-characterized tissue-specific enhancers, the traditional route to such a goal.

Dominant Negatives

Even where inactivation of a gene has not been accomplished by "site-selected" mutagenesis, homologous recombination or serendipity, it can be possible to study the effects of mutant genes. If a mutant (defective) gene is strongly overexpressed in an otherwise wild-type background, then the phenotype will tend toward the mutant form (a "dominant negative"), even if the full effect does not become apparent. Given that the full phenotype of an inactivating V-ATPase mutation is very likely to be rapid death, this has certain attractions for experimental study. The heat-shock promoter would be ideal for such purposes, as it drives very high levels of expression. This technique has already been applied to reduce protein kinase A activity by up to 400-fold in NIH 3T3 and Y1 adrenocortical tumor cells.[29]

THE *Drosophila* RENAL TUBULE—THE ULTIMATE EPITHELIUM?

Realizing that null mutants of genes as vital V-ATPase subunits were not likely to have any phenotype more informative than very early embryonic death, we have developed a Malpighian tubule fluid secretion assay in *Drosophila melanogaster*. As well as having a rich control repertoire (Fig. 4.9),[40,60,61] the tubules are exquisitely sensitive either to bafilomycin (the V-ATPase inhibitor) or to amiloride (the exchanger inhibitor).[40] Accordingly, if effects of mutations could be confined to the tubules, and preferably the tubules of the later larval and adult stages, then it will be possible to do epithelial transport physiology on these mutants, something so difficult as to be effectively impossible in any other organism.

Fortunately, the molecular genetic techniques outlined above make the generation of such conditional mutations relatively straightforward in *Drosophila melanogaster*, and we plan to employ the following tools to study the role of V-ATPases:

(1) Rescue of loss-of-function alleles. Although normally considered to be the paragon of molecular genetic virtuosity, we consider this to be the least informative of our approaches, for the reasons discussed above.

(2) Generation of dominant negative mutations by overexpression of mutated genes under control of the heat shock promoter. This has the advantage that, against a relatively normal background, huge amounts of mutant message can be overexpressed at a stage in the life cycle of the experimenter's choosing. The remaining background activity should help to maintain viability. In principle, these constructs would allow us to study the role of V-ATPase in any *Drosophila* tissue at any stage in the life cycle. Although heat-shock constructs have proved valuable in many studies of (for example) development, the main drawback is that there is some "breakthrough" expression of genes under heat shock promoter control even under normal conditions, and so might produce a phenotype. The long-term stability of the lines might therefore be compromised.

(3) Binary expression, driven by enhancer trap lines showing tubule-specific patterns of expression. This technology allows the enhancer trap line and the line containing the potentially deleterious construct to be propagated safely, as the mutated gene is only expressed in the progeny of the cross between the lines. To this end, we have screened around 1000 lines generated in the laboratory of Kim Kaiser in Glasgow (M. Y. Yang, J. D. Armstrong and K. Kaiser, unpublished), and have found several dozen which stain tubules or sub-regions within tubules (M.A. Sozen and J.A.T. Dow, unpublished observations). We plan therefore to characterize these lines more fully, and then to select appropriate lines to drive expression of constructs of our choice. Our impression is that this will provide a valuable and unique system for the study, not just of V-ATPases, but for any of a wide range of genes involved in ion transport or its control.

CONCLUSION

V-ATPases are important agents of epithelial work in insects, and they are proving useful models to extend our understanding of V-ATPase function in general. In the future, we hope to be able to address the "big questions":

- how are V-ATPases controlled?
- are plasma membrane V-ATPases different from endomembrane V-ATPases?
- what plasma-membrane transport processes are energized by gradients generated by the V-ATPase?

ACKNOWLEDGMENTS

I am most grateful to the Medical Research Council and the Nuffield Foundation for their support, and to John-Paul Sumner, Drs Simon Maddrell and Kim Kaiser, and Profs. Helmut Wieczorek and William Harvey for their critical reading of the manuscript.

REFERENCES

1. Harvey, W. R. and Nedergaard, S. (1964). Sodium-independent active transport of potassium in the isolated midgut of the cecropia silkworm. Proc.Natn.Acad.Sci.U.S.A. 51, 757-765.
2. Wood, J. L. and Harvey, W. R. (1975). Active transport of potassium by the Cecropia midgut; tracer kinetic theory and transport pool size. J.exp.Biol. 63, 301-311.
3. Wieczorek, H. (1992). The insect V-ATPase, a plasma-membrane proton pump energizing secondary active-transport - molecular analysis of electrogenic potassium-transport in the tobacco hornworm midgut. J.exp.Biol. 172, 335-343.
4. Wieczorek, H., Putzenlechner, M., Zeiske, W., and Klein, U. (1991). A vacuolar-type proton pump energizes K^+/H^+ antiport in an animal plasma membrane. J.Biol.Chem. 266, 15340-15347.
5. Wieczorek, H., Weerth, S., Schindlebeck, M., and Klein, U. (1989). A vacuolar-type proton pump in a vesicle fraction enriched with potassium transporting plasma membranes from tobacco hornworm midgut. J.Biol.Chem. 264, 11143-11148.
6. Dow, J. A. T. (1986). Insect midgut function. Adv. Insect Physiol. 19, 187-328.
7. Gluck, S. and Nelson, R. (1992). The role of the V-ATPase in renal epithelial H^+ transport. J.exp.Biol. 172, 205-218.
8. Harvey, B. J. (1992). Energization of sodium-absorption by the H^+-ATPase pump in mitochondria-rich cells of frog skin. J. exp. Biol. 172, 289-309.
9. Chatterjee, D., Chakraborty, M., Leit, M., Neff, L., Jamsakellokumpu, S., Fuchs, R., Bartkiewicz, M., Hernando, N., and Baron, R. (1992). The osteoclast proton pump differs in its pharmacology and catalytic subunits from other vacuolar H^+-ATPases. J.exp.Biol. 172, 193-204.
10. House, C. R. (1980). Physiology of invertebrate salivary glands. Biol.Rev. 55, 417-473.
11. Berridge, M. J. and Irvine, R. F. (1984). Inositol triphosphate, a novel second messenger in cellular signal transduction. Nature 312, 315-321.

12. Berridge, M. J., Lindley, B. D., and Prince, W. T. (1975). Stimulus-secretion coupling in an insect salivary gland: cell activation by elevated potassium concentrations. J.exp.Biol. 62, 629-636.

13. Just, F. and Walz, B. (1993). Immunocytochemical localization of H⁺-ATPase in salivary glands of the cockroach, Periplaneta americana. Proc. German Zool. Soc. 86, 28.

14. Klein, U. (1992). The insect V-ATPase, a plasma-membrane proton pump energizing secondary active transport - immunological evidence for the occurrence of a V-ATPase in insect ion-transporting epithelia. J.exp.Biol. 172, 345-354.

15. Harvey, W. R., Cioffi, M., and Wolfersberger, M. G. (1981). Portasomes as coupling factors in active ion transport and oxidative phosphorylation. Amer.Zool. 21, 775-791.

16. Thomas, M. V. and May, T. E. (1984a). Active potassium ion transport across the caterpillar midgut I. Tissue electrical properties and potassium ion transport inhibition. J.exp.Biol. 108, 273-291.

17. Thomas, M. V. and May, T. E. (1984b). Active potassium transport across the caterpillar midgut II. Intracellular microelectrode studies. J.exp.Biol. 108, 293-304.

18. Harvey, W. R. and Wolfersberger, M. G. (1979). Mechanism of inhibition of active potassium transport in isolated midgut of Manduca sexta by Bacillus thuringiensis endotoxin. J.exp.Biol. 83, 293-294.

19. Schweikl, H., Klein, U., Schindlebeck, M., and Wieczorek, H. (1989). A vacuolar-type ATPase, partially purified from potassium transporting plasma membranes of tobacco hornworm midgut. J.Biol.Chem. 264, 11136-11142.

20. Hakim, R. S., Baldwin, K. M., and Bayer, P. E. (1988). Cell differentiation in the embryonic midgut of the tobacco hornworm, Manduca sexta. Tissue & Cell 20, 51-62.

21. Baldwin, K. M. and Hakim, R. S. (1991). Growth and differentiation of the larval midgut epithelium during molting in the moth, Manduca sexta. Tissue & Cell 23, 411-422.

22. Wigglesworth, V.B. The principles of insect physiology. Seventh edition ed., London: Chapman and Hall, 1972.

23. Lawrence, P. A. The making of a fly. Oxford: Blackwell Scientific, 1993.

24. Flower, N. E. and Filshie, B. K. (1976). Goblet cell membrane differentiations in the midgut of a lepidopteran larva. J.Cell Sci. 20, 357-375.

25. Moffett, D. F. and Koch, A. (1992). Driving forces and pathways for H⁺ and K⁺ transport in insect midgut goblet cells. J. exp. Biol. 172, 403-415.

26. Moffett, D. F. and Koch, A. (1991). Lidocaine and barium distinguish separate routes of transbasal K⁺ uptake in the posterior midgut of the tobacco hornworm (Manduca sexta). J.exp.Biol. 157, 243-256.

27. Moffett, D. F. and Koch, A. R. (1988a). Electrophysiology of K⁺ transport by midgut epithelium of lepidopteran insect larvae I. The transbasal electrochemical gradient. J.exp.Biol. 135, 25-38.

28. Moffett, D. F. and Koch, A. R. (1988b). Electrophysiology of K⁺ transport by midgut epithelium of lepidoteran insect larvae II. The transapical electrochemical gradient. J.exp.Biol. 135, 39-49.

29. Clegg, C. H., Correll, L. A., Cadd, G. G., and McKnight, G. S. (1987). Inhibition of intracellular cAMP-dependent protein kinase using mutant genes of the regulatory type I subunit. J. Biol. Chem. 262, 13111-9.

30. Dow, J. A. T. (1992). pH gradients in lepidopteran midgut. J. exp. Biol. 172, 355-375.

31. Maddrell, S. H. P. and O'Donnell, M. J. (1992). Insect Malpighian tubules: V-ATPase action in ion and fluid transport. J.exp.Biol. 172, 417-429.

32. Harvey, B. J. and Dow, J. A. T. (1994). A high-conductance chloride channel in apical membranes of Drosophila melanogaster Malpighian tubule. In preparation

33. Brown, P. D., Greenwood, S. L., Robinson, J., and Boyd, R. (1993). Chloride channels of high conductance in the microvillous membrane of term human placenta. Placenta 14, 103-115.

34. Reeves, W. B. and Andreoli, T. E. (1992). Renal epithelial chloride channels. Annual Review Of Physiology 54, 29-50.

35. Phillips, J. E., Hanrahan, J., Chamberlin, M., and Thomson, B. (1986). Mechanisms and control of reabsorption in insect hindgut. Advances In Insect Physiology 19, 329-422.

36. Maddrell, S. H. P. and O'Donnell, M. J. (1993). Gramicidin switches transport in insect epithelia from potassium to sodium. J. exp. Biol. 177, 287-292.

37. Lebovitz, R. M., Takeyasu, K., and Fambrough, D. M. (1989). Molecular characterization and expression of the (Na$^+$ + K$^+$)-ATPase a-subunit in Drosophila melanogaster. EMBO J. 8, 193-202.

38. Anstee, J. H. and Bowler, K. (1979). Ouabain sensitivity of insect epithelial tissues. Comp.Biochem.Physiol. 62A, 763-769.

39. Bertram, G., Shleithoff, L., Zimmermann, P., and Wessing, A. (1991). Bafilomycin-A1 is a potent inhibitor of urine formation by Malpighian tubules of Drosophila hydei - is a vacuolar-type ATPase involved in ion and fluid secretion? J.Insect Physiol. 37, 201-209.

40. Dow, J. A. T., Maddrell, S. H. P., Görtz, A., Skaer, N. V., Brogan, S., and Kaiser, K. (1994b). The Malpighian tubules of Drosophila melanogaster: fluid secretion and its control. J. exp. Biol. 197, in press.

41. Maddrell, S. H. P. and Overton, J. A.(1988). Stimulation of sodium transport and fluid secretion by ouabain in an insect Malpighian tubule. J. exp. Biol. 137, 265-276.

42. Weltens, R., Leyssens, A., Zhang, A. L., Lohhrmann, E., Steels, P., and van Kerkhove, E. (1992). Unmasking of the apical electrogenic H pump in isolated Malpighian tubules (Formica polyctena) by the use of barium. Cell. Physiol. Biochem. 2, 101-116.

43. Ohya, Y., Umemoto, N., Tanida, I., Ohta, A., Iida, H., and Anraku, Y. (1991). Calcium-sensitive cls mutants of saccharomyces-cerevisiae showing a pet-phenotype are ascribable to defects of vacuolar membrane H$^+$-ATPase activity. J. Biol. Chem. 266, 13971-13977.

44. Maddrell, S. H. P. (1976). Excretion of alkaloids by Malpighian tubules of insects. J.exp.Biol. 64, 267-281.

45. Spring, J. H. and Phillips, J. E. (1980a). Studies on locust rectum I. Stimulants of electrogenic ion transport. J.exp.Biol. 86, 211-223.

46. Spring, J. H. and Phillips, J. E. (1980b). Studies on locust rectum II. Identification of specific ion transport processes regulated by corpora cardiaca and cyclic AMP. J.exp.Biol. 86, 225-236.

47. Klein, U., Timme, M., Novak, F. J. S., Lepier, A., Harvey, W. R., and

Wieczorek, H. (1994). In preparation

48. Thomson, R. B. and Phillips, J. E. (1992). Electrogenic proton secretion in the hindgut of the desert locust, Schistocerca gregaria. J. Membrane Biol. 125, 133-154.

49. Jan, Y. N. and Jan, L. Y. (1990). Genes required for specifying cell fates in Drosophila embryonic sensory nervous system. Trends In Neurosciences 13, 493-498.

50. Thurm, U. and Kuppers, J. (1980). Epithelial physiology of insect sensilla. In VBW80 - Insect biology in the future, (ed. M. Locke and D.S. Smith), pp. 735-76. London: Academic Press.

51. Wieczorek, H. (1982). A biochemical approach to the electrogenic potassium pump of insect sensilla: potassium sensitive ATPase in the labellum of the fly. J.comp.Physiol. 148A, 303-311.

52. Wieczorek, H. and Gnatzy, W. (1985). The electrogenic potassium pump of insect cuticular sensilla. Further characterisation of ouabain and azide-insensitive, K$^+$-stimulated ATPases in the labellum of the blowfly. Insect Biochem. 15, 225-232.

53. Bradley, T. J. (1984). Mitochondrial placement and function in insect ion-transporting cells. Amer.Zool. 24, 157-167.

54. Dow, J. A. T. and O'Donnell, M. J. (1990). Reversible alkalinization by Manduca sexta midgut. J.exp.Biol. 150, 247-256.

55. Gräf, R., Novak, F.J.S., Harvey, W.R., Wieczorek, H. (1992). Cloning and sequencing of cDNA encoding the putative insect plasma membrane V-ATPase subunit A. FEBS Letters 119-122.

56. Dow, J.A.T. and Harvey, W.R. (1988). The role of midgut electrogenic K$^+$ pump potential difference in regulating lumen K$^+$ and pH in larval lepidoptera. J.exp.Biol. 140, 455-463.

57. Dow, J. A. T. and Peacock, J. M. (1989). Microelectrode evidence for the electrical isolation of goblet cavities of the middle midgut of Manduca sexta. J.exp.Biol. 143, 101-114.

58. Giordana, B., Parenti, P., Hanozet, G. M., and Sacchi, V. F. (1985). Electrogenic K$^+$-basic amino acid cotransport in the midgut of lepidopteran larvae. J.Membrane Biol. 88,, 45-53.

59. Maddrell, S. H. P. (1991). The fastest fluid-secreting cell known: the upper Malpighian tubule cell of Rhodnius. BioEssays 13, 357-362.

60. Dow, J. A. T., Maddrell, S. H. P., Davies, S.-A., Skaer, N. J. V., and Kaiser, K. (1994a). A novel role for the nitric oxide/cyclic GMP signalling pathway: the control of fluid secretion in Drosophila. Amer.J.Physiol. 266, R1716-R1719.

61. Dow, J. A. T. and Maddrell, S. H. P. (1993). Fluid secretion by the Malpighian tubule of Drosophila melanogaster is stimulated by nitric oxide and cGMP. J. Physiol. 473, 97P.

62. Moriyama, Y., Takano, T., and Ohkuma, S. (1982). Acridine orange as a fluorescent probe for lysosomal proton pump. J. Biochem. 92, 1333-1336.

63. Russell, V. E. W., Klein, U., Reuveni, M., Spaeth, D. D., Wolfersberger, M. G., and Harvey, W. R. (1992). Antibodies to mammalian and plant V-ATPases cross react with the V-ATPase of insect cation-transporting plasma membranes. J.exp.Biol. 166, 131-143.

64. Gillespie, J., Ozanne, S., Percy, J., Warren, M., Haywood, J., and Apps, D. (1991). The vacuolar H$^+$-translocating ATPase of renal tubules con-

D. (1991). The vacuolar H$^+$-translocating ATPase of renal tubules contains a 115-kDa glycosylated subunit. FEBS Letters 282, 69-72.

65. Gräf, R., Lepier, A., Harvey, W. R., and Wieczorek, H. (1994). A novel 14-kDa V-ATPase subunit in the tobacco hornworm midgut. J.Biol.Chem. 269, in press.

66. Klein, U. and Zimmermann, B. (1991). The vacuolar-type ATPase from insect plasma membrane: immunocytochemical localization in insect sensilla. Cell Tissue Res. 266, 265-273.

67. Gill, S. S. and Ross, L. S. (1991). Molecular-cloning and characterization of the B-subunit of a vacuolar H$^+$-ATPase from the midgut and Malpighian tubules of Helicoverpa virescens. Arch.Biochem.Biophys. 291, 92-99.

68. Dow, J. A. T., Goodwin, S. F., and Kaiser, K. (1994). Analysis of the gene encoding the 57 kDa B-subunit of the V-ATPase in Drosophila melanogaster. In preparation

69. Dow, J. A. T., Goodwin, S. F., and Kaiser, K. (1992). Analysis of the gene encoding a 16-kDa proteolipid subunit of the vacuolar H$^+$-ATPase from Manduca sexta midgut and tubules. Gene 122, 355-360.

70. Meagher, L., McLean, P., and Finbow, M. E. (1990). Sequence of a cDNA from Drosophila coding for the 16 kD proteolipid component of the vacuolar H$^+$-ATPase. Nucleic Acids Res. 18, 6712.

71. Warner, A. E. (1988). The gap junction. J.Cell Sci. 89, 1-7.

72. Berdan, R. C. and Gilula, N. B. (1988). The arthropod gap junction and pseudo-gap junction: isolation and preliminary biochemical analysis. Cell Tissue Res. 251, 257-274.

73. Lane, N. J. and Finbow, M. (1988). Isolation of gap and septate junctions from arthropod tissues. J.Cell Biol. 107, 793a.

74. O'Donnell, M. J. (1992). A simple method for construction of flexible, subminiature ion-selective electrodes. J.exp.Biol. 162, 353-359.

75. Finbow, M. E. and Pitts, J. D. (1993). Is the gap junction channel - the connexon - made of connexin or ductin? J.Cell Sci. 106, 463-471.

76. Dermietzel, R., Volker, M., Hwang, T. K., Berzborn, R. J., and Meyer, H. E. (1989). A 16 kDa protein co-isolating with gap junctions from brain tissue belonging to the class of proteolipids of the vacuolar H$^+$-ATPases. FEBS Lett. 253, 1-5.

77. Ryerse, J. S. (1978). Ecdysterone switches off fluid secretion at pupation in insect Malpighian tubules. Nature 271, 745-746.

78. Ryerse, J. S. (1991). Gap junction protein tissue distribution and abundance in the adult brain of Drosophila. Tissue & Cell 23, 709-718.

79. Birman, S., Meunier, F.-M., Lesbats, B., Le Caer, J.-P., Rossier, J., and Israël, M. (1990). A 15 kD proteolipid found in mediatophore preparations from Torpedo electric organ presents high sequence homology with the bovine chromaffin granule protonophore. FEBS Lett. 261, 303-306.

80. Cioffi, M. and Wolfersberger, M. G. (1983). Isolation of separate apical, lateral and basal plasma membrane from cells of an insect epithelium. A procedure based on tissue organization and ultrastructure. Tissue & Cell 15, 781-803.

81. Bohrmann, J. (1993). Antisera against a channel-forming 16 kDa protein inhibit dye-coupling and bind to cell membranes in Drosophila ovarian follicles. J.Cell Sci. 105, 513-518.

Disposition and orientation of ductin (DCCD-reactive vacuolar H⁺-AT-Pase subunit) in mammalian membrane complexes. Exp. Cell Res. 207, 261-270.

83. Finbow, M. E., Eliopoulos, E. E., Jackson, P. J., Keen, J. N., Meagher, L., Thompson, P., Jones, P., and Findlay, J. B. C. (1992). Structure of a 16 kDa integral membrane protein that has identity to the putative proton channel of the vacuolar H⁺-ATPase. Protein Engineering 5, 7-15.

84. Finbow, M. E., Pitts, J. D., Goldstein, D. J., Schlegel, R., and Findlay, J. B. C. (1991). The E5 oncoprotein target: A 16-kDa channel-forming protein with diverse functions. Molecular Carcinogenesis 4, 441-444.

85. Rubin, G. M. (1988). Drosophila melanogaster as an experimental organism. Science 240, 1453-1459.

86. Lai, S. P., Watson, J. C., Hansen, J. N., and Sze, H. (1991). Molecular cloning and sequencing of cDNAs encoding the proteolipid subunit of the vacuolar H⁺-ATPase from a higher plant. J. Biol. Chem. 266, 16078-84.

87. Pietrantonio, P. V. and Gill, S. S. (1993). Sequence of a 17 kDa vacuolar H⁺-ATPase proteolipid subunit from insect midgut and Malpighian tubules. Insect Biochem. Molec. Biol. 23, 675-680.

88. Brown, D., Sabolic, I., and Gluck, S. (1992). Polarized targeting of V-ATPase in kidney epithelial cells. J.exp.Biol. 172, 231-243.

89. Gluck, S. L., Nelson, R. D., Lee, B. S., Wang, Z. Q., Guo, X. L., Fu, J. Y., and Zhang, K. (1992). Biochemistry of the renal V-ATPase. J.exp.Biol. 172, 219-229.

90. Brown, D., Sabolic, I., and Gluck, S. (1991). Colchicine-induced redistribution of proton pumps in kidney epithelial cells. Kidney International 40, S79-S83.

91. Bradley, T. J. (1989). Membrane dynamics in insect Malpighian tubules. Am.J.Physiol. 257, R967-R972.

92. Stone, D. K., Crider, B. P., and Xie, X.-s. (1990). Structural properties of vacuolar proton pumps. Kidney international 38, 649-653.

93. Butt, E., Geiger, J., Jarchau, T., Lohmann, S. M., and Walter, U. (1993). The cGMP-dependent protein-kinase - gene, protein, and function. Neurochemical Research 18, 27-42.

94. Moffett, D. F., Smith, C. J., and Green, J. M. (1983). Effects of caffeine, cAMP and A23187 on ion transport by the midgut of tobacco hornworm. Comp.Biochem.Physiol. 75C, 305-310.

95. Wolfersberger, M. G. and Giangiacomo, K. M. (1983). Active potassium transport by the isolated lepidopteran larval midgut: stimulation of net potassium flux and elimination of the slower phase decline of the short circuit current. J. exp. Biol. 102, 199-210.

96. Ryerse, J. S. (1980). The control of Malpighian tubule developmental physiology by 20-hydroxyecdysone and juvenile hormone. J. Insect Physiol. 26, 449-457.

97. Ryerse, J. S. (1989). Electron microscope immunolocation of gap junctions in Drosophila. Tisue & Cell 21, 835-839.

98. Cioffi, M. (1984). Comparative ultrastructure of arthropod transporting epithelia. Amer.Zool. 24, 139-156.

99. van Hille, B., Richener, H., Evans, D.B., Green, J. R. and Bilbe, G. (1993). Identification of two subunit A isoforms of the vacuolar

(1993). Identification of two subunit A isoforms of the vacuolar H$^+$-ATPase in human osteoclastoma. Journal of Biological Chemistry 268, 7075-80.

100. Nelson, H. and Nelson, N. (1990). Disruption of genes encoding subunits of yeast vacuolar H$^+$-ATPase causes conditional lethality. Proc. Natl Acad. Sci. U. S. A. 87, 3503-3507.

101. Noumi, T., Beltran, C., Nelson, H., and Nelson, N. (1991). Mutational analysis of yeast vacuolar H$^+$-ATPase. Proc. Natl Acad. Sci. U. S. A. 88, 1938-1942.

102. Lindsley, D. L. and Zimm, G. G. The genome of Drosophila melanogaster. San Diego: Academic Press, 1992.

103. Kaiser, K. and Goodwin, S. F. (1990). "Site-selected" transposon mutagenesis of Drosophila. Proc.Natn.Acad.Sci.U.S.A. 87, 1686-1690.

104. Spradling, A. C. and Rubin, G. M. (1982). Transposition of cloned P elements into Drosophila germ line chromosomes. Science 218, 341-347.

105. Gogarten, J. P., Fichmann, J., Braun, Y., Morgan, L., Styles, P., Taiz, S. L., DeLapp, K., and Taiz, L. (1992). The use of antisense mRNA to inhibit the tonoplast H$^+$-ATPase in carrot. Plant Cell 4, 851-64.

106. Zhao, J. J. and Pick, L. (1993). Generating loss-of-function phenotypes of the fushi tarazu gene with a targeted ribozyme in Drosophila. Nature 365, 448-451.

107. Knipple, D. C. and Marsella-Herrick, P. (1988). Versatile plasmid vectors for the construction, analysis and heat-inducible expression of hybrid genes in eukaryotic cells. Nucleic Acid Res. 16, 7748.

108. McGarry, T. J. and Lindquist, S. (1986). Inhibition of heat shock protein synthesis by heat-inducible antisense RNA. Proc.Natn.Acad.Sci.U.S.A. 83, 399-403.

109. Bellen, H. J., O'Kane, C., Wilson, C., Grossniklaus, U., Pearson, R. K., and Gehring, W. J. (1989). P-element-mediated enhancer detection: a versatile method to study development in Drosophila. Genes & Dev. 3, 1288-1300.

110. O'Kane, C. J. and Gehring, W. J. (1987). Detection in situ of genomic regulatory elements in Drosophila. Proc.Natl.Acad.Sci.U.S.A. 84, 9123-9127.

111. Kaiser, K. (1993). Second-generation enhancer traps. Current Biology 3, 560-562.

CHAPTER 5

MITOCHONDRIAL ATP SYNTHASE: STRUCTURE, BIOGENESIS AND PATHOLOGY

Howard T. Jacobs

INTRODUCTION

M itochondrial ATP synthase catalyses probably the most crucial enzymatic reaction for aerobic, eukaryotic cells, namely oxidative phosphorylation, driven by the proton gradient created by the mitochondrial electron transfer chain. The enzyme is also capable, under defined physiological circumstances, of catalyzing the reverse reaction, namely the energization of the inner mitochondrial membrane at the expense of ATP. For this reason, and because ATP hydrolysis is usually much easier to assay than proton-driven ATP synthesis, the enzyme is often referred to as the mitochondrial ATPase or H^+-ATPase, or as the F_1F_0 ATPase, after the names given to the two major sectors of the enzyme complex: the hydrophilic F_1 sector, with intrinsic ATPase activity, and the hydrophobic F_0 sector, which constitutes the proton channel through the inner mitochondrial membrane.

The enzyme is a complex hetero-oligomer, comprising at least 14 different polypeptide subunits in mammals,[1] probably a similar number (or slightly fewer) in plants,[2,3] and some 11 or 12 in fungi. It is also commonly denoted as complex V of the mitochondrial respiratory membrane, complexes I-IV comprising the enzymatic machinery directly responsible for electron transfer from NADH to molecular oxygen. In all eukaryotes studied, at least one, usually several subunits of mitochondrial ATP synthase are encoded in mitochondrial DNA, and the biogenesis of the enzyme therefore requires the co-operative expression of genes in two distinct compartments. The enzyme is structurally

Organellar Proton-ATPases, edited by Nathan Nelson; ©1994 R.G. Landes Company.

and functionally related to the ATP synthases of bacteria and chloroplasts, which are nevertheless simpler in terms of their subunit composition, and probably their propensity for regulation. The eubacterial
enzyme[4,5] nevertheless provides a convenient initial paradigm for studying
the properties of the mitochondrial enzyme. The present-day mitochondrial enzyme is believed to have originated in the endosymbiotic
event that created modern eukaryotes from an ancestral anaerobic protist,
that engulfed an aerobic bacterium to mutual advantage.

The features of mitochondrial ATP synthase and related proton-
ATPases have been extensively reviewed previously,[2,6-9] and it is my
purpose here firstly to bring these earlier treatments up to date, and
secondly to relate the properties of ATP synthase more specifically to
pathological states involving dysfunctions of the enzyme in a variety
of contexts.

SUBUNIT STRUCTURE AND FUNCTION

PURIFICATION AND SUB-FRACTIONATION

The mitochondrial ATP synthase is amongst the easiest of the
mitochondrial inner membrane redox complexes to isolate, and may
be conveniently prepared from mitoplasts by treatment with a mild
non-ionic detergent, such as n-heptyl β-thioglucoside, which releases
it into solution, and allows most of the remaining membrane components to be removed by centrifugation. Functionally active ATP synthase
may be recovered from such preparations by anion-exchange HPLC,[10]
reconstitution of the purified enzyme into proteoliposomes restores
proton-translocating activity. The F_1 ATPase may be easily released
from the F_0 sector by shaking with chloroform, or by sonication, although the latter treatment appears milder, and engenders less N-terminal proteolysis of F_1 subunits.[1] Walker and colleagues have recently
reported a modified procedure for the purification of the holoenzyme
in a form suitable for crystallization trials.[11]

SECTORAL SUB-STRUCTURE AND SUBUNIT COMPOSITION

The subunit composition of the mitochondrial ATP synthase from
various sources is summarized in Tables 5.1 and 5.2. It is worth pointing
out, at the outset, that in such an intensively studied area, which has
been investigated over a long period of time from a multidisciplinary
perspective, parallel nomenclatures have inevitably arisen, that are very
hard if not impossible to eliminate from the literature. In this presentation, I shall therefore attempt to use the most common names for
subunits of the complex, even though these are not necessarily the
most systematic or sensible. The fact that additional subunits have been
discovered or characterized only recently should argue against trying
to systematize the nomenclature prematurely. In addition, because the
enzyme is a member of a much larger family, no attempt to systematize the mitochondrial nomenclature can be undertaken without reference to other proton-ATPases and ATP synthases. Where alternative
names for subunits are in common use, or where a mitochondrial subunit is the structural or functional homologue of a known chloroplast

or bacterial subunit with a different name, I indicate this clearly at first usage, and the reader may wish to refer back frequently to Tables 5.1 and 5.2 for guidance.

The F_1 sector is peripherally located on the matrix side of the inner membrane, and consists universally (except in dicotyledonous plants) of five subunits (designated α, β, γ, δ and ε), in the stoichiometry $\alpha_3\beta_3\gamma\delta\varepsilon$. The molecular weight of beef heart F_1 as a whole, based on the amino acid sequences of its constituent subunits, is 371 kD (reference 12). Four of the subunits of mitochondrial F_1 are found in the F_1 sector of the *E. coli* enzyme. Some confusion may arise in that the *E. coli* F_1 also contains five subunits designated α, β, γ, δ and ε, but the *E. coli* and mitochondrial subunits δ do not correspond. Structural and molecular genetic analysis has confirmed that the fungal or mammalian mitochondrial δ subunit is the homologue of the bacterial (or chloroplast) ε subunit, and that the counterpart of the eubacterial δ subunit is found more loosely associated with the mitochondrial F_1 in the form of the so-called oligomycin sensitivity-conferring protein (OSCP). In plant mitochondria, the homologue of the bacterial ε subunit is designated δ' (see below).

OSCP is a member of a group of subunits which constitute a critical region of the enzyme, bridging the catalytic F_1 'headpiece' with the membranous proton channel of the F_0. This stalk region, sometimes referred to as the F_A sector, although regarded by some authors as a

---

Table 5.1. Subunits of mitochondrial ATP synthase

Sector	Subunit Name (Beef Heart F_1F_0)	Typical Mol Wt (kD)	Stoichiometry	Probable Function
F_1	α	56	3	Regulation of catalysis/coperativity/protein import
	β	52	3	Catalysis
	γ	35	1	Gating of proton channel
	δ	18	1	Anchoring to F_0
	ε	8	1	Anchoring to F_0
F_A/F_0	OSCP	21	1	Membrane attachment of F_1/Oligomycin sensitivity
	F_6	9	2	Membrane attachment of F_1
	IF_1	10	1	Regulation of ATP hydrolytic activity
F_0	6/a	23	?	Proton conductance
	b	25	2	Assembly/F_1 binding
	9/c	8	6-12	Proton conductance
	8/A6L	6	>1 ?	? (relic of ancient plasmid maintenance system?)
	d	19	1	Assembly/F_1 binding
	e	8	?	?

sub-domain of the F_0, contains subunits involved in the coupling function of ATP synthase.[13] In beef heart, the stalk region comprises, in addition to OSCP, the subunits designated as F_6 ('Coupling Factor 6') and IF_1 ('Inhibitor of F_1', also known as the Pullman and Monroy Inhibitor, or PMI). As far as is known, the latter two subunits have no counterparts in eubacteria or chloroplasts.

The F_0 proper contains at least 6 subunits in beef heart, three of which correspond with the well characterized subunits a, b and c of the *E. coli* enzyme, and three of which (d, e and subunit 8, also called A6L or chargerin II) have no counterparts in bacteria, and, as yet, no clearly proven function. It should be noted that, in the absence of genetic studies, or very exhaustive biochemical analyses to establish a direct function in ATP synthesis, the designation of several of the above subunits (or, for that matter, any additional proposed subunits) as bona fide constituents of mitochondrial ATP synthase remains unproven. However, the stoichiometric retention of the subunits mentioned, through many rounds of purification,[11] argues strongly that they are genuine components of the enzyme.

The beef heart enzyme is the most intensively studied, and may be taken as the prototypic mitochondrial ATP synthase. In the following discussion, the properties described may be taken to refer to the enzyme purified from this source, except where stated. To a lesser extent, the yeast and rat liver enzymes have proven invaluable for distinguishing those properties of the beef heart enzyme that are universal, from those constituting a tissue- or species-specific idiosyncrasy. Plant mitochondrial ATP synthases are less well characterized, especially in their F_0 sectors, but this field is rapidly catching up, aided by

---

Table 5.2. Subunit homologies between mitochondrial and other ATP synthases

Sector	Beef Heart mt F_1F_0	Plant (Dicot) mt F_1F_0	E. coli F_1F_0	Cyanobacteria F_1F_0	Chloroplast CF_1CF_0
F_1	α	α	α	α	α
	β	β	β	β	β
	γ	γ	γ	γ	γ
	δ	δ′	ε	ε	ε
	ε	ε	–	–	–
F_A/F_0	OSCP	δ	δ	δ	δ
	F_6	F_6	–	–	–
	IF_1	IF_1	–	–	–
F_0	6/a	6	a	a	IV
	b	b	b	b and b′	I and II
	9/c	9	c	c	III
	8/A6L	8	–	–	–
	d	d?	–	–	–
	e	?	–	–	–

the ease of gene cloning from heterologous sources using PCR technology, and the advent of routine peptide sequencing as an aid in gene cloning.

α AND β SUBUNITS: THE CATALYTIC CORE OF F_1 ATPASE

The F_1 sector may be dissociated and reconstituted, which has led to clear functional assignments to several of the subunits. The two largest subunits, α and β, together constitute a measurably active ATPase. In bacteria, re-constitution with the isolated γ subunit is sufficient to regenerate an enzyme with very similar properties to those of the entire F_1 (reference 14). The enzyme catalyses an essentially reversible reaction, and the physiological substrate for both ATPase and ATP synthase activities is the magnesium-complexed nucleotide. A number of lines of evidence support the view that the catalytic site is located within the β subunit. The isolated β subunit from bacterial and chloroplast sources has been shown to exhibit a low intrinsic ATPase activity,[15,16] and the high degree of conservation of the β subunit suggests that this property is likely to be general.

Six adenine nucleotide binding sites are present on the mitochondrial, as on the bacterial F_1, three of which readily exchange bound nucleotide for nucleotide in the medium, have a high K_d, and are classed as catalytic sites.[17,18] They exhibit strong negative co-operativity,[19] and bind a variety of adenine nucleotide analogues that block catalysis. The supposition that these sites are located wholly or mainly on the three β subunits is supported by cross-linking[20-23] and inhibitor studies,[24] and by the identification of highly conserved sequence motifs in the β subunit similar to those found in the nucleotide-binding sites of many purine nucleotide-binding proteins, including adenylate kinase, p21ras, and elongation factor Tu.[25,26] Most notable among these motifs is a glycine-rich loop, GX_4GKT (see Fig. 5.1). Recent mutational analysis of bacterially expressed rat liver β subunit has confirmed that this sequence is essential for ADP/ATP binding at the catalytic site, and that a second domain with sequence similarity to adenylate kinase is involved,[27] although several other such regions previously proposed as candidates for involvement, as well as a number of specific residues, were excluded. The glycine-rich sequence retains nucleotide-binding as a 50 amino acid peptide,[28] and binding studies suggest that it interacts with the pyrophosphate moiety of adenine nucleotides.[28]

The three additional sites fail to release bound nucleotide during ATP synthesis or hydrolysis, and are regarded as non-catalytic sites (see reference 9 for review). They are capable of binding GTP and pyrophosphate (PP_i), in addition to adenine nucleotides.[29] Inhibitor[24] and fluorescence-quenching[30] studies indicate that the catalytic and non-catalytic binding sites are in close proximity, almost certainly both involving residues on the β subunit. Cross-linking studies in the mitochondrial, bacterial and chloroplast enzyme have identified a pair of conserved tyrosine residues on the β subunit (Tyr-345 and Tyr-368 in the beef heart enzyme, corresponding with Tyr-331 and Tyr-354 in *E. coli*), that are involved in nucleotide binding. DCCD modification (see reference 9 for review) has identified two glutamate residues in a domain of the β subunit that are likely to be involved in catalysis: Glu-198 is modified in the beef heart subunit, and the nearby residue, equivalent to Glu-187 of beef heart, is modified in the thermophilic bacterium PS3. As the equilibrium constant on the surface of the enzyme is close to 1, it has been proposed that the free energy present in the electrochemical proton gradient ($\Delta\mu_H+$) promotes the release of bound ATP from the enzyme, by driving a reversible conformational change.

The α subunit, at least in the bacterial enzyme, also has adenine nucleotide binding capacity, and a putative nucleotide binding site can be discerned in the sequence of the mitochondrial α subunit, including a glycine-rich loop (see Fig. 5.1), leading many investigators to suggest that at least the non-catalytic nucleotide binding sites of the mitochondrial F_1 might be located at the interfaces of the α and β subunits. This view has received some support from cross-linking studies,

Bovine β	A	K	G	G	K	I	G	L	F	**G**	G	A	G	V	**G**	**K**	T	V	L	I	M	E	L	I	N	N
Maize β	Q	R	.	.	.	.	.	.	.	.	.	.	.	.	.	.	.	.	.	.	.	.	.	.	.	.
Yeast β	.	R	.	.	.	.	.	.	.	.	.	.	.	.	.	.	.	.	F	.	Q	.	.	.	.	.
E. coli β	.	.	.	.	.	.	V	.	.	.	.	.	.	.	.	.	.	.	N	M	.	.	.	.	R	.
Bovine α	G	R	G	Q	R	E	L	I	I	**G**	D	R	Q	T	**G**	**K**	T	S	I	A	I	D	T	I	I	N
Sea urchin α	.	.	.	.	.	.	.	.	.	.	.	.	.	.	.	.	.	A	V	.	.	.	.	.	.	.
Yeast α	.	.	.	.	.	.	.	.	.	.	.	.	.	.	.	.	.	A	V	.	L	.	.	.	L	.
Rice α	.	.	.	.	.	.	.	.	.	.	.	.	.	.	.	.	.	A	.	.	.	.	.	.	L	.
ATP synthase consensus	Ψ	r	G	Ψ	Π	x	x	Φ	Φ	**G**	x	x	Ψ	x	**G**	**K**	T	x	φ	x	φ	Δ	x	I	x	N
E. coli AK				M	R	I	I	L	L	**G**	A	P	G	A	**G**	**K**	G	T	Q	A	Q	F	I	M	E	K
Human p21ras	M	T	E	Y	K	L	V	V	V	**G**	A	G	G	V	**G**	**K**	S	A	L	T	I	Q	L	I	Q	N
Bacillus EF-Tu	K	S	H	A	N	I	G	T	I	**G**	H	V	D	H	**G**	**K**	T	T	L	T	A	A	I	T	T	V
Overall consensus	x	x	x	x	π	φ	x	φ	Φ	**G**	x	x	ψ	x	**G**	**K**	Ψ	ψ	x	ψ	x	x	φ	φ	x	x

Fig. 5.1. Comparison of amino acid sequences around the glycine-rich loop of the adenine-nucleotide binding site of mitochondrial ATP synthase α and β subunits from selected species, with those from the E. coli β subunit, and three other purine nucleotide-binding proteins: adenylate kinase (AK) from E. coli, p21ras (human), and Elongation Factor EF-Tu from Bacillus subtilis. Dots indicate residues conserved amongst the grouped α and β sequences. Consensus sequences are shown for ATP synthase (α and β subunits combined), and for the sequence set as a whole, with absolute identity denoted in upper case, and predominant amino acids (no more than one exception allowed) denoted in lower case, using the following ambiguity codes: Φ, φ - hydrophobic (ILVMFW); Δ - acidic (DE); Π, π - basic (RK), Ψ, ψ - neutral polar (GSTYQNPA). Dissimilarities are shown by a lower case x. The invariant residues of the core sequence (Gx_4GK) are boxed in bold, with the adjacent amino acids of conserved type shown boxed in dotted lines. Sequences were obtained from the GenBank-EMBL and NBRF databases, except the sea urchin (Strongylocentrotus purpuratus) α subunit sequence (A. Chisholm and H. T. Jacobs, unpublished data).

but a careful recent study of this type using the reagent FSBA (5'-p-fluorosulfonylbenzoyl-8-azido[³H]adenosine) has shown that the cross-linked residues both lie within the β subunit,[31] being His-427 and Tyr-345 of the beef heart polypeptide. Tyr-345 would appear to have been excluded, mutationally, from involvement with the catalytic binding site,[27] contradicting inferences from earlier cross-linking data.[23] Nevertheless, a role for the α subunit in regulatory nucleotide binding is strongly supported by mutational studies in fission yeast, in which a point mutation at Gln-173 in the putative nucleotide binding domain leads to a lowered enzyme affinity for ADP, and importantly, a loss of negative co-operativity.[32] Similar studies on the β subunit[33] have implicated the equivalent residue, Gln-170, in such co-operativity. The best current interpretation of the data is that non-catalytic adenine nucleotide binding, as well as substrate co-operativity, are likely to involve both of the major subunits. The precise roles of the nucleotide binding pockets and other residues at the interface(s) of the α and β subunit remain to be elucidated.

Co-operativity of the catalytic sites may be illustrated by the fact that substrate binding at one site accelerates the rate of product release from a second site by up to 5 orders of magnitude (see Ref. 7, 29). Recent binding studies on the beef heart F_1 ATPase, demonstrate, in addition, that occupancy of one catalytic site by inhibitory MgADP prevents binding to the others.[29]

In beef heart, the α subunit has an unusual N-terminus in the majority of chains, modified by the addition of pyrrolidonecarboxylic acid.[1] Recent genetic evidence from yeast, coupled with suggestive biochemical evidence from mammalian cells, has implied an additional, and most unexpected role for the α subunit, in mitochondrial protein import.[34-36] I shall discuss this in greater detail below (see 'Biogenesis'). In other respects, the α subunit exhibits a weak sequence similarity with the β subunit, mainly in the regions that have been implicated in nucleotide binding (see Fig. 5.1). It is plausible, though not proven, that the two subunits arose by duplication of an ancient gene.

The β subunit is the most highly conserved of the polypeptides of mitochondrial ATP synthase, exhibiting approximately 80% amino acid sequence identity in inter-kingdom comparisons, and about 75% identity with bacterial β subunits. The conservation extends over the length of the polypeptide, with the exception of most of the targeting N-terminal presequence, which, like all such elements, is highly variable even between relatively closely related species. The presequences of β subunits nevertheless fulfill all the criteria now regarded as 'standard' for such signals: the presence of many positively charged and hydroxyl-group amino acids, potentially adopting an amphiphilic structure, the almost complete absence of acidic residues, and the absence of runs of hydrophobic residues. An arginine at positions -2 and/or -10 is also found in most cases. The presequence of yeast F_1 β has, moreover, been used in a number of experimental contexts as a model to study targeting of mitochondrially destined proteins (see below, 'Biogenesis'). The only significant departure from the pattern of conservation of the β subunit is in the green alga *Chlamydomonas reinhardtii*, where a long

C-terminal domain is appended to an otherwise well conserved, typical β subunit.[37] The significance of this observation remains unclear, although one possible explanation may lie in the fact that exceptionally in this organism, mitochondrial and chloroplast targeting sequences are superficially rather similar, and the C-terminal extension may therefore be involved in some way in restricting the polypeptide to the mitochondrial destination.

The arrangement of the major subunits of the F_1 sector has been studied by low resolution X-ray diffraction, taking advantage of the cysteine residues found uniquely in the α subunit to identify it specifically by heavy atom labeling.[38] The α and β subunits each exist as trimeric arrays which are organized in two slightly offset, inter-digitated layers along the 3-fold axis. The layer of α subunits is located close to the axis, allowing them to interact with each other, while the β subunits are arrayed further from the axis, and interaction each with an α, but not another β subunit. At one end of the structure, part of the interface between the pairs of heterologous subunits encloses a solvent-accessible pocket, with the interfaces at the other end more open and exposed.

γ, δ AND ε SUBUNITS OF THE F_1 SECTOR

The three (stoichiometrically) unique subunits of the F_1 allow the catalytic centre formed by the α and β subunits to interact productively with the membranous proton-channel, via the 'stalk region' subunits (see below). The γ subunit is directly involved in coupling of proton flow to ATP synthesis, and appears to function specifically in the gating of the proton channel.[13,39] Based on thiol cross-linking studies, it interacts directly with the b subunit of the F_0. By analogy with results from *E. coli*, the γ subunit, especially its (highly conserved) carboxy-terminus, is assumed to have important roles both in catalysis and assembly of the F_1 (see Ref. 2). Residues in a conserved hydrophobic domain in the N-terminal portion of the subunit are also critical for assembly, and may be hypothesized as being important for inter-subunit interactions.

The δ and ε subunits form a tight complex, that is reconstitutable in vitro. It is generally believed that their role in the overall complex is essentially to anchor the catalytic portion of the F_1 to the F_0, and transmit the proton conductance signal in some way, as they do not appear to influence catalytic properties in vitro. The δ subunit has a pronounced hydrophobic character. In dicotyledonous (but not, apparently, in monocotyledonous) plants, the F_1 is isolated, exceptionally, as a six subunit complex. This complex contains an additional subunit homologous with the mammalian or yeast OSCP (see below), which in these species has (unfortunately) been designated as the δ subunit. The enzyme from these species does nevertheless contain a true homologue of the mammalian δ (or bacterial ε) subunit, which has been designated δ',[40] and also contains the ε subunit that is unique to the mitochondrial form of the enzyme.

In comparing the sequences of the mitochondrial γ, δ (in dicotyledonous plants δ') and ε genes between species, it is evident that all three are markedly less well conserved than the 'core' F_1 subunits α and β. This also holds up if the comparison is extended to the corresponding bacterial or chloroplast subunits (except of course for the ϵ subunit, which has no counterpart in the prokaryotic enzyme). Subunit γ is markedly better conserved than δ, especially within specific regions.

In yeast, amino acid sequencing has identified subunit ϵ as a 61-amino acid polypeptide, with net basic charge.[41] Similarity with the somewhat larger bovine ϵ subunit is relatively low (26% identity), although the mammalian and plant homologues are rather more similar, at least in their N-termini.[42] No convincing similarity with any other known protein has been established, therefore, the evolutionary origins of the subunit remain unknown. Unlike the majority of other ATP synthase subunits, subunit ϵ, at least in sweet potato,[43] is not made as a larger precursor with a cleavable N-terminal presequence; only the N-terminal methionine being removed to create the mature polypeptide. As regards its overall role in the complex, very little is known about the ϵ subunit, although a recent study, using phosphorescence of the single tryptophan residue of the bovine polypeptide, demonstrated a conformational change concomitant with Mg-ATP binding.[44] This is consistent with the view already put forward, that the single-chain subunits of the F_1 might be involved in transducing the proton conductance signal through the stalk/stem to the catalytic site(s), by means of a concerted series of conformational changes. The mitochondrial ϵ subunit is devoid of the inhibitor activity ascribed to the ϵ subunit of chloroplast ATPase: this is not surprising, in view of the fact that the mitochondrial and chloroplast ϵ subunits are not homologues.

Low resolution X-ray diffraction indicates that the γ, δ and ϵ subunits associate with a single pair of α and β subunits.[45] Recent studies to 6.5 Å resolution[46] have given some insight into the overall architecture of the F_1, and the nature of its asymmetry, contributed by the three unique subunits. From the main globular structure, 110 Å in diameter, protrudes a 40 Å stem, presumed to be part of the stalk that joins the F_1 to the membrane. The stem contains two α-helices in a coiled coil configuration, the longer of which extends through the centre of the structure for 90 Å, emerging on the other side into a 'dimple' 15 Å deep. A 'pit' next to the stem penetrates approximately 35 Å into the central particle. The α and β subunits of the globular particle are arranged as three/six fold-symmetry elements around the central axis, but with an inescapable overall asymmetry. The central-stem helices could provide the mechanism whereby conformational changes generated by proton conductance are transduced to the catalytic sites.

THE OLIGOMYCIN SENSITIVITY-CONFERRING PROTEIN OSCP

OSCP, a basic polypeptide in the 20-23 kD size range by SDS-PAGE, has been characterized from a number of species, including fungi[47] and mammals.[4,48] It has no intrinsic catalytic activity, and reconstitution studies have shown that it is not needed for association of the F_0 and F_1 sectors per se, but is absolutely required for the restoration of ATP synthesis in OSCP-depleted F_1-bound membrane preparations.[49,50] It is also needed to reconstitute ATP-driven reduction of NAD$^+$ by succinate, and oligomycin or dicyclohexylcarbodiimide (DCCD) sensitivity to the ATPase.[49] Bovine heart OSCP, cloned from a cDNA library, and expressed in *E. coli*, was able to mimic all these biological activities when added to OSCP-depleted ATP synthase complexes.[51] Although poorly conserved at the primary sequence level, OSCP shows a remarkably conserved secondary structure, even with the homologous (δ) subunit of chloroplast ATP synthase.[52] As expected, OSCP is accessible to proteolysis only on the matrix side of the complex.[53] Nested C-terminal deletion mutagenesis of the cloned protein[51] has shown that the extreme C-terminal 10 amino acids are essential for the biological activity of OSCP.

Evidence for the universality of OSCP in plant mitochondria is, thus far, somewhat indirect. Plant mitochondrial F_1 ATPases are commonly less oligomycin-sensitive than their mammalian counterparts, but in cross-reconstitution studies, plant F_1 productively interacts with the mammalian OSCP.[54] As already indicated, a structural homologue of OSCP has been isolated associated with the F_1 preparation in a number of species of dicotyledonous (but not monocotyledonous) plants, including pea[55] and sweet potato.[56] Reconstitution studies have yet to demonstrate that the protein functions in these species in an entirely analogous manner to mammalian or fungal OSCP. The fact that the OSCP-homologue associates intimately with the F_1 sector from at least some plants may indicate that the plant enzyme is more 'prokaryotic' in character than its fungal or metazoan counterparts. This recalls the observation, from ribosomal RNA sequencing, that the plant mitochondrial genetic system appears to have evolved much less far from its prokaryotic origins than the mitochondrial genomes of other taxonomic groups. It has even been suggested that plant mitochondria might have originated in a separate endosymbiotic event. On the other hand, plant mitochondrial ATP synthase possesses a number of clearly 'mitochondrial' features that distinguish it from the bacterial enzyme, including possession of a number of mitochondrial-specific subunits, and susceptibility to regulation by the IF_1 inhibitor.

THE 'INHIBITOR' SUBUNIT IF_1

An intrinsic, heat-stable polypeptide inhibitor of ATPase activity, designated IF_1 ('Inhibitor of F_1') is recovered from preparations of de-energized mitochondria from a variety of sources, and the polypeptide responsible has been purified from beef heart, rat liver, rat skeletal muscle, yeast and plants (see Refs. 2 and 57). The gene for it has been cloned from several sources, including plants, fungi and animals, revealing it to be a relatively well-conserved polypeptide of approximately 80 amino acids (e.g. 82 from rat[58,59]), varying in apparent molecular weight between 7.5-12.5 kD. The peptides from yeast and potato are cross-active, suggesting strong functional conservation. In vertebrates, IF_1 is present in fibroblasts, as well as in heart and skeletal muscle, but its reported stoichiometry in the enzyme appears to vary between tissues.[60] In some instances, such as so-called 'fast-rate' hearts, functional IF₁ may be absent. Such findings should be interpreted cautiously, however, as the amount of inhibitor isolated with a given preparation is likely to depend on the method used for purification, as well as the energetic state of the mitochondria prior to (or during) isolation. Inhibition can be effective at a stoichiometry of 1:1, under appropriate conditions.[61]

The ATPase inhibitor acts uniquely on what is traditionally regarded as the back reaction (ATP hydrolysis): oxidative phosphorylation is unaffected (discussed in Refs. 57 and 62), and indeed, evidence suggests that its binding to F₁ is dependent on the energetic state of mitochondria, and in particular, pH. The cloned and overexpressed rat inhibitor polypeptide exhibits a structural transition involving loss of α-helical content, in response to the lowering of pH, which correlates with activation of its inhibitory capacity.[59] Bacterially expressed bovine heart IF₁ undergoes a similar structural transition.[63] Data from studies on intact rabbit heart mitochondria have indicated two kinds of interaction of IF₁ with ATPase: a 'docking interaction', in which the subunit is loosely or non-functionally bound, and an 'inhibitory interaction' at a distinct regulatory site,[64] distinguishable by their response to Zn²⁺ ion. Recent evidence suggests that the 'docking' interaction is on a small membrane protein distinct from F₁.[65]

Cross-linking studies have located the binding site for the inhibitor subunit in the yeast enzyme at the interface between the α and β subunits.[62] In the bovine enzyme, mutational analysis has implicated residues 35-45 of the inhibitor polypeptide in activity.[66] This is confirmed by an analysis of short peptides from this region, that can function as active inhibitors.[67] This segment of IF₁ has some similarity to a portion of the β subunit near to the C-terminus,[68] which has led to the suggestion that it may interfere with communication between the α and β subunits,[66] through mimicry of an interfacial surface between them. This, in turn, would prevent the conformational changes required for nucleotide release from the catalytic site. In vitro studies confirm that the beef heart IF₁ traps nucleotide in one of the catalytic sites of the F₁ sector.[69] Two other polypeptides, of 9 and 15 kD respectively, apparently facilitate binding of the inhibitor subunit to F₁F₀ ATPase in yeast, although the possible counterparts of these factors in mammalian mitochondria have not been identified.

SUBUNIT F₆

The confusingly named subunit F₆ is the third component of the stalk region. It is a relatively well conserved, heat-stable but trypsin-sensitive, hydrophilic polypeptide of about 9 kD, found in both plant and animal mitochondria. Its presence in the fungal enzyme has not been demonstrated.[70] It is located uniquely on the matrix side of the inner membrane, is present in a stoichiometry of 2 with respect to the F₁ sector[53], and is absolutely required in the beef heart enzyme, both for binding of F₁ to the membrane[1], and for proton-driven ATP synthesis. The latter properties distinguish it from OSCP, in that its absence prevents stable interaction between the two sectors, whereas OSCP-depleted F₁ can still bind to the membrane.

Conserved Integral Subunits of the F_0 Proton Channel

The mitochondrial homologue of *E. coli* subunit a, usually designated subunit 6, is relatively poorly conserved at the primary sequence level, although it exhibits a well conserved hydropathy profile, and almost certainly adopts a highly conserved secondary structure. It is characterized by five trans-membrane helices, one of which contains seven residues conserved between the bovine mitochondrial and *E. coli* polypeptides. These are capable of forming an amphipathic face of the α helix, which has been suggested to form part of the proton pore[2] through the membrane. The yeast,[71] and perhaps plant[72] subunit 6 polypeptides are subject to N-terminal processing (see below, 'Genetics' and 'Biogenesis'). The mature polypeptide is generally about 23-25 kD in size, but often migrates faster on polyacrylamide gels, probably as a result of its hydrophobicity. Subunit 6 is not available to proteolysis from either side of the inner membrane, consistent with its proposed role as a core component of the proton channel.[53] The precise stoichiometry of subunit 6 in the complex has, for trivial technical reasons, proven difficult to assess. In *E. coli* it is present in a stoichiometry of 1, and it is probable that the mitochondrial enzyme is similar.

Subunit 9, also known as subunit c after its eubacterial homologue, alternatively as the 'DCCD-binding protein', or proteolipid (as a result of its extreme hydrophobicity and hence solubility in organic solvents), is generally regarded as the other essential component of the proton channel. Like subunit 6, it is proteolytically inaccessible from both sides of the inner membrane.[53] In yeast, six copies of subunit 9 are arrayed in the proton channel of the F_0 sector,[73] although estimates in mammals suggest a stoichiometry nearer to 10 or 12. The subunit is certainly, on a molar basis, one of the most abundant inner membrane proteins. Unlike subunit 6, it is a very highly conserved polypeptide, being identical in cattle, sheep and humans,[74] consistent with the idea that it constitutes the functional core of the F_0 proton channel. It is probably folded into a hairpin of two transmembrane α-helices, linked by a β-turn near the membrane surface.[75] A carboxyl group essential for proton conductance, which is also the site of reactivity with DCCD, is located in the C-terminal α-helix: in mitochondrial ATP synthases this is always a glutamate residue, although in *E. coli* it is an aspartate.[5,75] The C-terminal helical arm of the polypeptide, as well as glycine-18, are essential for the assembly of subunit 9 into the complex, in yeast.[76]

The 25 kD subunit b (previously denoted in yeast as subunit 4, and in mammalian cells by some authors as subunit F_0I or F_0I-PVP, based on its N-terminal sequence) is an amphiphilic protein[77] that has been shown to be involved in the binding of F_1 to the membranous part of the complex.[78] It is exposed on the matrix side of the membrane.[53] The yeast and beef heart polypeptides are structurally homologous with each other, and with the b subunit from *E. coli*,[79] as well as with the duplicated but diverged subunits I and II in chloroplasts.[2] It has also been characterized in plants.[35] The N-terminus carries the conserved sequence Pro-Val-Pro. In yeast the protein is critical for the correct assembly of the complex (see below, 'Biogenesis'), and its extreme C-terminus, based on studies in both yeast[80] and mammals,[81] also appears to be required for maintaining the integrity of the gating system for the proton channel. This is consistent with the topology of its bacterial homologue, whose C-terminus is in contact with the F_1 sector.[82] In mammalian mitochondria, subunit b is involved in both proton conduction and oligomycin-sensitivity, based on proteolytic cleavage and reconstitution experiments.[83,84]

'MITOCHONDRIAL-SPECIFIC' SUBUNITS OF F_0

As already stated, subunits d and e, as well as the mtDNA-encoded polypeptide designated subunit 8, are specific to the mitochondrial enzyme, having no counterparts in prokaryotes or chloroplasts. Genetic evidence from yeast[85] indicates clearly that subunit d (encoded by the ATP7 gene), which is 22% identical with its beef heart homologue,[79] is essential for functional, oligomycin-sensitive ATPase. The polypeptide isolated from various sources is of approximately 20 kD (173 amino acids in yeast), and, perhaps surprisingly, is mainly hydrophilic, suggesting a possible role in assembly and/or bridging to the F_1 sector. This is supported by topographic studies.[53] Its N-terminus is N-acetylated in yeast[85] and mammals[1], and apart from removal of the N-terminal methionine, no N-terminal processing of the kind that accompanies the import of most mitochondrial polypeptides from the cytosol has been detected.[85]

Subunit e has been purified and characterized only recently, from beef heart[1] and rat liver,[86] and it remains unclear how widespread is its occurrence, as it appears to be absent from plant and fungal ATP synthases. Based on amino acid sequencing and cDNA cloning, it is a basic, hydrophilic polypeptide of 70 amino acids (8 kD). Although the bovine and rat peptides are highly homologous with each other, they are completely unrelated to any of the previously characterized subunits of mitochondrial, chloroplast or bacterial ATP synthase. Subunit e has no known function. However, a significant stretch of sequence similarity is apparent, with members of the troponin T super-family. The conserved segment represents the Ca^{2+}-dependent tropomyosin-binding region of troponin T. Whether this indicates a site of interaction with a Ca^{2+}-dependent regulatory protein remains unknown, although regulation of ATP synthase by Ca^{2+} ion is well established (see below, 'Regulation').

Subunit 8, previously called A6L, because of its occurrence as an open reading frame upstream of the subunit 6 gene in mammalian mtDNAs, remains something of an enigma. Mutational studies in yeast identify it clearly as an essential component of functional ATP synthase, but its precise function, as well as its evolutionary origin, remain obscure. The protein is poorly conserved at the primary sequence level, except for the N-terminal tetrapeptide, which is almost universally Met-Pro-Gln-Leu, but exhibits a strikingly conserved hydropathy profile.[87,88] It also displays a considerable degree of length heterogeneity, ranging from 47 or 48 amino acids in most fungi, to over 70 in mammals. Overall, the protein is highly hydrophobic, but earlier suppositions that the extended hydrophobic segment in the middle of the polypeptide was a trans-membrane domain have been contradicted by recent mutational data in yeast (see below, Ref. 89), and to a lesser extent by the observation that the C-terminus is exposed on the matrix side of the membrane.[53] It is generally regarded as an abundant component of the F_0, but its precise stoichiometry remains in some doubt. The N-terminus of subunit 8, at least in bovine heart,[1] is N-formylated, hence escapes post-translational N-terminal processing.

Despite the absence of a workable DNA transformation system for mitochondria, the advent of nuclear ('allotopic') expression technology[90] has allowed a detailed structure/function analysis of subunit 8 to be conducted in yeast by Nagley's group,[89] and we have recently initiated a similar study in mouse cells. The use of N-terminal targeting information of different potencies has, in addition, allowed subunit 8 variants that are poorly assembled, but 'catalytically functional' to be distinguished from those that are truly inactive. Such experiments have revealed that the conserved N-terminus, as expected, is essential for function, whilst the (positively) charged carboxy-terminal region is required for efficient assembly of subunit 8 into the membrane-bound complex. Introduction of charged residues into the hydrophobic segment has surprisingly little effect on function, suggesting that it is unlikely to span the membrane. A suggested, revised topology, is that it lies in a hydrophobic pocket on the matrix surface.

Based on a statistically significant similarity of the hydropathy profile of subunit 8 to that of a family of plasmid-encoded, membrane-acting toxins involved in plasmid maintenance in bacteria, I have proposed that subunit 8 may originally have been acquired extraneously by the ATP synthase complex, from a plasmid maintenance system operating in the ancestral endosymbiont that evolved into present-day mitochondria.[88] I have further suggested that such a mechanism may have contributed to the (selfish) maintenance of mtDNA as a separate genome over evolutionary time, but that subsequently this system became unnecessary, due to the functional gulf that came to separate the two genetic systems of nucleus and mitochondrion.[88,91] In this view, subunit 8, which might initially have functioned as a poison of oxidative phosphorylation, could have gradually evolved a (putative) regulatory role in the functioning of the enzyme. Although this idea is extremely hard to test formally, it offers certain predictions regarding the likely function of the subunit that are now amenable to experimentation.

GENETICS

CODING LOCATION OF ATP SYNTHASE SUBUNIT GENES

As already mentioned, mitochondrial ATP synthase is encoded partially in nuclear DNA, and partially in mitochondrial DNA, in all eukaryotes studied. This situation, although highly unusual, is not unique, as it applies also to Complexes I, III and IV of the mitochondrial respiratory chain, as well as to a number of chloroplast enzymes, including its own ATP synthase. A fundamental problem in understanding the genetics of these enzymes is to explain the persistence of the organelle genomes, and the partition of genes between the two compartments. In this regard it is useful to consider the situation in regard to the family of F_1F_0 ATP synthases/ATPases, which is detailed in Table 5.3. Three generalities may be put forward to summarize the data. Firstly, the distribution of ATP synthase genes between nuclear and organellar genomes is clearly not invariant. Secondly, there is a preponderance of hydrophobic subunits encoded in organelle DNA. Thirdly, particular subunits, notably mitochondrial subunit 6 (equivalent to bacterial subunit

Table 5.3. Genomic locations of ATP synthase genes

Sector	Subunit	Animal mt	Plant mt	Fungal mt	Chloroplast Equivalent	Note
F_1	α	Nuclear	Mitochondrial	Nuclear	Chloroplast	
	β	Nuclear	Nuclear	Nuclear	Chloroplast	
	γ	Nuclear	Nuclear	Nuclear	Nuclear	
	δ	Nuclear	Nuclear	Nuclear	Chloroplast	ε Subunit in Chloroplasts
	ε	Nuclear	Nuclear	Nuclear	—	
F_A/F_O	OSCP	Nuclear	Nuclear	Nuclear	Nuclear	δ Subunit in Dicot Plants/ Chloroplasts
	F_6	Nuclear	Nuclear ?	?	—	
	IF_1	Nuclear	Nuclear	Nuclear	—	
F_O	6/a	Mitochondrial	Mitochondrial	Mitochondrial	Chloroplast	Possibly Nuclear in Plasmodium
	b	Nuclear	Nuclear	Nuclear	1 Nuclear 1 Chloroplast	
	9/c	Nuclear	Mitochondrial	Nuclear and/or Mitochondrial	Chloroplast	
	8/A6L	Mitochondrial	Mitochondrial	Mitochondrial	—	Possibly Nuclear in Nematodes
	d	Nuclear	? Nuclear	Nuclear	—	
	e	Nuclear	?	—	—	

a or chloroplast subunit IV), are almost universally organelle encoded. Each of these statements must nevertheless be qualified.

Despite the absence of a universally applicable pattern of gene partition, the coding capacity of organelle DNA is relatively static in each of the major groups of eukaryotes. For example, ATP synthase subunits 6 and 8, but not subunit 9, are mtDNA-encoded in all metazoans, except nematodes, where subunit 8 has not been identified.[92] This is consistent with the idea that during the early period of eukaryotic evolution, prior to the emergence of the major modern taxonomic groups, the coding capacity of mitochondrial and chloroplast DNAs remained relatively fluid, with continuing transfer of genes from organelle to nuclear genomes. Subsequently, the gene contents of organelle DNAs are presumed to have become effectively fixed, as a result of the increasing genetic distance, hence functional isolation, between the genetic systems operating in the two compartments. This appears to have entrained certain conserved features of the biogenetic programme for ATP synthase and other organellar redox complexes, which have become established and effectively immutable in the major taxa. These questions will be returned to below (see 'Biogenesis').

The hydrophobic bias of organelle-encoded subunits is also not absolute, as a number of hydrophobic polypeptides of ATP synthase are nuclear-encoded in at least some contexts, most notably mitochondrial subunit 9 in most fungi and animals. The possible case of subunit 8 in nematodes, where no mitochondrial gene has yet been found,[92] may also be mentioned. Nuclear-coded subunit b is also an integral membrane protein with a hydrophobic domain. On the other hand, several hydrophilic subunits, including most of the F_1 sector in chloroplast ATP synthase, and notably subunit α in plant mitochondria, are organelle-encoded. This indicates that although hydrophobicity of the encoded protein seems to have disfavoured gene transfer to the nucleus, presumably as a result of the concomitant problems in solubilizing the cytosolic product for its journey to and through the organellar membrane system, the problem has been solved a number of times in eukaryote evolution, and that, furthermore, hydrophobicity cannot be advanced as the sole reason for the retention of particular genes in organelle DNA.

As far as subunit 6 is concerned, the malaria parasite *Plasmodium* presents an interesting exception. *Plasmodium* and closely related groups of protozoa are highly unusual amongst non-photosynthetic organisms, in possessing two organelle genomes.[93] A linear (concatemerized) genome of approximately 6 kb encodes several core components of the mitochondrial respiratory chain, as well as modularized ribosomal RNA genes ('rRNA genes in pieces'), whereas a 35 kb circular genome encodes a separate set of rRNAs, as well as number of polypeptides related to those found in the chloroplast DNAs of higher plants. No genes for subunits of ATP synthase have yet been identified in either of these genomes (J. E. Feagin, personal communication), although it remains possible that they are present, but encrypted in an unrecognizable form that requires extensive RNA editing to reveal. Alternatively, subunit 6 (and perhaps subunit 8, if it is present) might be nuclear-encoded in this organism. Apart from this case, the near-universality of an organellar genomic location for the subunit 6 gene, especially in view of the relative fluidity of organelle gene content in respect of other subunits of the enzyme, suggests that some features of this polypeptide, or the manner of its assembly into F_0, require its synthesis at the site of its incorporation into the inner membrane.

An unusual situation in respect of subunit 9 pertains in fungi. In yeasts such as *Saccharomyces cerevisiae* or *Candida parapsilosis*, the gene is mitochondrial,[94] but in filamentous fungi such as *Neurospora crassa* or *Aspergillus nidulans* a functional gene for the subunit is present in nuclear DNA, that encodes the polypeptide incorporated into the complex during vegetative growth.[95,96] Nevertheless, these fungi contain, in their mitochondrial genomes, an unexpressed copy of the subunit 9 gene, that resembles the corresponding mitochondrial gene from yeast.[97,98] Whether this additional gene is ever expressed, for example under physiological conditions not yet reproduced in the laboratory, or whether it is a true pseudogene, remains to be determined. This mitochondrial copy of the gene is absent entirely in *Podospora anserina*.[99]

COPY NUMBER AND ORGANIZATION
OF NUCLEAR ATP SYNTHASE GENES

Nuclear genes for the α and β subunits of F_1 have been characterized from a wide diversity of species, including many animals, plants, fungi and protozoa. Almost invariably, the genes are unlinked, and in some instances, such as the fungus *Neurospora crassa*, both genes are single copy.[100] In organisms where highly expressed genes are distinguished by a specific pattern of codon usage, such as in filamentous fungi, the α and β subunit genes fall into this class.[100] Comparing these genes in different species, intron number and position are extremely diverse, confounding the generally held view that intron-exon structure reflects the division of proteins into functional domains.

The exact number of α and β subunit genes in different metazoan species remains to be definitively documented, although it appears that β is a single-copy gene in most, if not all true diploid animals, whereas α is encoded by at least two genes in the cow, one of which is highly expressed in liver, the other in heart.[101,102] The situation is completely different in plants, where, as already indicated, the α subunit is encoded in mtDNA (in some species in multiple, usually identical copies), but the β subunit is encoded by a multi-gene family, at least in tobacco,[103] where two different isoforms are expressed, respectively, in leaves and in developing pollen. It is not yet known if this reflects any functional specialization of the ATP synthase in relation to gametogenesis, or merely a duplication of functionally equivalent polypeptides expressed in different tissues under distinct regulatory information. Lomax and Grossman[104] have argued that the existence of tissue-specific isoforms of various subunits of the mitochondrial redox complexes could be one mechanism for generating complex, tissue-dependent patterns of regulation of oxidative ATP synthesis.

Two subunit 9 genes were initially isolated from bovine cDNA libraries[105] and (unlinked) homologues of each gene have subsequently been cloned from human nuclear DNA.[106] The amino acid sequences of the two gene products, designated P1 and P2, show complete identity over the mature protein segments,[105] but considerable sequence divergence within the targeting presequences. Whether these genes are functionally interchangeable at the precursor protein level remains to be demonstrated, but they nevertheless exhibit a striking pattern of tissue-specific expression.[105] The two genes are expressed at similar levels in tissues of mesodermal origin, but the P2 gene is expressed at a 3-fold greater level than P1, in tissues of endodermal origin. This suggests that the tissue-specificity of subunit 9 gene expression probably arises very early in development. The results of sequence analysis imply that the subunit 9 gene duplicated between 100-200 million years ago, in the mammalian lineage.

Complementary DNAs corresponding with the human and bovine δ subunits have been cloned and characterized.[107,108] At the genomic level, the gene appears to be single-copy in both cases, although in the case of human, two classes of cDNA encoding identical polypeptides were isolated, one of which contained a 296 bp insert in the 3' untranslated region. It remains unclear whether this indicates a second copy of the gene that has somehow escaped detection, an alternative splicing pattern, or an inter-individual polymorphism.

Most of the remaining ATP synthase genes that have thus far been characterized appear to be single-copy, and unlinked, although the data on this are rather scant: the genes for most subunits have been studied closely at the genomic level in no more than a handful of species. Many genes have been characterized only through single cDNAs in one or a few disparate organisms. The existence of pseudogenes for some subunits, for example subunit 9 in mammals,[74] also complicates the analysis. Therefore, there is certainly much to learn about the genomic constitution and evolution of the various gene families that are required for the biosynthesis of mitochondrial ATP synthase.

ORGANIZATION OF MITOCHONDRIALLY-ENCODED ATP SYNTHASE GENES

Although the mitochondrial genome is always present in multicopy, genes within it are generally present only once per molecule. The notable exception is in plants, where mtDNA commonly contains multiple copies of a number of sequences, that function as recombinogenic sites for genomic rearrangements. Not surprisingly, ATP synthase genes are frequently involved in these, for example the *atpA* gene of soybean, encoding subunit α (Ref. 109). In addition, plant mtDNA contains a considerable and variable amount of non-coding sequence, probably generated during such recombination events, and a number of pseudogenes, such as the partial copy of ATP synthase subunit 9 in potato.[110]

In all animals and fungi thus far investigated, except nematodes where the subunit 8 gene is missing,[92] the genes for mitochondrially encoded ATP synthase subunits 8 and 6 are adjacent, co-transcribed, and form a single, di-cistronic mRNA. In all metazoans, the two polypeptides are encoded in different reading frames that overlap, often by just 1-2 nucleotides as in sea urchins.[111] The large number of re-arrangements of the mitochondrial genome that have occurred since the dawn of eukaryote evolution, even within the kingdoms, mean that this continued association is most unlikely to be due to chance, and must reflect some requirement for the synthesis of the two subunits to be closely coupled or co-regulated (see below, 'Biogenesis'). Unfortunately, the absence of subunit 8 in bacteria means that it is not possible to search in prokaryotes for a precedent to explain or investigate this phenomenon.

In plants, genes for ATP synthase subunit 6 show a peculiar organization. Commonly, especially in cereals such as maize[112] or sorghum,[72] they are present in multi-copy, with identical sequences of the 'core polypeptide' but highly divergent N-terminal extensions, typically of 100-200 codons, specifying peptides of unknown provenance or function.

EVOLUTIONARY CONSIDERATIONS

The mitochondrial location of several genes for ATP synthase subunits, coupled with the idiosyncrasies of mitochondrial gene expression and sequence evolution, place the evolution of these genes under constraints that do not apply to the nuclear-encoded subunits. Two striking examples are the effects of RNA editing on the amino acid sequence of the subunit 6 gene in trypanosomes,[113,114] and the effects of codon bias on the evolution of the subunit 6 and 8 genes in insects.[115] ATP synthase subunit 6 is a highly encrypted gene in a number of kinetoplastid (trypanosome) species whose mitochondrial genomes have thus far been characterized.[114] In such cases, a significant fraction of the coding sequence is created by RNA editing of the primary transcript, via insertion and deletion of U residues, for which reason the genes were not even initially recognized. Polypeptides encoded by such genes tend to exhibit a highly unusual amino acid composition compared with that of the homologous proteins from other taxa, reflecting the tendency to include runs of U residues at editing sites. Although overall protein structure, and the amino acid sequence of regions critical for function, are conserved, these subunits are much more diverged than expected, compared with other mtDNA-encoded polypeptides that are not subject to radical editing.

In insect mtDNA, the extreme (A+T) bias imposes an unusual pattern of sequence evolution, in which third base variability is actually lower than that at the first or second base positions, in comparing long diverged species, such as *Drosophila* and the honeybee *Apis*.[115] In addition, the content of amino acids with A or T in the first and second position is far above expectation in *Apis*, compared with gene sequences from other taxa which do not show such a compositional bias. Consistent with this view, the *Apis* genes, where the (A+T) bias is even greater than in *Drosophila*, appear to have evolved significantly faster than their dipteran homologues. Furthermore, detailed analysis of the patterns of sequence evolution of the ATP synthase subunit 6 gene in various *Drosophila* species has shown evidence for apparent inconsistencies in the rates of change at synonymous and non-synonymous sites between species.[116] The cause of these discrepancies is debatable, and codon bias is only one of several possible explanations, but some caution is signalled in attempting to place too great a reliance on one particular type of sequence information in phylogenetic analysis. The effects of such phenomena on the evolution of nuclear genes for ATP synthase subunits have also not been investigated systematically, and it is worth pointing out that because of the likelihood of co-evolution, such effects might render the use of nuclear-coded mitochondrial gene sequences for phylogenetic analysis equally hazardous, for organisms with extreme codon bias in mtDNA.

BIOGENESIS

MITOCHONDRIAL GENE EXPRESSION
AND ATP SYNTHASE BIOGENESIS

The mitochondrial genomic location of a number of ATP synthase genes in all, or almost all eukaryotes, necessitates an elaborate program of organellar genomic maintenance and organellar protein synthesis, to ensure the appropriate supply of the relevant subunits in response to physiological or developmental requirements. This operates at several levels, a number of which are potentially subject to regulation. A complete discussion of mitochondrial DNA and its expression machinery is beyond the scope of this work: I shall therefore refer only to those aspects of particular relevance to ATP synthase. An important, and unresolved question is the degree to which the supply of mtDNA-encoded subunits is limiting for the assembly of functional redox complexes in the inner mitochondrial membrane. The mtDNA-encoded subunits of ATP synthase, like those of the other redox complexes, are core constituents of the enzyme, and essential for its assembly, suggesting an important, and perhaps regulatory role in the overall biogenetic programme. Evidence has been presented to support such a hypothesis, in the case of cytochrome oxidase in yeast.[117] However, the validity of this idea, and its applicability to ATP synthase, remain unproven.

Studies of mtDNA content in different mammalian tissues indicates that the modulation of genomic copy number is the principal mechanism for regulating the rate of synthesis of mtDNA-encoded gene products in development.[118,119] Exactly how this is achieved remains mysterious, although the step in mtDNA synthesis most likely to be regulated is the decision to extend or terminate the growing H-strand, at the terminus of the D-loop. In bovine cells this has recently been shown to involve a sequence-specific template-binding protein with contra-helicase activity,[120] and proteins with binding sites at replication pause sites close to the origins for each strand have also been characterized in sea urchins.[121-123] How these activities are regulated in development and during the cell-cycle remains to be elucidated, but it is likely that they represent a key system for controlling the biosynthesis of ATP synthase, as well as the other redox complexes partially encoded in mtDNA.

Mitochondrial transcription is multicistronic in most organisms, but fungi employ multiple promoters, whereas in mammals, each strand of the mitochondrial genome is transcribed in its entirety, from just two promoters. In plants, most mitochondrial transcription units contain only one, or a few genes. In vitro transcriptional characterization of the maize *atp1* gene, encoding subunit α, has identified an 11 nucleotide sequence required for efficient initiation.[124] This core promoter appears to be shared with many maize mitochondrial genes, suggesting that if transcription is regulated, such regulation does not operate in a gene-specific manner.

In all cases, a complex and unusual method of RNA processing is required, to generate mature mRNAs. In animals, this involves the so-called punctuation model,[125] in which tRNA sequences directly flank most mRNAs, potentially requiring single cleavages to create simultaneously the termini of both a tRNA and the adjacent mRNA. Several exceptions may be noted, however, especially in organisms such as sea urchins, where most mitochondrial tRNA genes are clustered.[111] Strikingly, the ATP synthase subunit 8/6 mRNA is, in almost all metazoans, not flanked at its 3' end by a tRNA gene, suggesting a special mechanism for its maturation. In principle, this could allow the biosynthesis of ATP synthase to be independently regulated from that of the other mitochondrial redox complexes, although there is no evidence that this occurs. Three possibilities have been suggested: (i) the involvement of a cryptic, tRNA-like structure in the downstream COIII gene, that directs the processing machinery (in this case presumed to be RNase P) to the appropriate site;[126] (ii) the use of a transcriptional attenuator resembling bacterial *rho*-independent terminators, which might involve a hairpin structure at the gene boundary;[127] (iii) the involvement of a template binding protein that promotes polymerase dissociation, by analogy with the mechanism of attenuation beyond the mitochondrial rRNA genes in mammalian mtDNA. No convincing evidence has been presented to support any of these models, although a DNA-binding protein with a binding site exactly at the gene boundary has been characterized in sea urchins.[123] The 3' terminus of the mRNA appears to be precise, however (unpublished data), suggesting that specific RNA processing is more likely.

The maturation of yeast mitochondrial mRNAs involves an entirely different mechanism, with 3' cleavage at a conserved dodecamer sequence,[128] and a requirement for splicing in a number of genes (not including those coding for ATP synthase subunits). A yeast (nuclear) mutant has been isolated, which is partially deficient in subunits 6 and 8, and in which the mRNA for these subunits is present at reduced levels, despite the fact that the primary transcript covering the transcription unit in which these genes lie is actually elevated with respect to wild-type.[129] Transcripts of other mitochondrial genes are unaffected. This strongly suggests a specific deficiency in subunit 8/6 RNA processing, although the primary lesion could, conceivably, be in translation or mRNA stability. A distinct nuclear gene (NAM1/MTF2), originally identified by virtue of a defect in splicing of several mitochondrial pre-mRNAs in the mutant strain, is required specifically for the stabilization (or processing) of the *atp6* mRNA, even in intron-lacking strains.[130] In the intronless strain, other transcripts are unaffected, but the subunit 8/6 mRNA is strongly under-represented, compared with wild-type. The *NAM1/MTF2* gene has been proposed to encode an RNA-binding protein capable of interacting with stem-loop regions, thus contributing to their stability. Maturation of the mitochondrial subunit 9 mRNA in yeast appears to be under the control of a further nuclear gene.[131] Why so many different nuclear gene products should be required for the expression of so few, individual genes in mtDNA, remains a mystery. It would seem logical that this could allow the different redox complexes to be independently regulated, but such an idea is hard to apply to the subunits of a single enzyme, such as ATP synthase.

In at least three groups of organisms, namely kinetoplastids (trypanosomes), slime molds and higher plants, mitochondrial RNA synthesis is complicated by the requirements for RNA editing.[113] An entirely distinct mode of editing appears to operate in each of these taxa, and it remains unclear to what extent editing, which is certainly developmentally regulated in trypanosomes, is a significant factor in controlling the rate of synthesis of specific mtDNA-encoded gene products. In plant mitochondria, transcripts of ATP synthase subunits 6, 9 and α from a wide range of species, including tobacco,[132] wheat,[133] potato,[108] the evening primrose *Oenothera*,[134] sorghum[135,136] and petunia[137] are extensively edited, resulting in the creation of reading frames that more closely resemble those of the corresponding genes from other taxa. Editing in plants involves C to U and, exceptionally, U to C modifications, that commonly create functional stop codons, both at the 3' ends of the relevant genes, and often just upstream of start codons, where they interrupt upstream continuations of the reading frame. The importance of such changes is unknown, but it appears to be an efficient process, as unedited, partially edited or mis-edited transcripts are present only at very low levels (for review, see Ref. 138), except in pathological states. The extent of editing required for a given mRNA varies greatly between different genes of a single organism, and between the same gene in different organisms. For example, the 1533 bp open reading frame of the subunit α gene in *Oenothera* is edited at only four sites, whereas the much shorter subunit 6 mRNA is edited in the same species at 21 sites.[134] The advent of an in vitro system in which editing is supported[139] should help unravel its mechanism, and the basis of its specificity. RNA editing in trypanosomes occurs by an entirely different mechanism, involving guide RNAs, encoded in both the maxicircle and minicircle kinetoplast (mitochondrial) DNAs, which provide templating information for the insertion and deletion of runs of U residues.[113]

Mitochondrial translation is also unusual in a number of respects. In yeast, multiple mRNA-specific translation factors have been identified, initially by mutational studies, that interact with sequences in the 5' untranslated regions of mitochondrial mRNAs.[140] No such factors have yet been characterized for the subunit 8/6 mRNA, but at least two nuclear-coded gene products are probably required specifically for the translation and/or stabilization of the mRNA for ATP synthase subunit 9. In strains mutated for the *ATP13* gene, subunit 9 is absent from mitochondria, and is not synthesized in pulse-labeling studies.[141] The mRNA for subunit 9 is present in the mutant, however, but at a significantly reduced level, with concomitant accumulation of precursor transcripts. Two other mutants, designated aep1 and *aep2*, also fail to accumulate subunit 9.[142] *aep2* strains have much reduced levels of subunit 9 mRNA, and accumulation of precursor transcripts,[143] but it is unclear whether the primary lesion is in RNA processing, stabilization or translation. The *aep2* gene has been cloned,[143] and is unrelated to *ATP13*: both genes, however, represent previously unknown gene families.

Metazoan mitochondrial mRNAs contain no 5' untranslated sequences, although it has been proposed that mRNA-specific factors that interact with the 5' regions of coding sequence[144] might mediate translational regulation, by analogy with yeast. The unusual arrangement of the subunit 8/6 mRNA strongly suggests translational coupling of some kind, the purpose of which might be to ensure that the two subunits are always produced concomitantly, and/or with a fixed stoichiometry. The possibility that one or both polypeptides might directly regulate translation of the common mRNA, in a positive or negative feedback loop, must also be considered, although there is currently no evidence to support this idea. The strong implication from the conserved gene arrangement is that unregulated expression of subunit 8 and/or 6 is in some way deleterious: recent preliminary experiments in my own laboratory, in which mouse subunits 8 and 6 have been engineered for nuclear expression, suggest that over-expression of the two subunits has only a modest effect on glucose-grown cells, but seriously impairs growth and survival on a substrate such as galactose, which requires metabolization via mitochondrial oxidative phosphorylation to ensure cell viability (L. Sutherland and H. T. Jacobs, unpublished data).

In general, the mtDNA-encoded polypeptides are not subjected to N-terminal processing of any kind, and commonly start with the formyl-methionine contributed by the initiator tRNA, although a clear exception is subunit 6, at least in yeast,[71] where processing creates the terminus Ser-Pro-Leu-Asp. Examination of plant mitochondrial subunit 6 genes revealed this or a very similar sequence flanking the 'core polypeptide' of 249-252 amino acids, and delineating it from the variable, and lengthy N-terminal extensions not seen in other taxa.[72] It is therefore highly likely that precursor polypeptides in plant mitochondria are cleaved at these sites, to generate mature subunit 6 with a conserved N-terminus and 'standard' overall length.

MITOCHONDRIAL IMPORT AND ASSEMBLY OF ATP SYNTHASE

As already mentioned, recent data has implicated the F_1 α subunit in mitochondrial import and/or processing, mainly in yeast. A yeast mutant deleted for the nuclear gene for the α subunit exhibits a widespread deficiency of mitochondrial function, characterized by the inability of isolated mitochondria from the mutant to import a wide variety of precursor proteins.[34] In the mutant, the import of all N-terminal presequence-containing precursors, except that of F_1 α itself, appears to be blocked at a step after binding, but prior to proteolytic processing. The import of the adenine nucleotide translocator protein, which lacks an N-terminal presequence, and is imported via a distinct pathway,[35] is not affected. In vivo pulse-labeling studies reveal a particularly reduced rate of import of the HSP60 chaperonin, a key cofactor in the assembly of the mitochondrial redox complexes. By contrast, a yeast strain deleted for the β subunit gene exhibits a defect only in ATP synthase, with none of the above pleiotropic consequences.

The significance of these observations is far from clear, but the simplest interpretation is that the F_1 α subunit has some generalized role in facilitating the import of presequence-containing precursors (except itself) to mitochondria. A weak similarity between sequences within the α subunit and members of the HSP60 family of protein chaperones has previously been noted,[36] and in mammalian cells, the α subunit is, indeed, a heat-shock protein. Based on such data, coupled with suggestions that the protein might be partitioned to peroxisomes as well as to mitochondria, a role has been proposed for it as a molecular chaperone involved in protein import and/or multi-subunit assembly in more than one organelle.[36] It is still unclear whether the role of the subunit in mitochondrial protein import is direct, or indirect (e.g. via a specific role in the import of the HSP60 protein, which is in turn required more generally). The degree to which these various inferences are generalizable is not known, nor can their impact on ATP synthase assembly and function be assessed at present.

No such complications attend the β subunit, and, as indicated earlier, studies of protein import by mitochondria have frequently used the ATP synthase β subunit as a model, as its import pathway is in virtually every respect typical for a nuclear-coded mitochondrial polypeptide. Like the remaining nuclear-coded subunits of ATP synthase, it is, as far as is known, synthesized on free cytosolic ribosomes and imported post-translationally into mitochondria at membrane contact sites, via a surface receptor system (for review see Ref. 35). The extent to which polysomes actively translating such nascent polypeptides associate with mitochondria in vivo remains controversial.[145] Precursors of most of the nuclear-coded ATP synthase subunits contain conventional N-terminal presequences, typically of 20-50 amino acids, that are cleaved in the matrix by an efficient metalloprotease, revealing the mature N-terminus. Precursor polypeptides are present at very low steady-state concentrations, but can be detected by 2-D polyacrylamide gel electrophoresis and immunoblotting, where sensitive and specific antisera are available (e.g. see Ref. 146 for a recent characterization of the β subunit and its precursor in rat liver mitochondria).

Physical studies of the β subunit presequence indicate that it adopts the stereotypic structure predicted for mitochondrial targeting peptides, namely an amphiphilic α helix, when placed in structure-promoting solvent environments.[147] The transition to the α helical configuration initiates at two positions within the presequence, consistent with studies implying redundancy of targeting information within it.

Subunit 9 has also attracted some interest in its mechanism of import, due to its extreme hydrophobicity. In this case, the problem of cytosolic solubility appears to have been solved by the evolution of a very long, hydrophilic N-terminal presequence,[95] which is postulated to wrap around the hydrophobic core of the protein (the so-called 'life-belt' hypothesis). The yeast subunit 9 presequence, especially when duplicated, has proved an effective way of artificially delivering otherwise inefficiently targeted proteins to mitochondria.[89]

Two subunits of the ATP synthase complex, namely subunit ε of the F_1, and subunit d of the of F_0 sector, are synthesized in the cytosol without a cleavable amino-terminal presequence, at least in those organisms where the metabolism of the two polypeptides has been elucidated.[43,85] It is not known where in the polypeptides resides the information to target them to mitochondria, nor whether they enter the organelle through a distinct pathway from that traversed by polypeptides with 'conventional' targeting presequences. This property is not unique to these subunits: a number of other mitochondrial membrane polypeptides appear to lack cleavable presequences, including, for example, the adenine nucleotide translocase, and several of the smaller, nuclear-coded subunits of complexes III and IV. Subunits ε and d are absent from eubacterial ATP synthase, but they are not unique in this regard. They are also rather small, but some equally small or smaller subunits, such as subunits e or F_6, are manufactured with conventional presequences.

Some of the main features of the assembly pathway for ATP synthase in yeast have been elucidated,[76] using various mt-mutants (point mutants in mtDNA), together with monoclonal antibody probes for the F_1 α and β subunits. As is the case for the other mitochondrial redox complexes, the mtDNA-encoded subunits appear to play a key role in the assembly of the functional holoenzyme. However, in yeast mutants in which the synthesis of all three mtDNA-encoded subunits is abolished, the F_1 sector is nevertheless assembled into a catalytically active, membrane-associated complex. This contrasts with the reported assembly pathway in *E. coli*, where the incorporation of F_1 subunit β into the membrane is an early step, requiring F_0 subunits a and/or c. In yeast, assembly proceeds with the sequential addition of the mtDNA-encoded subunits 9, 8 and 6 to a membrane-bound complex, that subsequently interacts with a membrane-associated, pre-assembled F_1 sector. Incorporation of two polypeptides of 18 and 25 kD that were not specifically identified in the study cited, but which may represent OSCP and/or subunits b or d, appears to be a late step in assembly.

At least two nuclear-coded subunits of the F_0 are needed for the assembly of the mitochondrially encoded polypeptides into the membrane sector: specifically, the incorporation of subunit 6 into the complex is completely dependent on both subunits b and d, based on the properties of the mutant strains lacking these subunits.[78,80,85] If these correspond with the 18 and 25 kD polypeptides mentioned above, then this may indicate that subunits 6, b and d assemble as a pre-complex, before insertion into the membrane. Site-directed deletion mutagenesis of subunit b has identified the 10 C-terminal amino acids as critical for its ability to recruit subunit 6 into the complex.[85] Exactly how the F_1 sector could associate with the membrane in the absence of the F_0, including subunits b and d, is unclear, but this need not involve a strictly physiological mechanism. Indeed, the presence of non-coupled ATPase in mitochondria may be highly deleterious. Alternatively, there might be a transient 'docking' site on the membrane, for fully assembled F_1 moieties to be stored in an inactive state, awaiting assembly of the corresponding F_0 sector.

The products of two genes, *ATP11* and *ATP12*, which are not themselves subunits of ATP synthase, are required for a late step in the assembly of F_1. In strains mutant for either gene, unincorporated α and β subunits accumulate in an inactive state in mitochondria.[148] Both genes have been cloned,[149,150] and neither, based on sequence, is a member of a previously recognized gene family; both appear to be targeted to mitochondria via conventional N-terminal presequences. The bacterially expressed *ATP11* protein has been shown to bind to the α and β subunits, using affinity chromatography.[149] *ATP12* appears to be localized to the mitochondrial matrix, or loosely bound to the inner membrane, and is present in vivo, based on its sedimentation properties, as part of a larger complex.[150]

The features of the import/assembly pathway described above are put together in the model shown in Figure 5.2. Although a number of aspects of this model are fanciful, it is consistent with the published data from yeast, and forms a working hypothesis for further investigation. The extent to which it applies in detail in other organisms is presently not known, although the broad principles are almost certainly conserved. Mention should be made at this point of the special situation that applies in plants, where two entirely separate extranuclear genetic systems operate, in mitochondria and chloroplasts respectively, each of which directs the synthesis of an ATP synthase, among other functions. Nuclear-coded chloroplast polypeptides are also targeted, in general, via cleavable presequences, although their properties differ somewhat from those of mitochondrial targeting presequences. Exactly how the plant cell recognizes and sorts mito-chondrially destined subunits of the F_1F_0 ATP synthase from chloroplast-destined subunits of CF_1CF_0 is the subject of intensive ongoing research.

Developmental Regulation of ATP Synthase Biosynthesis

The regulation of mitochondrial function at the level of both nuclear and mitochondrial gene expression must necessarily operate in a great diversity of physiological and developmental situations, and it will not be surprising to learn that different strategies are prominent in different organisms. Several examples of regulated mitochondrial 'differentiation' involving ATP synthase may be cited. Bloodstream forms of *Trypanosoma brucei*, for example, are essentially glycolytic: only a partial electron transfer chain is operative, via the so-called (non-proton pumping) alternative oxidase. Nevertheless, oligomycin-sensitive ATPase is present, and is essential for the creation and maintenance, at the expense of ATP, of the mitochondrial membrane potential,[151] which is presumably used to drive ion, metabolite and protein transport across the mitochondrial membrane system. In this context, genes for the ATPase complex, both nuclear and mitochondrial, must perforce be independently regulated from those of the respiratory chain. Nevertheless, the ATP synthase complex is quantitatively regulated in the trypanosome life cycle, being induced at least three-fold in the procyclic compared with the early bloodstream form, based on both enzymatic and immunochemical measurements.[152]

In an entirely different context, mitochondrial oxidative function in mammalian skeletal muscle responds rapidly to the demands of physical activity, by enhanced expression of genes for mitochondrial components. This results in a co-ordinate increase in the content of enzymes of mitochondrial energy metabolism, and elevated capacity for mitochondrial ATP production.[153] A similar co-ordinated elevation of mitochondrial oxidative phosphorylation capacity occurs developmentally, as shown by studies in the neonatal rat liver,[154] as the organism adapts to aerobic life outside of the uterine environment. A further important regulator of mitochondrial biogenetic activity in mammals is thyroid hormone. However, in this case the effects are not co-ordinate. Rat cardiomyocytes respond to prior thyroxine treatment by a specific accumulation of ATP synthase activity, in the absence of elevated respiratory capacity,[155] and the enzyme also shows altered regulatory properties (see below 'Regulation').

In early animal development in many species, a pronounced accumulation of mitochondrial redox capacity is well documented. ATP synthase specific activity, for example, rises at least two-fold during cleavage and gastrulation of the sea urchin embryo.[156] Circumstantial evidence supports the view that interference with mitochondrial energetic function during embryonic development in this organism, for example by inhibiting mitochondrial protein synthesis with chloramphenicol, may have specific morphogenetic consequences.[157] In addition, classical histochemical staining experiments imply a gradient of oxidative activity in sea urchin embryos.[158]

Although there is no convincing evidence for differential mitochondrial function in vertebrate embryos, the mRNA for the α subunit of mitochondrial ATP synthase has been identified as a localized maternal message in *Xenopus* eggs.[159] In situ hybridization has confirmed that it is some 20-fold enriched in the animal pole region of the *Xenopus* egg, and is selectively inherited by presumptive ectoderm cells into which this cytoplasmic domain is partitioned during cleavage. Although mitochondria are also slightly enriched in the animal pole cytoplasm, this is quantitatively modest, suggesting that a specific differentiation of embryonic mitochondria is thus programmed. It is currently unknown whether other mRNAs for ATP synthase subunits or other mitochondrial proteins are similarly localized. Other animal pole-localized mRNAs include a number of putative regulatory molecules, such as transcription factors, proteases, and RNA-binding proteins. A family of mRNAs localized to the opposite (vegetal) pole of the egg encode members of the TGF-β family of growth factors, that are implicated in mesoderm induction. These observations do not necessarily imply a role for mitochondrial ATP synthase in morphogenesis in

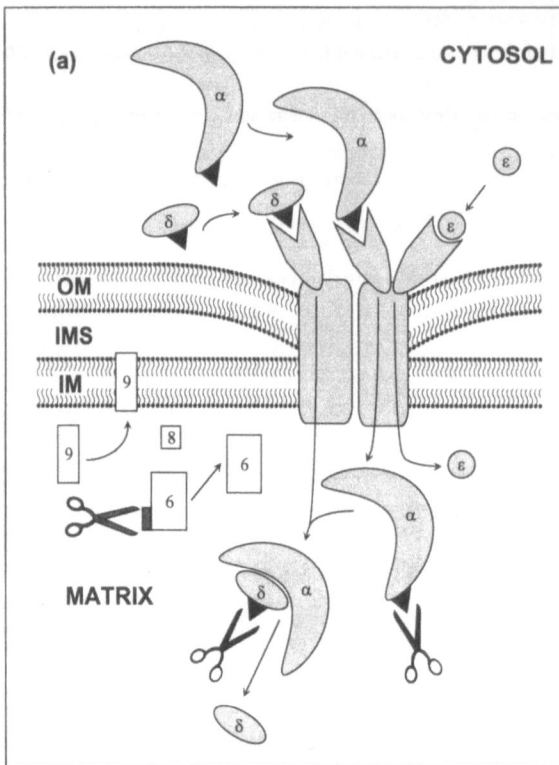

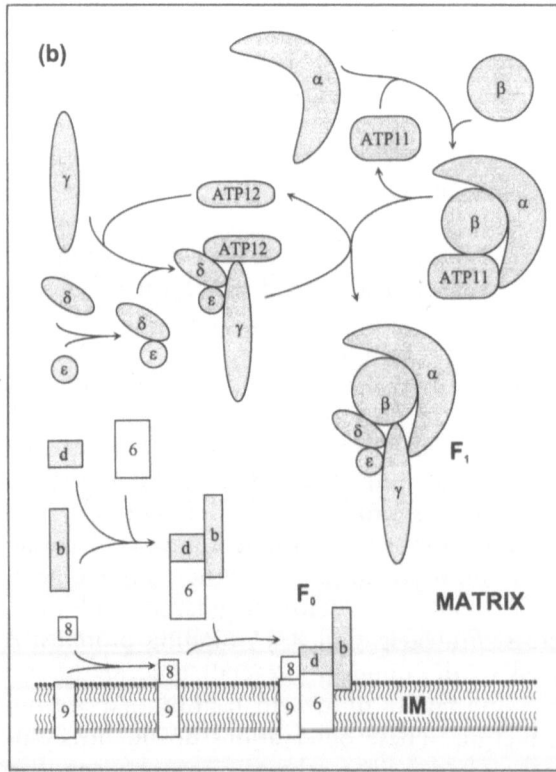

Fig. 5.2 Model scheme for the import, processing and assembly of yeast ATP synthase, based on the current literature. Many details are conjectural. Polypeptides are denoted as follows: mtDNA-encoded polypeptides: open symbols, with dark shading indicating the cleaved N-terminus of subunit 6; nuclear-coded polypeptides: lightly shaded symbols, with black triangles indicating N-terminal presequences; rectangles: subunits of the F_0 sector, ellipses/circles; subunits of the F_1 sector, except subunit α, which is denoted by a special symbol to indicate its putative molecular chaperone role; rounded rectangles: stalk subunits and other polypeptides; irregular symbols: receptors for mitochondrial protein import; scissors: matrix proteases. The compartments corresponding to cytosol, outer (OM) and inner (IM) membranes, the intermembrane space (IMS) and mitochondrial matrix are as shown. For clarity, subunit stoichiometries are not shown: for details see text. (a) Import of nuclear-coded subunits, and processing of nuclear and mtDNA-encoded subunits. Only nuclear-coded subunits α, δ and ε are shown: subunit d probably follows the same pathway as subunit ε; all other nuclear-coded subunits contain N-terminal presequences and follow the pathway shown for subunit δ. All import receptors appear to interact with a 'common import complex', whose exact polypeptide composition is not yet established: such complexes are located at contact sites between inner and outer membranes. Subunit α is

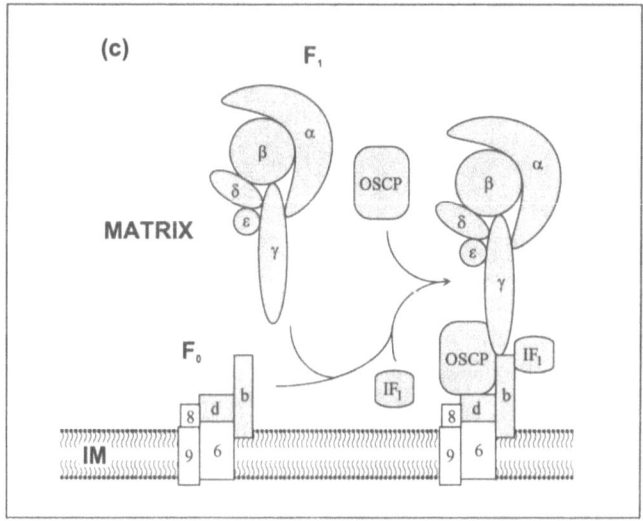

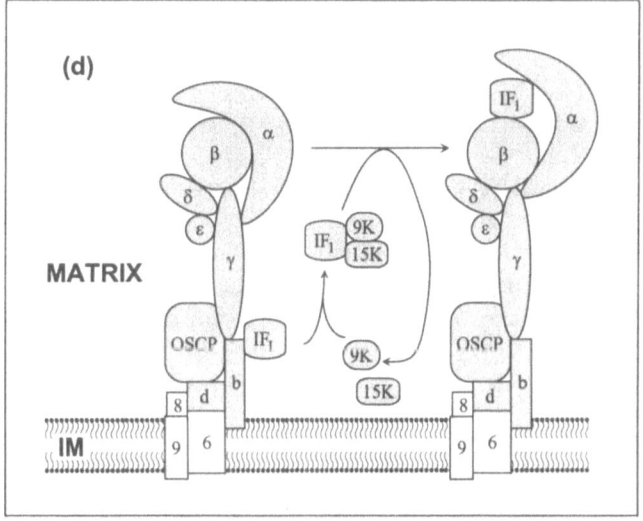

required for the import and/or processing of all other presequence-containing subunits. Other molecular chaperones, especially the heat-shock proteins of the HSP70 and HSP60 families, are also required for import, but for clarity are not shown. The mtDNA-encoded polypeptides follow distinct fates: subunit 6 is proteolytically processed, and subunit 9 almost certainly associates spontaneously with the inner membrane. (b) Assembly of F_0 and F_1 sectors. The exact nature of the complex formed by ATP12 is unknown: the one shown is only one of several possibilities. Although the diagram depicts the assembly of the F_0 and F_1 sectors as separate entities prior to their being brought together in the holoenzyme, subunits b and γ, which interact directly, may bind to each other before the assembly of the F_0 and F_1 sectors is complete. (c) Assembly of F_0 and F_1 sectors with each other, and with the stalk subunits OSCP and IF_1, forming the fully functional ATP synthase complex The exact steps at which OSCP and IF_1 are incorporated, as well as their topology in the overall complex are not known. The docking site for IF_1 is assumed to be on the F_0, although this is not certain. (d) Re-organization of the complex consequent on activation of the ATPase inhibitor subunit IF_1, which associates with the F_1 sector at the interface between the α and β subunits. This appears to be mediated by two polypeptides of 9 and 15 kD (designated 9K and 15K), whose exact nature and role is unclear.

the frog, but nevertheless indicate that the biosynthesis of the enzyme is regulated in early development in a highly unusual manner that must reflect some requirement of the developing ectoderm of the embryo. The possible role of subunit α as a molecular chaperone involved in other aspects of mitochondrial biogenesis may be relevant here.

REGULATED EXPRESSION OF NUCLEAR GENES FOR ATP SYNTHASE

As indicated above, nuclear genes for ATP synthase are likely to be regulated in many contexts, by a multiplicity of factors. For example, all, or almost all genes for mitochondrial products are expected to respond positively to signals promoting growth, both in lower eukaryotes such as yeast, as well as higher organisms. In lower eukaryotes, nutritional source is also of great importance in determining the importance of mitochondrial proton-driven ATP synthesis. In metazoans, oxidative phosphorylation genes would be expected to be up-regulated in tissues where respiratory ATP generation is important, such as cardiac and skeletal muscle, and the central nervous system. A considerable body of evidence indicates that the primary mechanism in most of these cases is transcriptional regulation.

The evidence is particularly convincing in relation to the generalized induction of mitochondrial biogenesis that accompanies growth in mammalian cells. As expected, ATP synthase genes are key targets for regulation of this type. Transcriptional activation is not, however, the result of a single 'master' transcription factor, acting on the whole set of nuclear-coded mitochondrial genes. Rather, an 'intersecting set' of common transcription factors acts on the 5' DNA of the various genes involved, to bring about co-ordinate induction.

Bovine and human subunit α genes that have been cloned and characterized are flanked at their 5' end by an array of putative binding sites for known and hypothesized transcriptional regulators.[160] In the case of the bovine gene these include ubiquitous factors, such as Sp1, Ap-1 and cAMP-response elements, as well as factors specifically implicated in the regulation of mitochondrial function. The latter include the elements Mt1, Mt3 and Mt4, found in the promoters of a number of genes for subunits of complex III, as well as in the mtDNA D-loop region,[161] and the factor NRF-2 (Nuclear Respiratory Factor-2, Ref. 162). It may be hypothesized that one or many of these factors is involved in co-ordinating the up- or down-regulation of ATP synthase and probably other mitochondrial redox protein genes, in a variety of contexts. The promoter of a human α subunit gene, although lacking consensus TATA and CCAAT signals, also contains Sp-1 binding sites, as well as binding-sites for factors with an ETS domain, shared with a number of complex IV genes as well as the human β subunit gene. Binding-sites are also present for the transcription factors GFI and GFII, implicated in co-ordinate up-regulation of a large number of nuclear genes for mitochondrial proteins[163] in both yeast and higher eukaryotes.

Reporter gene assays have been used to define an enhancer region in the 5' flanking DNA of the human β subunit gene, which includes sequences related to the Mt3 and Mt4 elements.[164] The transcription factor designated NRF-1 (Nuclear Respiratory Factor-1), that has been shown to enhance the transcription of a large number of nuclear genes for mitochondrial proteins,[165] binds to and activates the promoter of the human γ subunit gene.[166]

How is this modulation of nuclear transcriptional activity co-ordinated with mitochondrial gene activation? An attractive idea that has been put forward is that one or more key regulatory components of the transcriptional apparatus is partitioned between nucleus and mitochondria, and act(s) in some way to co-ordinate transcriptional activity in the two compartments. This idea has received some support from studies in yeast, which showed that many nuclear loci involved in mitochondrial biogenesis respond transcriptionally to alterations in the mitochondrial genotype.[167,168] Studies in mammalian cells, however, appear to rule out a simple model of this kind involving the major mtDNA transcription factor mtTF1: mtDNA promoter sequences containing binding sites for this factor have no significant effect on the transcription of nuclear reporter genes in a variety of human cell-lines, including *rho⁰* cells lacking mtDNA.[169] This does not rule out the possibility that other factors, binding to mtDNA elements such as Mt3 or Mt4, might be involved in such control circuitry, although, there is no convincing evidence for the intra-mitochondrial presence of any such DNA-binding proteins.

Such a 'general transcriptional induction' is clearly not the whole story, however, as a number of the redox complex polypeptides, including at least two subunits of ATP synthase (namely subunits α and 9) are encoded by multi-gene families expressed in a tissue-specific fashion, as has already been indicated. In the case of the subunit 9 genes isolated from human DNA,[106] one (P1) contains conventional 5' promoter elements, including TATA and CCAAT sequences, whereas the other (P2) does not, despite the presence of a CpG island indicative of a transcriptional start. The significance of this observation is presently unknown, though it suggests that the two genes might be transcribed as members of entirely distinct gene sets.

Tissue-specific, as opposed to general enhancement of ATP synthase biosynthesis is also illustrated by studies of the effects of muscle-specific transcription factors on genes of the oxidative phosphorylation system. Analysis of the promoter regions of human genes for the muscle-specific isoform of the adenine nucleotide translocase and the β subunit of ATP synthase, by reporter gene methodology and protein-binding assays in vitro, has revealed two overlapping regulatory elements with possible roles in the transcriptional control of the biogenesis of the oxidative phosphorylation apparatus.[170] The region in which these elements lie functions as a positive transcriptional regulator in a muscle-specific fashion.[171] One of the two elements, designated the REBOX (8 bp), binds to a ubiquitous factor, whose activity appears to be sensitive to NADH and thyroxine, whereas the overlapping 13 bp OXBOX binds to a factor found only in myogenic cells. These findings indicate that oxidative phosphorylation genes, notably for the nuclear-coded subunits of ATP synthase, are likely to be regulated transcriptionally in response to both physiological and developmental cues, and that factors binding (competitively?) to REBOX and OXBOX elements might be involved in co-ordinating the activity of many such genes.

It is important to stress that despite the evidence for 'co-ordinate induction', the transcriptional response to a given physiological signal of nuclear genes coding for different mitochondrial proteins is not necessarily uniform, as shown by the differential effects of thyroid hormone on mitochondrial and various nuclear genes.[172,173] Mitochondrial protein synthesis is uniformly up-regulated, but only a subset of nuclear-coded mRNAs for mitochondrial redox complex polypeptides is induced. Some discrepancy is evident, however, in comparing the results obtained in different laboratories, in regard to the regulation of the ATP synthase β subunit mRNA in response to this hormone. Cuezva and colleagues have demonstrated, using nuclear 'run-on' transcription, that a rapid, marked elevation in the level of this mRNA seen in the liver of thyroid hormone-treated newborn rats is due to a specific increase in the rate of its transcription[174]. Nelson's laboratory, however, found no difference in the level of the β subunit mRNA, after 24 hours of thyroid hormone treatment.[173] It is unclear whether this reflects age or strain differences, or the timing of hormone treatment.

In addition, nuclear genes for ATP synthase are regulated post-transcriptionally, offering a possible route whereby genes for this particular enzyme complex might be differentially regulated with respect to the other mitochondrial redox complexes. One particularly intriguing example is of the (single-copy) γ subunit gene, transcripts of which have been shown to be alternately spliced in a tissue-specific fashion[175] in both humans and cows. The alternate splice gives rise to two distinct isoforms of the polypeptide, that differ by the addition of a single C-terminal aspartate residue in the liver isoform, that is absent from the heart isoform. Given the likely importance of the γ subunit in gating of the proton channel, and evidence that the C-terminus of the polypeptide is of functional significance, the alternate splice is highly likely to modulate the properties of the enzyme in a tissue-specific fashion.

Regulation also operates downstream of RNA processing. F_1 β subunit mRNA is translationally down-regulated in brown adipose tissue (BAT), leading to specific depletion of ATP synthase in this cell-type.[176] The F_1 β mRNA from BAT is translated in vitro in the rabbit reticulocyte system with 2-3 fold lower efficiency than the corresponding mRNA from other tissues, implying that a covalent modification of the RNA is responsible, perhaps poly(A) length, which has elsewhere been shown to correlate with translational competence. The half-life of F_1 β subunit mRNA in BAT is also 3-7 fold shorter than in liver or heart muscle. Translational enhancement has similarly been invoked to account for the up-regulation of nuclear genes for mitochondrial biogenesis, specifically the β subunit of ATP synthase, in the neonatal rat liver.[177]

REGULATION

IMPORTANCE OF REGULATED SYNTHESIS OF ATP

The regulation of mitochondrial respiration and ATP synthesis has for decades been one of the key areas studied by bioenergeticists, perhaps not surprisingly, in view of the central importance of these processes in the metabolism of all higher organisms. For an excellent recent review, see Ref. 178. Although capable of catalysing a reversible reaction, mitochondrial ATP synthase under normal physiological conditions is effectively a unidirectional enzyme, converting at least the major portion of the free energy conserved by the electron transfer chain in the form of the electrochemical proton gradient ($\Delta\mu_H+$) into ATP, by phosphorylation of ADP. ATP synthase competes with at least two other processes for the utilization of the proton gradient: (i) the pumping of substrates and other ions, including phosphate anion, one of the substrates of the ATP synthase, across the inner mitochondrial membrane, and (ii) proton leak, which in some animal tissues, notably brown adipose tissue, is a precisely regulated thermogenic process, mediated by a specific inner membrane proton channel formed by the uncoupler protein UCP (reviewed in Ref. 179). In turn, the ATP produced by mitochondrial ATP synthase is utilized by a variety of competing pathways in a highly tissue-specific manner, in mammals notably by the actomyosin ATPase, which produces contractile energy in muscle, and the Na^+/K^+ ATPase, of greatest importance in the nervous system and kidney. Biosynthesis in general, and protein synthesis in particular, is also a major user of ATP in all cells, and urea synthesis and gluconeogenesis are especially important in the liver.

To a first approximation, the rate of oxidative phosphorylation is driven in all cells and tissues by the rate of ATP utilization, with a concomitant feedback effect on the respiratory chain. The validity of this 'respiratory control' hypothesis, originally formulated by Chance and Williams in the 1950s,[180] is supported by a large body of data on the properties of the individually dissected components of mitochondrial energy metabolism. Nevertheless, a more modern view is that control over respiration and ATP synthesis is exercised jointly by a number of reactions and pathways, that are susceptible to subtle, and tissue-specific variations in different metabolic circumstances.[181,182] ATP synthesis, in some circumstances[178] can actually regulate the rate of ATP utilization by some of the major pathways described above. Because of the interdependence of the components of the system, it is therefore extremely difficult to model a complete picture of the competing influences on each individual enzyme.

The great diversity of rates and types of ATP usage in different eukaryotic cell-types, the competing uses of the proton gradient, and the varied rate of ATP production by other pathways, notably glycolysis under defined substrate conditions, have led to a plethora of views regarding the regulation of mitochondrial ATP synthase. Rather than attempt a general model, I shall focus on the results obtained with studies on specific systems, principally the mammalian heart, and attempt to draw only the simplest of general conclusions. The difficul-

inherent in measuring substrate levels and flux in a highly compartmentalized system such as mitochondria, compounded by the inaccessibility and complexity of the enzyme, renders potentially unsafe any conclusions based solely upon in vitro studies. For similar reasons, definitive in vivo measurements are equally hard to perform.

REGULATORY MECHANISMS INFLUENCING ATP SYNTHESIS AND HYDROLYSIS

In principle, the activity of ATP synthase, like that of any enzyme, may be said to depend on four factors: firstly, the availability of substrate, which is the primary determinant of flux through most enzymes; secondly, the action of effector molecules binding at catalytic or regulatory sites in the enzyme, and influencing its kinetic properties in response to physiological demand; thirdly, the structural modification of the enzyme, leading to more permanent or tissue-specific differences in its properties; and fourthly, alteration in the steady-state level of the enzyme or its constituent subunits, in response to physiological or developmental signals. The fourth of these is already discussed above under 'Biogenesis'. I shall deal with the remaining levels in turn.

Considerations of the regulation of ATP synthase activity by substrate and product availability are complicated by the fact that there are at least two types of 'substrate' for the enzyme, namely the small molecules ADP, inorganic phosphate (P_i) and ATP which it interconverts, and the proton gradient which drives the 'forward' reaction. Because ADP and P_i are provided from, and ATP supplied to an extramitochondrial location, the transport of these molecules must be considered to be an integral part of the overall reaction. In the case of ADP and ATP, this transport is effected by a translocase enzyme that is not proton-driven, and for P_i by a related carrier that is. The overall stoichiometry of the reaction may be summarized as: ADP(out) + P_i(out) + 4H$^+$(out) = ATP(out) + 4H$^+$(in). Note that in considering substrate levels, their location is of critical importance. Considering this overall reaction, the balance of evidence[57] suggests that short-term alterations in substrate and product levels are insufficient to explain the fluctuations in ATP synthase (and ATPase) activity observed in different physiological circumstances, implying that regulation at other levels must be operating, at least in some tissues.

Data are probably most convincing in this regard in the mammalian heart (see Ref. 57 for a recent review). Measurements of cytoplasmic ADP and P_i levels by freeze clamping and ^{31}P-nmr[183,184] and of $\Delta\mu_{H^+}$ by TPMP probe distribution[185] indicated that all three substrates for the forward reaction are non-saturating under physiological conditions, hence capable of regulating activity. Nevertheless, stimulation of the perfused rat or dog heart with isoprenaline, producing a 2- to 4-fold elevation of ATP synthase activity, is accompanied by no significant changes in ADP or P_i levels.[183,186] Measurements of the flux across ATP synthase in both perfused heart[187] and isolated heart mitochondria[185] indicate a considerable imbalance between forward and back reactions, precluding an explanation for the above result based on the assumption that the enzyme is operating close to equilibrium. In addition, direct measurements of $\Delta\mu_{H^+}$ indicate that if anything it falls slightly rather than rises, in response to increased heart work rate.[188] These findings indicate that the enzyme must be subject to direct regulation by one or more of the other mechanisms implied above.

ATP synthase regulation in the heart appears to involve at least two distinct mechanisms: down-regulation in response to ischemia,[189,190] and up-regulation in response to electrical or hormonal stimuli indicating increased work-rate.[191,192] The most likely mechanism for the former involves the inhibitor subunit IF$_1$ already described, which, based on its properties in vitro, is postulated to associate rapidly with the enzyme in conditions of de-energization, as a result of lowered pH. This has the effect of limiting ATP losses by hydrolysis, which would otherwise occur in circumstances of transient interruption of electron flow. Re-energization of mitochondria is accompanied by a brief lag (5-30 s) in attainment of maximal rates of ATP synthesis (reviewed in ref. 57), which probably reflects the structural reorganization of the enzyme, involving loss of the inhibitor subunit. A simpler form of allosteric regulation involving a small molecule seems to be precluded by the observation that the F$_1$ fragment itself can be isolated from de-energized mitochondria in an inactive state.[193]

Although a critical role for IF$_1$ in protecting against catastrophic loss of ATP in the heart under ischemic conditions is well supported by the in vitro data, its applicability in vivo depends upon assumptions regarding the effect of ischemia on $\Delta\mu_{H^+}$, and the primary contribution of mitochondrial ATP synthase to ATP hydrolysis under these conditions. Neither assumption is supported unequivocally. Recent studies on isolated ventricular cardiomyocytes indicate that a state of deep hypoxia, involving a full aerobic-anaerobic metabolic transition, may be induced without loss of mitochondrial membrane potential.[194] Furthermore, recent measurements of net myocardial ATP loss in total ischemia[195] attribute a significant fraction, but not the majority, to the mitochondrial ATP synthase/ATPase enzyme. It has alternatively been suggested[57] that the most important protective role of IF$_1$ may be during recovery from ischemia, when most tissue damage occurs, rather than during ischemia per se. A transient loss of $\Delta\mu_{H^+}$ may accompany recovery,[196] supporting this view.

Up-regulation appears to be mediated by the concentration of intra-mitochondrial Ca^{2+} ion,[192] the only second messenger known to enter mitochondria, but is again indirect, as activation survives dilution of the enzyme. The involvement of a protein kinase/phosphatase cascade is unlikely, as no phosphorylated forms of ATP synthase subunits have been demonstrated. Another possible mechanism is redox-mediated modification of the enzyme, by analogy with chloroplast ATP synthase, where the γ subunit is reversibly reduced through the action of thioredoxin.[197,198] However, physiological changes occur in mitochondrial ATP synthase activity without any significant alteration in the redox state, therefore it is likely that this is a specific adaptation of chloroplasts, where it operates as a more general regulatory mechanism.

The best candidate molecule for involvement in Ca^{2+}-mediated up-regulation is a second heat-stable polypeptide of 6.3 kD, designated CaBI (calcium binding inhibitor), which is structurally unrelated to IF_1.[199] CaBI appears to act as a dimer, which associates with ATP synthase in the absence of Ca^{2+} ion, but is monomerized and stripped from the complex by Ca^{2+} levels of 1 μM or above.[200] The level of CaBI in different tissues varies considerably,[62] with the ratio of CaBI to IF_1 being highest in heart and skeletal muscle, of the tissues where it has been measured. In addition to ATP synthase, calcium levels regulate a number of key mitochondrial enzymes in the heart, including pyruvate and Krebs cycle dehydrogenases, but interestingly, via a multiplicity of mechanisms. The reason why peptide inhibitors, rather than small molecules, are involved in ATP synthase regulation is unclear. One suggestion that has been put forward[57] is that they might have more than one target, allowing them to mediate a rapid switch between competing metabolic pathways. Overall, the regulatory system provided by CaBI allows the rate of ATP production to be adjusted to the rate of ATP utilization, thereby maintaining adenine nucleotide levels at roughly constant values over a wide range of work rates, which may be important in cellular homeostasis. In turn, increased flux through ATP synthase entrains the stimulation of electron transfer and substrate utilization.

It is important to stress that whilst both of the inhibitory polypeptides described above are widespread in animal tissues, IF_1 being properly regarded as an integral subunit of the complex, their activity is far from universal. Thus, thermogenic brown adipose tissue (BAT) from cold-acclimated rats contains equivalent levels of the IF_1 and CaBI polypeptides as non-thermogenic tissue from control rats.[201] Nevertheless, mitochondria from non-thermogenic BAT exhibit high ATP synthase and low ATPase activity, both of which respond to calcium regulation, whereas those from thermogenic BAT, by contrast, exhibit high ATPase and low ATP synthase activity, with minimal calcium sensitivity. This suggests that the inhibitors do not function in this tissue, at least in the uncoupled state, in the same way as they do in heart or skeletal muscle.

Regulation of mitochondrial ATP synthase by inhibitory peptides is not confined to mammals. In yeast, mutational studies have confirmed that the yeast homologue of the IF_1 inhibitor subunit is required for the suppression of ATP hydrolysis under conditions where the electrochemical gradient of the inner mitochondrial membrane is dissipated.[62] In plants, two forms of ATP synthase have been inferred to be present, respectively, in resting organs (such as potato tuber) and in metabolically active organs (such as pea leaves), distinguishable by their response to de-energization. Loss of membrane potential in pea leaf mitochondria results in a high rate of ATP hydrolysis, which is almost undetectable when potato tuber mitochondria are treated similarly.[202] ATPase activity can be induced in potato tuber mitochondria, however, by detergent treatments known to remove the inhibitor protein.[203,204] These findings suggest that the inhibitor function is either absent or compromised in active tissues.

Because the phosphate/proton synporter and the adenine nucleotide antiporter are important components of the overall reaction catalysed by ATP synthase, their possible regulatory roles in mitochondrial ATP synthesis must also be considered. The phosphate carrier operates close to equilibrium, and its measured capacity is far in excess of the requirements of ATP synthesis, at least in liver,[205] indicating that it is unlikely to be a site of regulation. However, the data for the ADP/ATP translocase are more equivocal. Several investigators have found the activity of this carrier to be limiting for respiration in isolated mitochondria from a variety of tissues, although this is contradicted by in vivo data (reviewed in Ref. 178). In vitro, fatty acyl-CoA can inhibit the carrier from liver,[206] but the available evidence suggests that this does not limit the rate of synthesis in vivo, and in rat heart, the effect may even be reversed. Overall, the question of an indirect regulatory role over ATP synthase exercised by the adenine nucleotide carrier remains unresolved.

PATHOLOGY

As expected for an enzyme central to metabolism, defects in the synthesis, function or regulation of mitochondrial ATP synthase are associated with pathological states in a variety of organisms, notably plants and humans. I shall deal with plants first, as they provide an interesting model.

CYTOPLASMIC MALE STERILITY: A PLANT MITOCHONDRIAL CYTOPATHY

Plants, like humans, appear to suffer from specific cytopathic effects resulting from mtDNA mutations, most notably the phenomenon of cytoplasmic male sterility (CMS: see Ref. 207), which is of economic importance. Plant mtDNA undergoes promiscuous re-arrangements, and is also subject to invasion by plasmids, both of which make it difficult to assign precisely the molecular lesions giving rise to CMS, as mtDNAs from male-sterile strains of many plants differ in multiple sites from those of their male-fertile counterparts.

In a number of specific instances, the phenotype has been mapped genetically, or by in vitro genetic manipulation, to a novel open reading frame created by mtDNA re-arrangements. The evidence for this is most definitive in the case of the male-sterile T-race of maize, where a novel 13 kD polypeptide is synthesized in the male-sterile plant, and is also responsible for sensitivity to several membrane-active fungal toxins.[207,208] Another well characterized example in petunia[209] involves an abnormal fusion gene comprising the 5' portion of ATP synthase subunit 9, joined to an aberrant segment of the gene for cytochrome oxidase subunit II (COX II). In such cases, it is presumed that segments of the fusion gene that are related to domains of bona fide mitochondrial gene products interact with partner proteins in the corresponding respiratory complex(es), in this example ATP synthase (and/or cytochrome oxidase), and disrupt the assembly or function of the complexes. Conceivably, such fusion proteins might also affect the biosynthesis of the wild-type subunit, by an interference with feedback inhibition.

In other cases, mtDNA re-arrangements, usually partial duplications, place mitochondrial genes, including open reading frames of unknown (and possibly selfish) function under the control of information associated with bona fide mitochondrial transcription units. Inevitably, ATP synthase subunit genes are frequently involved in many of these phenomena. One such case is in rapeseed (*Brassica napus L.*), where, in the male-sterile strain, a novel open reading frame of 105 amino acids is found upstream of the *atp6* gene, with which it is probably co-expressed.[210] It remains unclear whether it is the expression of the novel polypeptide per se, or an interference with the expression of the ATP synthase subunit 6 gene, that leads to CMS in this plant. A similar case in sunflower involves re-arrangements close to or within the subunit α (*atpA*) gene, that lead to a male-sterile-specific pattern of transcription of the *atpA* region [211] A male-sterile strain of sugar beet is characterized by a subunit α transcript with completely different 5' information (but the same coding sequence) from that found in male-fertile plants.[212]

In rice, a duplicated copy of the subunit 6 gene is present in male-sterile plants, with a different 3' untranslated sequence from that of the 'wild-type' copy. In this case, the phenotype of a nuclear fertility restorer mutant (*Rf-1*) strongly suggests the involvement of the extra subunit 6 gene in CMS, because the transcript of the duplicated gene is differentially processed and edited in the *Rf-1* background.[213] The CMS phenotype in this case appears again to be due to the synthesis of an abnormal polypeptide, here resulting from the inappropriate translation of an unedited or partially edited mRNA for ATP synthase subunit 6, whose editing is impaired by the inability of the wild-type strain to process the transcript of the extra *atp6* gene correctly. A similar situation has been engineered experimentally, by transgenic expression of the unedited wheat ATP synthase subunit 9 gene in tobacco,[214] providing strong support for this hypothesis.

The overall conclusion from such studies is that CMS can result from a variety of genetic lesions, many of which appear to result in the production of an abnormal polypeptide, usually related to a subunit of ATP synthase. This provides an interesting model for some human disorders, which are discussed below.

To date, CMS has only been reported in plants. However, the fact that sperm are highly dependent on mitochondrial function and appear to contain numerous specific isoforms of mitochondrial proteins, suggests that a similar phenomenon could also occur in animals. One intriguing finding in the moth *Heliothis* is the report of a sperm-specific protein associated with the mitochondrial ATP synthase, whose presence is correlated with a hybrid sterility phenotype.[248] This is highly reminiscent of the CMS-associated synthesis of abnormal mitochondrial proteins, in the many cases reviewed above.

PATHOLOGICAL MITOCHONDRIAL MUTATIONS AFFECTING ATP SYNTHASE GENES

Mitochondrial ATP synthase deficiency per se has only rarely been reported in the clinical literature.[215,216] This could indicate that defects

tile, late infantile, juvenile, and adult (Kufs' disease) forms of the disorder are known, and there is an ovine model.[236,237] Linkage analysis has mapped genes for the infantile and juvenile forms of the disease to distinct loci within the human genome. As the disease progresses, neuronal function is compromised by the lysosomal accumulation of large amounts of protein, comprising solely or mainly intact subunit 9 (subunit c) of mitochondrial ATP synthase, in both the sheep[236] and human late-infantile, juvenile and adult[238,239] forms of the disease. Specific lysosomal accumulation of this polypeptide is diagnostic for the disorder, and does not occur in other lysosomal storage diseases. It also accumulates to some degree in other tissues, such as cultured skin fibroblasts[239] from Batten's disease patients, but not controls. Cloning of the subunit 9 genes from sheep with the disorder has confirmed the absence of any mutation in the coding sequence (including the presequences), and shown that expression levels in diseased animals are no different from those in unaffected controls.[74] The generally accepted current hypothesis is that the disease results from a specific defect in the metabolism of subunit 9 of ATP synthase, rather than a lysosomal defect per se. It is quite easy to imagine that an abundant polypeptide of such extreme hydrophobicity might depend upon rather special mechanisms for its turnover. Cloning of the nuclear genes mapping at the disease loci will obviously be crucial in unravelling the exact disease mechanism.

ATP Synthase Defects Consequent on External Causes

As well as sporadic or inherited defects in genes for ATP synthase and its biosynthesis/regulation, other genetic or physiological defects may have secondary consequences mediated through ATP synthase. Chronic hypertension or hyperthyroidism, for example, have a number of effects on cardiac ATP synthase, which may be of pathological significance. Most relevant studies have been carried out in the rat. Studies of rat cardiomyocytes from hypertensive or hyperthyroid animals indicate a rise in the basal level of ATP synthase activity, no effect on down-regulation by anoxia, but abolition of up-regulation by energy demand.[155,240] Antihypertensive therapy or removal of thyroxine reverses these effects. These findings have been interpreted to suggest that heart failure resulting from an inability to regulate ATP synthase correctly might be a pathological consequence of these conditions.

The effect of chronic alcohol consumption on mitochondrial ATP synthase has also been extensively investigated. Liver mitochondria from ethanol-fed rats show a consistent depression of enzyme activity, and this has recently been suggested[241] to be a specific consequence of a drop in synthesis of the mtDNA-encoded subunit 6 and, to a lesser extent, subunit 8. Levels of nuclear-encoded ATP synthase polypeptides are only modestly affected, if at all, but the deficiency of the mtDNA-encoded subunits of the complex results in a failure to assemble a significant proportion of the nuclear-coded subunits into functional ATP synthase. Similarly, studies of mitochondrial ATP synthase activity in cardiomyocytes of alcohol-fed rats indicates a loss of the ability to up-regulate the enzyme.[242] Although unproven in relation to humans, these findings suggest a possible direct interference with mitochondrial biogenesis in key organs, resulting from chronic alcohol consumption, and may explain some of the phenotypes of alcohol-induced disease, such as cardiomyopathy.

In addition to regulation during normal development, mitochondrial oxidative phosphorylation is subject to a marked, age-related decline, which involves a dramatic loss of ATP synthase capacity.[243] Recent attention has focused on the possible role of mitochondrial genetic lesions and of oxidative damage by free oxygen radicals in this process.[244] It has been proposed that mtDNA damage might promote a vicious cycle, by virtue of the enhanced rate of production of reactive oxygen species resulting from respiratory impairment. Damage to the ATP synthase by reactive oxygen species is well documented,[245] and may be of critical importance to cardiomyocytes during ischemia.[246] Recent evidence indicates that this inactivation is mediated by iron ions not tightly bound in the enzyme.[247] The possibility that aging and degenerative disease in a more general sense might be attributable to the compromise of mitochondrial ATP synthesis by such a mechanism is obviously both attractive and potentially testable.

CONCLUSIONS

As will be clear from the foregoing, the study of mitochondrial ATP synthase is an extremely eclectic endeavour. This reflects its status as a 'crossroads' enzyme. In fact, a more appropriate metaphor might be a multi-layered highway intersection, for the enzyme and its major product, ATP, are simultaneously at a metabolic intersection between catabolism and biosynthesis, and a genetic intersection between the nuclear and mitochondrial genomes and their respective expression machineries. In addition, by virtue of the endosymbiotic event that brought the modern, aerobic, eukaryotic cell into being, the enzyme represents a key evolutionary intersection in the history of life. These features of the enzyme, taken together with the large number of unanswered questions regarding its mechanism, biogenesis and regulation, mean that our understanding of it is certain to expand enormously in the coming years. Finally, the great complexity of the enzyme, and its biosynthetic and regulatory apparatus, imply that dysfunctions of oxidative phosphorylation are likely to be involved in a considerable range of pathological states, of which the manifestations described above are only the most obvious. Research in this area is likely therefore to have a significant impact on human health, the nature of which is not easy to predict.

ACKNOWLEDGEMENTS

I am grateful to my research colleagues for their constant intellectual stimulation, as well as for permission to mention their unpublished data. I am also indebted to Ian Holt and Joanna Poulton for their input to the ideas contained in this chapter, and to BD for his invaluable secretarial assistance in carrying out the literature survey. My research on ATP synthase and mitochondrial disease is currently supported by MRC and the Wellcome Trust, whom I thank for their support.

REFERENCES

1. Walker JE, Lutter R, Dupuis A et al. (1991) Identification of the subunits of F_1F_0-ATPase from bovine heart mitochondria. Biochemistry 30, 5369-5378.

2. Glaser E, Norling B (1991) Chloroplast and plant mitochondrial ATP synthases. Curr Topics Bioenerget 16, 223-263

3. Hamasur B, Glaser E (1992) Plant mitochondrial F_0F_1 ATP synthase. Identification of the individual subunits and properties of the purified spinach leaf mitochondrial ATP synthase. Eur J Biochem 205, 409-416.

4. Walker JE, Saraste M, Gay NT (1984) The unc operon: nucleotide sequence, regulation and structure of ATP synthase. Biochim Biophys Acta 768, 164-200.

5. Hoppe J, Sebald W (1984) The proton-conducting F_0 part of bacterial ATP synthases. Biochim Biophys Acta 768, 1-27

6. Nagley P (1988) Eukaryotic membrane genetics: The F_0 sector of mitochondrial ATP synthase. Trends in Genet 4, 46-52.

7. Penefsky HS, Cross RL (1991) Structure and mechanism of F_0F_1-type ATP synthases and ATPases. Adv Enzymol 64, 183-214.

8. Cox GB, Devenish RJ, Gibson F et al. (1992) The structure and assembly of ATP synthase. In Molecular Mechanisms in Bioenergetics (Ernster L, ed.), Elsevier, Amsterdam. pp. 283-315.

9. Senior AE (1988) ATP synthesis by oxidative phosphorylation. Physiol Rev 68, 177-231

10. Yoshihara Y, Nagase H, Yamame T et al. (1991) H^+-ATP synthase from rat liver mitochondria. A simple, rapid purification method of the functional complex and its characterization. Biochemistry 30, 6854-6860.

11. Lutter R, Saraste M, van Walraven HS et al. (1993) F_1F_0-ATP synthase from bovine heart mitochondria: development of the purification of a monodisperse oligomycin-sensitive ATPase. Biochem J 295, 799-806.

12. Walker JE, Fearnley IM, Gay NT et al. (1985) Primary structure and subunit stoichiometry of F_1-ATPase from bovine mitochondria. J Mol Biol 184, 677-701.

13. Papa S, Guerrieri F, Zanotti Z et al. (1992) F_0 and F_1 subunits involved in the gate and coupling function of mitochondrial H^+ ATP synthase. Ann New York Acad Sci 671, 345-358.

14. Kagawa Y, Nikiwa N (1981) Conversion of stable ATPase to labile ATPase by acetylation, and the $\alpha\beta$ and $\alpha\gamma$ subunit complexes during its reconstitution. Biochem Biophys Res Commun 100, 1370-1376.

15. Harris DA, Boork J, Baltscheffsky M (1985) Hydrolysis of adenosine 5'-triphosphate by the isolated catalytic subunit of the coupling ATPase from Rhodospirillum rubrum. Biochemistry 24, 3876-3883.

16. Frasch WD, Green J, Cagniat J et al. (1989) ATP hydrolysis catalysed by a β-subunit preparation purified from the chloroplast energy transducing complex CF_0CF_1. J Biol Chem 264, 5065-5069.

17. Cross RL, Nalin CM (1982) Adenine nucleotide binding-sites on beef heart F_1-ATPase. Evidence for 3 exchangeable sites that are distinct from 3 non-catalytic sites. J Biol Chem 257, 2874-2881.

18. Wise JG, Duncan TM, Latchney LR et al. (1983) The properties of F_1-ATPase from the uncD412 mutant of Escherichia coli. Biochem J 215, 343-350.

19. Kayalar KC, Rosing J, Boyer PD (1977) An alternative site sequence for oxidative phosphorylation suggested by measurement of substrate binding patterns and exchange reaction inhibitions. J Biol Chem 252, 2486-2491.

20. Garin J, Boulay F, Issartel JP et al. (1986) Identification of amino acid residues photolabeled with 2-azido[α^{32}P] adenosine diphosphate in the β-subunit of beef heart F_1-ATPase. Biochemistry 25, 4431-4437.

21. Cross RL, Cunningham D, Miller CG et al. (1987) Adenine nucleotide binding sites on beef heart F_1-ATPase. Photoaffinity labeling of β-subunit Tyr-368 at a noncatalytic site and β-Tyr-345 at a catalytic site. Proc Natl Acad Sci USA 84, 5715-5719.

22. Esch FS, Allison WS (1978) Identification of a tyrosine residue at a nucleotide binding site in the β subunit of the mitochondrial ATPase with p-fluorosulfonyl[^{14}C]-benzoyl-5'-adenosine. J Biol Chem 253, 6100-6106.

23. Bullough DA, Allison WS (1986) Inactivation of the bovine heart mitochondrial F_1-ATPase by 5'-para-fluorosulfonylbenzoyl[^{3}H]inosine is accompanied by modification of Tyrosine-345 in a single β-subunit. J Biol Chem 261, 4171-4177.

24. chondrial F_1-ATPase. J Biol Chem 266, 6101-6105.

25. Duncan TM, Parsonage D, Senior AE (1986) Structure of the nucleotide binding site in the β-subunit of Escherichia coli F_1-ATPase. FEBS Lett 208, 1-6.

26. Fry DC, Kuby SA, Mildvan AS (1986) ATP binding site of adenylate kinase. Mechanistic implications of its homology with ras-encoded p21, F_1-ATPase and other nucleotide-binding proteins. Proc Natl Acad Sci USA 83, 907-911.

27. Thomas PJ, Garboczi DN, Pedersen PL (1992) Mutational analysis of the consensus nucleotide binding sequences in the rat liver mitochondrial ATP synthase β-subunit. J Biol Chem 267, 20331-20338.

28. Thomas PJ, Garboczi, Pedersen PL (1992) Mitochondrial F-type ATPases: the glycine-rich loop of the β-subunit is a pyrophosphate binding domain. Acta Physiol Scand 146, 23-29.

29. Milgrom YM, Cross RL (1993) Nucleotide binding sites on beef heart mitochondrial F_1-ATPase. Co-operative interaction between sites and specificity of noncatalytic sites. J Biol Chem 268, 23179-23185.

30. Divita G, Goody RS, Gautheron DC et al. (1993) Structural mapping of catalytic site with respect to α subunit and noncatalytic site in yeast mitochondrial F_1-ATPase using fluorescence resonance energy transfer. J Biol Chem 268, 13178-13186.

31. Zhuo S, Garrod S, Miller P et al. (1992) Irradiation of the bovine mitochondrial F_1-ATPase previously inactivated with 5'-p-fluorosulfonylbenzoyl-8-azido[^{3}H]adenosine cross-links His-β4277 to Tyr-β345 within the same β subunit. J Biol Chem 267, 12916-12927

32. Falson P, Maffey L, Conrath K et al. (1991) α subunit of mitochondrial F_1-ATPase from the fission yeast. J Biol Chem 266, 287-293

33. Falson P, Leterme S, Capiau C et al. (1991) β subunit of mitochondrial F_1-ATPase from the fission yeast. Deduced sequence of the wild type protein and identification of a mutation that increases nucleotide binding.

Eur J Biochem 200, 61-67.

34. Yuan HB, Douglas MG (1992) The mitochondrial F_1-ATPase α subunit is necessary for efficient import of mitochondrial protein precursors. J Biol Chem 267, 14697-14702.

35. Neupert W (1993) Mitochondrial protein import: specific recognition and membrane translocation of preproteins. J Membrane Biol 135, 191-207.

36. Luis AM, Alconada A, Cuezva JM (1990) The α regulatory subunit of the mitochondrial F_1-ATPase is a heat-shock protein. J Biol Chem 265, 7713-7716.

37. Franzén L-G, Falk G (1992) Nucleotide sequence of cDNA clones encoding the β subunit of mitochondrial ATP synthase from the green alga Chlamydomonas reinhardtii: the precursor protein encoded by the cDNA contains both an N-terminal presequence and a C-terminal extension. Plant Mol Biol 19, 771-780.

38. Bianchet M, Ysern X, Hullihen J et al. (1991) Mitochondrial ATP synthase. Quaternary structure of the F_1 moiety at 3.6 Å determined by X-ray diffraction analysis. J Biol Chem 266, 21197-21201.

39. Zanotti F, Guerrieri F, Capozza G et al. (1992) Role of F_0 and F_1 subunits in the gating and coupling function of mitochondrial H^+-ATP synthase. Eur J Biochem 208, 9-16.

40. Morikami A, Aiso K, Asahi T et al. (1992) The δ' subunit of higher plant 6-subunit mitochondrial F_1-ATPase is homologous to the δ subunit of animal mitochondrial F_1-ATPase. J Biol Chem 267, 72-76.

41. Arselin G, Gandar J-C, Guérin G et al. (1991) Isolation and complete amino acid sequence of the mitochondrial ATP synthase ε-subunit of the yeast Saccharomyces cerevisiae. J Biol Chem 266, 723-727.

42. Kimura T, Nakamura K, Koyiura H et al. (1989) Correspondence of minor subunits of plant mitochondrial F_1-ATPase to the F_1F_0-ATPase subunits of other organisms. J Biol Chem 264, 3183-3186.

43. Morikami A, Ehara G, Yuuki K et al. (1993) Molecular cloning and characterization of cDNAs for the γ- and ε-subunits of mitochondrial F_1F_0 ATP synthase from the sweet potato. J Biol Chem 268, 17205-17210.

44. Solaini G, Baracca A, Castelli GP et al. (1993) Tryptophan phosphorescence as a structural probe of mitochondrial F_1-ATPase ε-subunit. Eur J Biochem 214, 729-734.

45. Amzel LM, Pedersen PL (1978) Adenosine triphosphatase from rat liver mitochondria. Crystallization and X-ray diffraction studies of the F_1-component of the enzyme. J Biol Chem 253, 2067-2069.

46. Abrahams JP, Lutter R, Todd RJ et al. (1993) Inherent asymmetry of the structure of F_1-ATPase from bovine heart mitochondria at 6.5 Å resolution. EMBO J 12, 1775-1780.

47. Uh M, Jones D, Mueller DM (1990) The gene coding for the yeast oligomycin sensitivity-conferring protein. J Biol Chem 265, 19047-19052.

48. Ovchinnikov YA, Modyanov NN, Grinkevich VA et al. (1984) Oligomycin sensitivity-conferring protein (OSCP) of beef heart mitochondria. Internal sequence homology and structural relationship with other proteins. FEBS Lett 250, 625-628.

49. Senior AE (1971) Bioenergetics 2, 141-150.

50. Joshi S, Huang Y (1991) ATP synthase complex from bovine heart mitochondria: the oligomycin sensitivity conferring protein is essential for

dicyclohexylcarbodiimide-sensitive ATPase. Biochim Biophys Acta 1067, 255-258.

51. Joshi S, Javed AA, Gibbs LC (1992) Oligomycin sensitivity-conferring protein (OSCP) of mitochondrial ATP synthase. J Biol Chem 267, 12860-12867.

52. Engelbrecht S, Reed J, Penin F et al. (1991) Subunit δ of chloroplast F_0 F_1-ATPase and OSCP of mitochondrial F_0 F_1-ATPase: a comparison by CD-spectroscopy. Z Naturforsch 46c, 759-764.

53. Hekman C, Tomich JM, Hatefi Y (1991) Mitochondrial ATP synthase complex. membrane topography and stoichiometry of the F_0 subunits. J Biol Chem 266, 13564-13571.

54. Norling B, Tourikas C, Hamasur B et al. (1990) On the subunit composition of plant mitochondrial ATP synthase. Biochim Biophys Acta 1015, 49-52.

55. Horak A, Horak H, Dunbar B et al. (1989) Plant mitochondrial F_1-ATPase. The presence of the oligomycin sensitivity-conferring protein (OSCP). Biochem J 263, 301-304.

56. Kimura T, Takeda S, Asahi T et al. (1990) The primary structure of a precursor for the δ-subunit of sweet potato mitochondrial F_1-ATPase deduced from full-length cDNA. J Biol Chem 265, 16079-6085.

57. Harris DA, Das AM (1991) Control of mitochondrial ATP synthase in the heart. Biochem J 280, 561-573.

58. Higuti T, Kuroiwa K, Kawamura Y et al. (1993) Molecular cloning and sequencing of cDNAs for the import precursors of oligomycin sensitivity conferring protein, ATPase inhibitor protein, and subunit c of H·-ATP synthase in rat mitochondria. Biochim Biophys Acta 1172, 311-314.

59. Lebowitz MS, Pedersen PL (1993) Regulation of the mitochondrial ATP synthase/ATPase complex: cDNA cloning, sequence, overexpression and secondary structural characterization of a functional protein inhibitor. Arch Biochem Biophys 301, 64-70.

60. Yamada EW, Huzel NJ (1992) Distribution of the ATPase inhibitor proteins of mitochondria in mammalian tissues including fibroblasts from a patient with Luft's disease. Biochim Biophys Acta 1139, 143-147.

61. Gomez-Fernandez JC, Harris DA (1978) Thermodynamic analysis of the interaction between the coupling ATPase from ox heart mitochondria and its naturally occurring inhibitor protein. Biochem J 176, 967-975.

62. Kimura H, Hashimoto T, Yoshida Y et al. (1993) Binding of an intrinsic ATPase inhibitor to the interface between α- and β-subunits of F_1F_0ATPase upon de-energization of mitochondria. J Biochem 113, 350-354.

63. Van Heeke G, Deforce L, Schnizer RA et al. (1993) Recombinant bovine heart mitochondrial F_1-ATPase inhibitor protein: overproduction in Escherichia coli, purification and structural studies. Biochemistry 32, 10140-10149.

64. Rouslin W, Broge CW, Chernyak BV (1993) Effects of Zn^{2+} on the activity and binding of the mitochondrial ATPase inhibitor protein, IF_1. J Bioenerget Biomembranes 25, 297-306.

65. Lopez-Media Villa C, Vigny H, Godinot C (1993) Docking the mitochondrial inhibitor protein IF_1 to a membrane receptor different from the F_1-ATPase β subunit. Eur J Biochem 215, 487-496.

66. Schnizer R, Shaw R, Couton J et al. (1993) Identification of critical amino

acid residues of the bovine mitochondrial F_1-ATPase inhibitor protein by alanine-scanning mutagenesis. Protein Engineering 6, 49.

67. Stout JS, Partridge BE, Dibbern DA et al. (1993) Peptide analogs of the beef heart mitochondrial F_1-ATPase inhibitor protein. Biochemistry 32, 7496-7502.

68. Jackson PJ, Harris DA (1988)The mitochondrial ATP synthase inhibitor protein binds near the C-terminus of the F_1 β-subunit. FEBS Lett 229, 224-228.

69. Milgrom YM (1991) When beef heart mitochondrial F_1-ATPase is inhibited by inhibitor protein a nucleotide is trapped in one of the catalytic sites. Eur J Biochem 200, 789-795.

70. Fang J, Jacobs JW, Kanner BI et al. (1984) Amino acid sequence of bovine heart coupling factor 6. Proc Natl Acad Sci USA 81, 6603-6607.

71. Michon T, Galante M, Velours J (1988) NH_2-terminal sequence of the isolated yeast ATP synthase subunit 6 reveals post-translational cleavage. Eur J Biochem 172, 621-625.

72. Mullen JA, Pring DR, Kempken F et al. (1992) Sorghum mitochondrial atp6: divergent amino extensions to a conserved core polypeptide. Plant Mol Biol 20, 71-79.

73. Sebald W, Graf T, Lukins HB (1979) The dicyclohexylcarbodiimide-binding protein of the mitochondrial ATPase complex from Neurospora crassa and Saccharomyces cervisiae. Identification and isolation. Eur J Biochem 93, 587-599.

74. Medd SM, Walker JE, Jolly RD (1993) Characterization of the expressed genes for subunit c of mitochondrial ATP synthase in sheep with ceroid lipofuscinosis. Biochem J 293, 65-73

75. Sebald W, Hoppe J (1981) On the structure and genetics of the proteolipid subunit of the ATP synthase complex. Curr Topics Bioenerget 12, 2-64.

76. Hadikusomo RG, Meltzer S, Choo WM et al. (1988) The definition of mitochondrial H^+-ATPase assembly defects in mit⁻ mutants of Saccharomyces cerevisiae with a monoclonal antibody to the enzyme complex as an assembly probe. Biochim Biophys Acta 933, 212-222.

77. Velours J, Durrens P, Aigle M et al. (1988) ATP4, the structural gene for yeast F_0F_1-ATPase subunit 4. Eur J Biochem 170, 634-642.

78. Paul MF, Velours J, de Chateau Bodeau GA et al. (1989)The role of subunit 4, a nuclear-encoded protein of the F_0 sector of yeast mitochondrial ATP synthase in the assembly of the whole complex. Eur J Biochem 185, 163-171.

79. Walker JE, Runswick MJ, Poulter I (1987) ATP synthase from bovine mitochondria. The characterization and sequence analysis of 2 membrane associated subunits and of the corresponding cDNAs. J Mol Biol 197, 89-100.

80. Paul M-F, Guerin B, Velours J (1992) The C-terminal region of subunit 4 (subunit b) is essential for assembly of the F_0 portion of yeast mitochondrial ATP synthase. Eur J Biochem 205, 163-172.

81. Guerrieri F, Zanotti F, Capozza G, Colaianni G, Ronchi S, Papa S (1991) Structural and functional characterization of subunits of the F_0 sector of the mitochondrial F_0F_1-ATP synthase. Biochim Biophys Acta 1059, 348-354.

82. Hoppe J, Brunner J, Jorgensen BB (1984) Structure of the membrane-

embedded F_0 part of F_0F_1-ATP synthase from Escherichia coli as inferred from labeling with 3-(trifluoromethyl)-3-(M-[^{125}I]iodophenyl)diazirine. Biochemistry 23, 5610-5616.

83. Houstek J, Kopecky J, Zanotti F et al. (1988) Topological and functional characterization of the F_0I subunit of the membrane moiety of the mitochondrial H$^+$-ATP synthase. Eur J Biochem 173, 1-8.

84. Zanotti F, Guerrieri F, Capozza G et al. (1988) Identification of nucleus-encoded F_0I protein of bovine heart mitochondrial H$^+$-ATPase as a functional part of the F_0 moiety. FEBS Lett 237, 9-14.

85. Norais N, Promé D, Velours J (1991) ATP synthase of yeast mitochondria. Characterization of subunit d and sequence analysis of the structural gene ATP7. J Biol Chem 266, 16541-16549.

86. Higuti T, Kuroiwa K, Kawamura Y et al. (1992) Complete amino acid sequence of subunit e of rat liver H$^+$-ATP synthase. Biochemistry 31, 12451-12454.

87. Clary DO, Wolstenholme D (1985) The mitochondrial DNA molecule of Drosophila yakuba: nucleotide sequence, gene organization and genetic code. J Mol Evol 22, 116-125.

88. Jacobs HT (1991) Structural similarities between a mitochondrially encoded polypeptide and a family of prokaryotic respiratory proteins involved in plasmid maintenance suggest a novel mechanism for the evolutionary maintenance of mitochondrial DNA. J. Mol. Evol. 32, 333-339

89. Devenish RJ, Papakonstantinou T, Galanis M et al. (1992) Structure/function analysis of yeast mitochondrial ATP synthase subunit 8. Ann New York Acad Sci 671, 403-414.

90. Nagley P, Devenish RJ (1989) Leading organellar proteins along new pathways: the relocation of mitochondrial and chloroplast genes to the nucleus. Trends in Biochem Sci 14, 31-35.

91. Jacobs H T (1991) Phylogenetic implications of the plasmid-like features of mitochondrial DNA. In: The Unity of Evolutionary Biology (E C Dudley, ed.), Dioscorides Press, Portland, Oregon. pp. 838-851.

92. Okimoto R, Macfarlane JL, Clary DO et al. (1992) The mitochondrial genome of two nematodes, Caenorhabditis elegans and Ascaris suum. Genetics 130, 471-498.

93. Howe CJ (1992) Plastid origin of an extrachromosomal DNA molecule from Plasmodium, the causative agent of malaria. J Theor Biol 158, 199-205.

94. Hensgens LAM, Grivell LA, Borst P et al. (1979) Nucleotide sequence of the mitochondrial structural gene for subunit 9 of yeast ATPase complex. Proc Natl Acad Sci USA 76, 1663-1667.

95. Viebrock A, Perz A, Sebald W (1982) The imported preprotein of the proteolipid subunit of the mitochondrial ATP synthase from Neurospora crassa. Molecular cloning and sequence of the messenger RNA. EMBO J 5, 565-571.

96. Ward M, Turner G (1986) The ATP synthase subunit 9 gene of Aspergillus nidulans: sequence and transcription. Mol Gen Genet 295, 331-338.

97. Van den Boogaart P, Samallo J, Agsteribbe E (1982) Similar genes for a mitochondrial ATPase subunit in the nuclear and mitochondrial genomes of Neurospora crassa. Nature 298, 187-189.

98. Brown TA, Ray JA, Waring RB et al. (1984) A mitochondrial reading frame which may code for a second form of ATPase subunit 9 in Aspergillus nidulans. Curr Genet 8, 489-492.

99. Ridder R, Künkele K-P, Osiewacz HD (1991) Sequence of the nuclear ATP synthase subunit 9 gene of Podospora anserina: lack of similarity to the mitochondrial genome. Curr Genet 20, 349-351.

100. Bowman EJ, Knock TE (1992) Structure of the genes encoding the α and β subunits of the Neurospora crassa mitochondrial ATP. synthase. Gene 114, 157-163.

101. Breen GAM (1988) Bovine liver cDNA clones encoding a precursor of the subunit of the mitochondrial ATP synthase complex. Biochem Biophys Res Commun 152, 264-269.

102. Walker JE, Powell SJ, Vinas O et al. (1989) ATP synthase from bovine mitochondria. A complementary DNA sequence of the import precursor of a heart isoform of the α subunit. Biochemistry 28, 4702-4708

103. De Paepe R, Forchioni A, Chétrit P et al. (1993) Specific mitochondrial proteins in pollen: presence of an additional ATP synthase β subunit. Proc Natl Acad Sci USA 90, 5934-5938.

104. Lomax MI, Grossman LI (1989) Tissue-specific genes for respiratory proteins. Trends in Biochem Sci 14, 501-503.

105. Gay NT, Walker JE (1985) Two genes encoding the bovine mitochondrial ATP synthase proteolipid specify precursors with different import sequences and are expressed in a tissue-specific manner. EMBO J 4, 3519-3524.

106. Dyer MR, Walker JE (1993) Sequences of members of the human gene family for the c subunit of mitochondrial ATP synthase. Biochem J 293, 51-64.

107. Jordan EM, Breen GAM (1992) Molecular cloning of an import precursor of the β-subunit of the human mitochondrial ATP synthase complex. Biochim Biophys Acta 1130, 123-126.

108. Runswick MJ, Medd SM, Walker JE (1990) The δ-subunit of ATP synthase from bovine mitochondria. Biochem J 266, 421-426.

109. Chanut FA, Grabau EA, Gesteland RF (1991) Complex organization of the soybean mitochondrial genome: recombination repeats and multiple transcripts at the atpA loci. Curr Genet 23, 234-247.

110. Dell'Orto P, Moenne A, Graves PV, Jordana X (1993) The potato mitochondrial ATP synthase subunit 9: gene structure, RNA editing and partial protein sequence. Plant Sci 88, 45-53.

111. Jacobs HT, Elliott DJ, Math VB et al. (1988) Nucleotide sequence and gene organization of sea urchin mitochondrial DNA. J Mol Biol 202, 185-2177.

112. Dewey RE, Levings CS III, Timothy DH (1985) Nucleotide sequence of ATPase subunit 6 gene of maize mitochondria. Plant Physiol 79, 914-919.

113. Stuart K (1991) RNA editing in mitochondrial mRNA of trypanosomes. Trends in Biochem Sci 16, 68-72.

114. Bhat GJ, Koslowsky DJ, Feagin JE et al. (1990) An extensively edited mitochondrial transcript in kinetoplastids encodes a protein homologous to ATPase subunit 6. Cell 61, 885-894.

115. Crozier RH, Crozier YC (1992) The cytochrome b and ATPase genes of honeybee mitochondrial DNA. Mol Biol Evol 9, 474-482.

116. Kaneko M, Satta Y, Matsuura ET et al. (1993) Evolution of the mito-chondrial ATPase 6 gene in Drosophila: unusually high level of polymor-phism in D. melanogaster. Genet Res 61, 195-204.
117. Black-Schaefer CL, McCourt JD, Poyton RO et al. (1991) Mitochondrial gene expression in Saccharomyces cerevisiae. Proteolysis of nascent chains in isolated mitochondria optimized for protein synthesis. Biochem J 274, 199-205.
118. Williams RS (1986) Mitochondrial gene expression in mammalian stri-ated muscle. Evidence that variation in gene dosage is the major regula-tory event. J Biol Chem 261, 2345-2349.
119. Robin ED, Wong R (1989) Mitochondrial DNA molecules and virtual number of mitochondria per cell in mammalian cells. J Cell Physiol 136, 507-513.
120. Madsen CS, Givizzani SC, Hauswirth WW (1993) Protein binding to a single termination-associated sequence of the mitochondrial DNA D-loop region. Mol Cell Biol 13, 2162-2171.
121. Roberti M, Mustich A, Gadaleta MN et al. (1991) Identification of two homologous mitochondrial DNA sequences which bind strongly and spe-cifically to a mitochondrial protein of Paracentrotus lividus. Nucl Acids Res 19, 6249-6254.
122. Qureshi SA, Jacobs HT (1993) Characterization of a high-affinity bind-ing site for a DNA-binding protein from sea urchin embryo mitochon-dria. Nucl Acids Res 21, 811-816.
123. Qureshi SA, Jacobs HT (1993) Two distinct, sequence-specific DNA bind-ing proteins interact independently with the major replication pause re-gion of sea urchin mitochondrial DNA. Nucl Acids Res 21, 2801-2808.
124. Rapp WD, Stern DB (1992) A conserved 11 nucleotide sequence con-tains an essential promoter element of the maize mitochondrial atp1 gene. EMBO J 11, 1065-1073.
125. Ojala D, Montoya J, Attardi G (1981) tRNA punctuation model of RNA processing in human mitochondria. Nature 290, 470-474.
126. Attardi G (1985) Animal mitochondrial DNA: an extreme example of genetic economy. Int Rev Cytol 93, 93-145.
127. Cantatore P, Roberti M, Rainaldi G et al. (1988) Clustering of tRNA genes in Paracentrotus lividus mitochondrial DNA. Curr Genet 13, 91-96.
128. Butow RA, Zhu H, Perlman P et al. (1989) The role of a conserved dodecamer sequence in yeast mitochondrial gene expression. Genome 31, 757-760.
129. Pelissier PP, Camougrand NM, Manon ST et al. (1992) Regulation by nuclear genes of the mitochondrial synthesis of subunits 6 and 8 of the ATP synthase of Saccharomyces cerevisiae. J Biol Chem 267, 24467-24473.
130. Groudinsky O, Bousquet I, Wallis MG et al. (1993) The NAM1/MTF2 nuclear gene product is selectively required for the stability and/or pro-cessing of mitochondrial transcripts of the atp6 and of the mosaic, cox1 and cytb genes in Saccharomyces cerevisiae. Mol Gen Genet 240, 419-427.
131. Ziaja K, Michaelis G, Lisowsky T (1993) Nuclear control of the messen-ger RNA expression for mitochondrial ATPase subunit 9 in a new yeast mutant. J Mol Biol 229, 909-916.
132. Hernould M, Mouras A, Litvak S et al. (1992) RNA editing of the mito-chondrial atp9 transcript from tobacco. Nucl Acids Res 20, 1809.

133. Bégu D, Graves P-V, Domec C, Arselin G et al. (1990) RNA editing of wheat mitochondrial ATP synthase subunit 9: direct protein and cDNA sequencing. The Plant Cell 2, 1283-1290.

134. Schuster W, Brennicke A (1991) RNA editing in ATPase subunit 6 mRNAs in Oenothera mitochondria. a new termination codon shortens the reading frame by 35 amino acids. FEBS Lett 295, 97-101.

135. Salazar RA, Pring DR, Kempken F (1991) Editing of mitochondrial atp9 transcripts from two sorghum lines. Curr Genet 20, 483-486.

136. Kempken F, Mullen JA, Pring DR et al. (1991) RNA editing of sorghum mitochondrial atp6 transcripts changes 15 amino acids and generates a carboxy-terminus identical to yeast. Curr Genet 20, 417-422.

137. Wintz H, Hanson MR (1991) A termination codon is created by RNA editing in the petunia mitochondrial atp9 gene transcript. Curr Genet 19, 61-64.

138. Wissinger B (1992) Regenerating good sense: RNA editing and trans-splicing in plant mitochondria. Trends in Genet 8, 322-328.

139. Araya A, Domec C, Begu D et al. (1992) An in vitro system for the editing of ATP synthase subunit 9 mRNA using wheat mitochondrial extracts. Proc Natl Acad Sci USA 89, 1040-1044.

140. Fox TD, Costanzo MC, Strick CA et al. (1988) Translational regulation of mitochondrial gene expression by nuclear genes of Saccharomyces cerevisiae. Phil Trans Roy Soc Lond B 319, 97-105.

141. Ackerman SH, Gatti DL, Gellefors P et al. (1991) ATP13, a nuclear gene of Saccharomyces cerevisiae essential for the expression of subunit 9 of the mitochondrial ATPase. FEBS Lett 278, 234-238.

142. Payne MJ, Schweizer E, Lukins HB (1991) Properties of two nuclear pet mutants affecting expression of the mitochondrial oli1 gene of Saccharomyces cerevisiae. Curr Genet 19, 343-351.

143. Finnegan PM, Payne MJ, Keramidaris E et al. (1991) Characterization of a yeast nuclear gene, AEP2, required for accumulation of mitochondrial mRNA encoding subunit 9 of the ATP synthase. Curr Genet 20, 53-61.

144. Denslow ND, Michaels GS, Montoya J et al. (1989) Mechanism of messenger RNA binding to bovine mitochondrial ribosomes. J Biol Chem 264, 8328-8338.

145. Verner K (1993) Co-translational protein import into mitochondria: an alternative view. Trends in Biochem Sci 18, 366-371.

146. Sanatarén JF, Alconada A, Cuezva JM (1993) Examination of processing of the rat liver mitochondrial F1-ATPase β subunit precursor protein by high resolution 2D-gel electrophoresis. J Biochem 113, 129-131.

147. Bruch MD, Hoyt DW (1992) Conformational analysis of a mitochondrial presequence derived from the F_1-ATPase β-subunit by CD and NMR spectroscopy. Biochim Biophys Acta 1159, 81-93.

148. Ackerman SH, Tzagoloff A (1990) Identification of 2 nuclear genes (ATP11, ATP12) required for the assembly of the yeast F_1-ATPase. Proc Natl Acad Sci USA 87, 4986-4990.

149. Ackerman SH, Martin J, Tzagoloff A (1992) Characterization of ATP11 and detection of the encoded protein in mitochondria of Saccharomyces cerevisiae. J Biol Chem 267, 7386-7394

150. Bowman S, Ackerman SH, Griffiths DE et al. (1991) Characterization of ATP12, a yeast nuclear gene required for the assembly of the mitochon-

drial F₁ ATPase. J Biol Chem 266, 7517-7523.

151. Nolan DP, Voorheis P (1992) The mitochondrion in bloodstream forms of Trypanosoma brucei is energized by the electrogenic pumping of protons catalysed by the F_1F_0-ATPase. Eur J Biochem 209, 207-216.

152. Williams N, Choi SY-W, Ruyechan WT et al. (1991) The mitochondrial ATP synthase of Trypanosoma brucei: developmental regulation throughout the life cycle. Arch Biochem Biophys 288, 509-515.

153. Wibom R, Hultman E, Johansson M et al. (1992) Adaptation of mitochondrial ATP production in human skeletal muscle to endurance training and detraining. J Appl Physiol 73, 2004-2010.

154. Valcarce C, Navarreta P, Encabo E et al. (1988) Postnatal development of rat liver mitochondrial functions. The roles of protein synthesis and of adenine nucleotides. J Biol Chem 263, 7767-7775.

155. Das AM, Harris DA (1991) Control of mitochondrial ATP synthase in rat cardiomyocytes: effects of thyroid hormone. Biochim Biophys Acta 1096, 284-290.

156. Devlin RB (1982) Biogenesis of the mitochondrial ATPase from sea urchin embryos. J Biol Chem 257, 9711-9716.

157. Lallier R (1975) Animalization and vegetalization. In: Czihak G, ed. The Sea Urchin Embryo. Berlin: Springer-Verlag, pp. 473-509.

158. Hörstadius, S (1973) Experimental Embryology of Echinoderms. Oxford: Clarendon Press.

159. Weeks DL, Melton DA (1987) A maternal messenger RNA localized to the animal pole of Xenopus eggs encodes a subunit of mitochondrial ATPase. Proc Natl Acad Sci USA 84, 2798-2802.

160. Pierce DJ, Jordan EM, Breen GAM (1992) Structural organization of a nuclear gene for the α subunit of the bovine mitochondrial ATP synthase complex. Biochim Biophys Acta 1132, 265-275.

161. Suzuki H, Hosokawa Y, Nishikimi M et al. (1991) Existence of common homologous elements in the transcriptional regulatory regions of human nuclear genes and mitochondrial gene for the oxidative phosphorylation system. J Biol Chem 266, 2333-2338.

162. Virbasius JV, Virbasius CMA, Scarpulla RC (1993) Identity of GABP with NRF-2, a multisubunit activator of cytochrome oxidase expression, reveals a cellular role for an ETS domain activator of viral proteins. Genes & Development 7, 380-392.

163. Jordan EM, Breen GAM (1993) Upstream region of a nuclear gene encoding the α subunit of the human mitochondrial F_0F_1 ATP synthase. Biochim Biophys Acta 1173, 115-117.

164. Tomura H, Endo H, Kagawa Y et al. (1990) Novel regulatory enhancer in the nuclear gene of the human mitochondrial ATP synthase β subunit. J Biol Chem 265, 6525-6527.

165. Evans MJ, Scarpulla RC (1990) NRF-1, a transactivator of nuclear-coded respiratory genes in animal cells. Genes & Development 4, 1023-1034.

166. Chau CMS, Evans MJ, Scarpulla RC (1992) Nuclear respiratory factor 1 activation in genes encoding the γ subunit of ATP synthase, eukaryotic initiation factor 2α, and tyrosine aminotransferase. Specific interaction of purified NRF-1 with multiple target genes. J Biol Chem 267, 6999-7006.

167. Parikh VS, Morgan MM, Scott R et al. (1987) The mitochondrial genotype can influence nuclear gene expression in yeast. Science 235, 576-580.

168. Marczynski GT, Schultz, PW, Jaehning JA (1989) Use of yeast nuclear DNA sequences to define the mitochondrial RNA polymerase promoter in vitro. Mol Cell Biol 9, 3193-3202.

169. Sewards R, Wiseman B, Jacobs HT (1994) Apparent functional independence of the nuclear and mitochondrial transcription systems in mammalian cells. Mol Gen Genet (in press).

170. Chung AG, Stepien G, Haraguchi Y et al. (1992) Transcriptional control of nuclear genes for the mitochondrial muscle ADP/ATP translocator and the ATP synthase β subunit. J Biol Chem 267, 21154-21161.

171. Li K, Hodge JA, Wallace DC (1990) OXBOX, a positive transcriptional regulator of the heart-skeletal muscle ADP/ATP translocator gene. J Biol Chem 265, 20585-20588.

172. Nelson BD, Mutvei A, Joste V (1984) Regulation of biosynthesis of the rat liver inner mitochondrial membrane by thyroid hormone. Arch Biochem Biophys 228, 41-48.

173. Joste V, Goitom Z, Nelson BD (1989) Thyroid hormone regulation of nuclear-coded mitochondrial inner membrane polypeptides of the liver. Eur J Biochem 184, 255-260.

174. Izquierdo JM, Cuezva JM (1993) Thyroid hormones promote transcriptional activation of the nuclear gene coding for mitochondrial β-F_1-ATPase in rat liver. FEBS Lett 323, 109-112.

175. Matsuda C, Endo H, Hirata H et al. (1993) Tissue-specific isoforms of the bovine mitochondrial γ-subunit. FEBS Lett 325, 281-284.

176. Tvrdík P, Kuzela S, Houstek J (1992) Low translational efficiency of the F_1-ATPase β subunit mRNA largely accounts for the decreased ATPase content in brown adipose tissue mitochondria. FEBS Lett 313, 23-26.

177. Luis AM, Izquierdo JM, Ostronoff LK et al. (1993) Translational regulation of mitochondrial differentiation in neonatal rat liver. Specific increase in the translational efficiency of the nuclear-encoded mitochondrial β F_1 ATPase messenger RNA. J Biol Chem 268, 1868-1875.

178. Brown GC (1992) Control of respiration and ATP synthesis in mammalian mitochondria and cells. Biochem J 284, 1-13.

179. Nicholls DG, Locke RM (1984) Thermogenic mechanisms in brown fat. Physiol Rev 64, 1-64

180. Chance B, Williams GR (1956) The respiratory chain and oxidative phosphorylation. Adv Enzymol 17, 65-134.

181. Tager JM, Wanders AJA, Groen AK et al. (1983) Control of mitochondrial respiration FEBS Lett 151, 1-9.

182. Groen AK, Wanders RJA, Westerhoff HG et al. (1982) Quantification of the contribution of various steps to the control of mitochondrial respiration. J Biol Chem 257, 27544-2757.

183. Katz LA, Swain JA, Portman MA et al. (1989) Relation between phosphate metabolites and oxygen consumption of heart. Am J Physiol 255, H189-H19.

184. From AHL, Petein MA, Michurski SP et al. (1986) ^{31}P-N.M.R. studies of respiratory regulation in the intact myocardium. FEBS Lett 206, 257-261.

185. LaNoue KF, Jeffries FMH, Radda GK (1985) Kinetic control of mitochondrial ATP synthesis. Biochemistry 25, 7667-7675.

186. Unitt JF, McCormack JG, Reid D et al. (1989) Direct evidence for a

role of intramitochondrial Ca^{2+} in the regulation of oxidative phosphorylation in the stimulated rat heart. Studies using ^{31}P-N.M.R. and ruthenium red. Biochem J 262, 393-401.

187. Matthews PM, Bland JL, Gadian DG et al. (1981) The steady-state rate of ATP synthesis in the perfused rat heart measured by ^{31}P-N.M.R. saturation transfer. Biochem Biophys Res Commun 103, 1052-1059.

188. Kauppinen R (1983) Proton electrochemical potential of the inner mitochondrial membrane in isolated perfused rat hearts as measured by exogenous probes. Biochim Biophys Acta 725, 131-137.

189. Rouslin W (1983) Protonic inhibition of the mitochondrial oligomycin-sensitive adenosine 5'-triphosphatase in ischemic and autolyzing cardiac muscle. Possible mechanisms for the mitigation of ATP hydrolysis under nonenergizing conditions. J Biol Chem 258, 9657-9661.

190. Rouslin W (1987) Persistence of mitochondrial competence during myocardial autolysis. Am J Physiol 252, H633-H627.

191. Das AM, Harris DA (1989) Reversible modulation of the mitochondrial ATP synthase with energy demand in cultured rat cardiomyocytes. FEBS Lett 256, 97-100.

192. Das AM, Harris DA (1990) Control of mitochondrial ATP synthase in heart cells. Inactive to active transitions caused by beating or positive inotropic agents. Cardiovasc Res 24, 411-417.

193. Adolfsen R, McClung JA, Moudrianakis EN (1975) Electrophoretic microheterogeneity and subunit composition of the 13S coupling factors of oxidative and photosynthetic phosphorylation. Biochemistry 14, 1727-1735.

194. Noll T, Koop A, Piper HM (1992) Mitochondrial ATP synthase activity in cardiomycetes after aerobic-anaerobic metabolic transition. Am J Physiol 262, C1297-C1303.

195. Jennings RB, Reimer KA, Steenbergen C (1991) Effects of inhibition of the mitochondrial ATPase on net myocardial ATP in total ischemia. J Mol Cell Cardiol 23, 1383-1395.

196. Crompton M, Costi A (1990) A heart mitochondrial Ca^{2+}-dependent pore of possible relevance to re-perfusion-induced injury. Evidence that ADP facilitates pore inter-conversion between the closed and open states. Biochem J 266, 33-39.

197. Mills JD, Mitchell P (1982) Modulation of coupling factor ATPase activity in intact chloroplasts. Reversal of thiol modulation in the dark. Biochim Biophys Acta 679, 75-83.

198. Junesch U, Graber P (1987) Influence of the redox state and the activation of the chloroplast ATP synthase on proton transport-coupled ATP synthesis and hydrolysis. Biochim Biophys Acta 893, 275-288.

199. Yamada EW, Huzel NJ (1988) The calcium-binding ATPase inhibitor protein from bovine heart mitochondria. Purification and properties. J Biol Chem 263, 11498-11503.

200. Yamada EW, Huzel NJ, Dickison JC (1981) Reversal, by uncouplers of oxidative phosphorylation and by Ca^{2+}, of the inhibition of mitochondrial ATPase by the ATPase inhibitor protein of rat skeletal muscle. J Biol Chem 256, 10203-10207.

201. Yamada EW, Huzel NJ, Bose R et al. (1992) ATPase inhibitor proteins of brown-adipose-tissue mitochondria from warm- and cold-acclimated rats. Biochem J 287, 151-157.

202. Valerio M, Haraux F, Gardeström P et al. (1993) Tissue specificity of the regulation of ATP hydrolysis by isolated plant mitochondria. FEBS Lett 318, 113-117.
203. Grubmeyer G, Spencer M (1980) ATPase activity of pea cotyledon submitochondrial particles. Plant Physiol 65, 281-285.
204. Jung DW, Laties GG (1976) Trypsin-induced ATPase activity in potato mitochondria. Plant Phsyiol 57, 583-588.
205. Ligeti E, Brandolin G, Dupont Y et al. (1985) Substrate-induced modifications of the intrinsic fluorescence of the isolated adenine nucleotide carrier protein. Demonstration of distinct conformational states. Biochemistry 24, 4423-4428.
206. Vignais PV (1976) Molecular and physiological aspects of adenine nucleotide transport in mitochondria. Biochim Biophys Acta 456, 1-38.
207. Nair CKK (1993) Mitochondrial genome organization and cytoplasmic male sterility in plants. J Biosciences 18, 407-422.
208. Huang J, Lee SH, Lin C et al. (1990) Expression in yeast of the T-urf13 protein from Texas male-sterile maize mitochondria confers sensitivity to methomyl and to Texas-cytoplasm-specific fungal toxins. EMBO J 9, 339-347.
209. Young EG, Hanson MR (1987) A fused mitochondrial gene associated with cytoplasmic male-sterility is developmentally regulated. Cell 50, 41-49.
210. Handa H, Nakajima K (1992) Different organization and altered transcription of the mitochondrial atp6 gene in the male sterile cytoplasm of rapeseed (Brassica napus L.) Curr Genet 21, 153-159.
211. Siculella L, Palmer JD (1988) Physical and gene organization of mitochondrial DNA in fertile and male-sterile sunflower. CMS-associated alterations in structure and transcription of the atpA gene. Nucl Acids Res 16, 3787-3799.
212. Sena M, Mikami T, Kinoshita T (1993) The sugar beet mitochondrial gene for the ATPase alpha-subunit: sequence, transcription and rearrangements in cytoplasmic male-sterile plants. Curr Genet 24, 164-170.
213. Iwabuchi M, Kyozuka J, Shimamoto K (1993) Processing followed by complete editing of an altered mitochondrial atp6 RNA restores fertility of cytoplasmic male sterile rice. EMBO J 12, 1437-1446.
214. Hernould M, Suharsono S, Litvak S et al. (1993) Male-sterility induction in transgenic tobacco plants with an unedited atp9 mitochondrial gene from wheat. Proc Natl Acad Sci USA 90, 2370-2374.
215. Schotland DL, DiMauro S, Bonilla E et al. (1976) Neuromuscular disorder associated with a defect in mitochondrial energy supply. Arch Neurol 33, 475-479.
216. Holme E, Greter J, Jacobson C-E, Larsson N-G et al. (1992) Mitochondrial ATP-synthase deficiency in a child with 3-methylglutaconic aciduria. Pediatr Res 32, 731-735.
217. Holt IJ, Harding AE, Petty RKH et al. (1990) A new mitochondrial disease asssociated with mtDNA heteroplasmy. Am J Hum Genet 46, 428-433.
218. Tatuch Y, Chistodoulou J, Feigenbaum A et al. (1992) Heteroplasmic mtDNA mutation 8993 (T—> G) can cause Leigh disease when the percentage of abnormal mtDNA is high. Am J Hum Genet 50, 852-858.
219. Santorelli FM, Shanske S, Macaya A et al.(1993) The mutation at nt-

8993 is a common cause of Leigh syndrome. Ann Neurol 34, 827-834.

220. de Vries D, van Engelen BGM, Gabreëls FJM et al. (1993) A second missense mutation in the mitochondrial ATPase 6 gene in Leigh's Syndrome. Ann Neurol 34, 410-412.

221. Tatuch Y, Robinson BH (1993) The mitochondrial DNA mutation at 8993 associated with NARP slows the rate of ATP synthesis in isolated lymphoblast mitochondria. Biochem Biophys Res Commun 192, 124-128.

222. Trounce I, Neil S, Wallace DC (1993) Cytoplasmic transfer of the mitochondrial DNA 8993 $^{T \to G}$ (ATP6) point mutation associated with Leigh's disease into mtDNA-less cells shows co-segregation of ATP synthase defect. Am J Hum Genet 53, 157.

223. Hartzog PE, Cain BD (1993) The aleu207 —> arg mutation in F_1F_0-ATP synthase from Escherichia coli. A model for human mitochondrial disease. J Biol Chem 268, 12250-12252.

224. Wallace DC (1992) Diseases of the mitochondrial DNA. Ann Rev Biochem 61, 1175-1212.

225. Poulton J, Deadman ME, Bindoff L et al. (1993) Families of mtDNA rearrangements can be detected in patients with mtDNA deletions: duplications may be a transient intermediate form. Human Molecular Genetics 2, 23-30.

226. Heddi A, Lestienne P, Wallace DC et al. (1993) Mitochondrial DNA expression in mitochondrial myopathies and coordinated expression of nuclear genes involved in ATP production. J Biol Chem 268, 12156-12163.

227. Kelley RI, Cheatham JP, Clark BJ et al. (1991) X-linked dilated cardiomyopathy with neutropenia, growth retardation and 3-methylglutaconic aciduria. J Pediatr 119, 738-747.

228. Bolhuis PA, Hensels GW, Hulsebos TJM et al. (1991) Mapping of the locus for X-linked cardioskeletal myopathy with neutropenia and abnormal mitochondria (Barth syndrome) to Xq28. Am J Hum Gent 48, 481-485.

229. Barth PG, Scholte HR, Berden JA et al. (1983) An X-linked mitochondrial disease affecting cardiac muscle, skeletal muscle and neutrophil leucocytes. J Neurol Sci 62, 327-355.

230. DiMauro S, Bonilla E, Zeviani M et al. (1987) Mitochondrial myopathies. J Inh Metab Dis 10, Suppl 1, 113-128.

231. Luft R, Ikkos D, Palmieri G et al. (1962) A case of severe hypermetabolism of nonthyroid origin with a defect in the maintenance of mitochondrial respiratory control: a correlated clinical, biochemical and morphological study. J Clin Invest 41, 1776-1804.

232. DiMauro S, Bonilla E, Lee CP et al (1976) Luft's disease. Further biochemical and ultrastructural studies of skeletal muscle in a second case. J Neurol Sci 27, 217-232.

233. Lake BD (1992) Lysosomal and peroxisomal disorders. In: Hume-Adams J, Duchen LW, eds. Greenfield's Neuropathology. 5th ed. London: Edward Arnold, pp. 709-810.

234. Rider AJ, Rider DL (1985) Batten's disease: neuronal ceroid lipofuscinosis. Arch Biol 96, 369-370.

235. Berkovic SF, Carpenter S, Anderman F et al. (1988) Kufs' disease: a critical reappraisal. Brain 111, 27 62.

236. Palmer DN, Martinus RD, Cooper SM et al. (1989) Ovine ceroid

lipofuscinosis. The major lipopigment protein and the lipid-binding sub-unit of mitochondrial ATP synthase have the same NH$_2$-terminal sequence. J Biol Chem 264, 5736-5740.

237. Fearnley IM, Walker JE, Martinus RD et al. (1990) The sequence of the major protein stored in ovine ceroid lipofuscinosis is identical with that of the dicyclohexylcarbodiimide-binding proteoplipid of mitochondrial ATP synthase. Biochem J 268, 751-758.

238. Hall NA, Lake BD, Dewji NN et al. (1991) Lysosomal storage of sub-unit c of mitochondrial ATP synthase in Batten's disease (ceroid-lipofuscinosis). Biochem J 275, 269-272.

239. Kominami E, Ezaki J, Muno D et al. (1992) Specific storage of subunit c of mitochondrial ATP synthase in lysosomes of neuronal ceroid lipofuscinosis (Batten's disease). J Biochem 111, 278-282.

240. Das AM, Harris DA (1990) Defects in regulation of mitochondrial ATP synthase in cardiomyocytes from spontaneously hypertensive rats. Am J Physiol 259, H1264-H1269.

241. Coleman WB, Spach P, Cunningham C (1991) Effect of chronic ethanol consumption on the subunit structure of the mitochondrial ATP synthase. FASEB J, 5, A845.

242. Das AM, Harris DA (1993) Regulation of the mitochondrial ATP synthase is defective in rat heart during alcohol-induced cardiomyopathy. Biochim Biophys Acta 1181, 295-299.

243. Guerrieri F, Capozza G, Kalous M et al. (1992) Age-related changes of mitochondrial F$_1$F$_0$ ATP synthase. Ann New York Acad Sci 671, 395-402.

244. Miquel J (1992) An update on the mitochondrial DNA mutation hy-pothesis of cell aging. Mutation Res 275, 209-216.

245. Malis CD, Bonventre JV (1986) Mechanism of potentiation of oxygen free-radical injury to renal mitochondria. A model for post-ischemic and toxic mitochondrial damage. J Biol Chem 261, 4201-4208.

246. Das AM, Harris DA (1990) Regulation of the mitochondrial ATP synthase in intact rat cardiomyocytes. Biochem J 266, 355-361.

247. Lippe G, Comelli M, Mazzilis D et al. (1991) The inactivation of mito-chondrial F$_1$ ATPase by H$_2$O$_2$ is mediated by iron ions not tightly bound in the protein. Biochem Biophys Res Commun 181, 764-770.

248. MillerSG (1987) Association of a sperm-specific protein with the F$_1$F$_0$ ATPase in *Heliothis*. Implications of sterility in Heliothis virescens x Heliothis subflexa backcross hybrids. Insect Biochem 17, 417.

CHAPTER 6

PROPERTIES OF KIDNEY PLASMA MEMBRANE VACUOLAR H⁺-ATPASES: PROTON PUMPS RESPONSIBLE FOR BICARBONATE TRANSPORT, URINARY ACIDIFICATION, AND ACID-BASE HOMEOSTASIS

Stephen L. Gluck, Raoul D. Nelson, Beth S.M. Lee, L. Shannon Holliday, and Masahiro Iyori

FUNCTION AND PROPERTIES OF THE RENAL VACUOLAR H⁺-ATPASE IN BICARBONATE TRANSPORT AND URINARY ACIDIFICATION

SITES OF H⁺ TRANSPORT BY VACUOLAR H⁺-ATPASES IN THE NEPHRON

To maintain acid-base homeostasis, the kidney must reabsorb all of the 4500 millimoles of bicarbonate filtered by the glomerulus and regenerate the approximately 70 millimoles of bicarbonate consumed by daily metabolism.[1,2] The kidney accomplishes both bicarbonate reabsorption and regeneration by using hydrogen ion secretion,[1,2] and vacuolar H⁺-ATPases residing on the plasma membrane have an important or essential role in these processes in several nephron segments.[3-9]

Organellar Proton-ATPases, edited by Nathan Nelson; ©1994 R.G. Landes Company.

The plasma membrane vacuolar H+-ATPases differ from vacuolar H+-ATPases of intracellular organelles in several ways. The plasma membrane vacuolar H+-ATPase in hydrogen ion-transporting renal epithelial cells reside at high densities,[10] and they have a polarized distribution[11,12] that allows for vectorial secretion of hydrogen ion. The plasma membrane vacuolar H+-ATPases of the nephron are also subject to physiologic regulation[3,8,13] that allows the kidney to preserve acid-base balance.

The proximal tubule reabsorbs 80% of the filtered bicarbonate;[14] 60-70% of this 3500 millimoles of bicarbonate is reabsorbed by proton secretion driven by the Na+/H+ antiporter of the proximal brush border, and 30-40% is attributable to proton secretion by a brush border membrane vacuolar H+-ATPase.[5,15]

The thick ascending limb reabsorbs 10-15% of the filtered bicarbonate.[16] At least 80% of the proton secretion in this segment comes from a Na+/H+ antiporter in the luminal membrane, and up to 20% originates from a luminal membrane vacuolar H+-ATPase.[17-20] The distal convoluted tubule normal reabsorbs only 1-2% of the filtered bicarbonate; 65-80 % of the bicarbonate reabsorption is from Na+/H+ antiport.[21] The remainder occurs by an amiloride-insensitive mechanism that may arise from a vacuolar H+-ATPase.[12,21]

The collecting duct is responsible for reabsorption of the remaining bicarbonate in the tubular fluid and for the additional secretion of hydrogen ion that results in the addition of bicarbonate to the blood. The cortical portion of the collecting duct contains several segments that are morphologically and functionally distinct: the connecting tubule, the initial collecting tubule, and the cortical collecting tubule. These segments are a mosaic of two major cell types: the principal cells, comprising about 60% of the cells, and the intercalated cells. The intercalated cells are responsible for most or all of the hydrogen ion transport in these segments.[1] In these two segments, there are both acid-secreting (α or type A) and bicarbonate-secreting (β or type B) intercalated cells[2] that differ in the type and arrangement of ion transporters that they possess. These properties are discussed in greater detail below. The majority of hydrogen ion transport in intercalated cells arises from a vacuolar H+-ATPase, which is abundant in the intercalated cells. The vacuolar H+-ATPase is probably the enzyme responsible for regulation of net acid secretion in the distal nephron.[26,27] Intercalated cells also appear to have a H+/K+-ATPase, that secretes H+ in exchange for potassium reabsorption.[28] This enzyme probably has an important function in active reabsorption of potassium from the urine during potassium depletion; its role in urinary acidification and acid-base homeostasis is still uncertain.

---

[1] With the exception of the outer medulla in the rabbit kidney, in which the principal cells also have a luminal membrane vacuolar H+-ATPase, and therefore may contribute to H+ secretion.[22]

[2] Studies indicate that there are more than two different phenotypic forms of intercalated cells.[23-25]

The outer medullary collecting tubule has two functionally distinct segments, the outer and inner stripes. Like the cortical collecting tubule, the outer medullary collecting tubule is composed of principal and intercalated cells; however, these segments have only α (acid-secreting) intercalated cells, and no bicarbonate-secreting cells (reviewed in ref. 29). The inner stripe has the highest rate of H⁺ secretion of the collecting duct segments.[30]

The inner medullary collecting duct has a few α intercalated cells in the first third of its length, but is otherwise composed of a single cell type, the inner medullary collecting duct cell. These cells secrete hydrogen ion by an unknown mechanism, possibly an H⁺/K⁺-ATPase. The importance of this segment to overall net acid secretion is unresolved.

PROPERTIES OF THE VACUOLAR H⁺-ATPase IN KIDNEY

TRANSPORT PROPERTIES

Transport Properties of Vacuolar H⁺-ATPase in Turtle Urinary Bladder

General Properties
Many of the physiologic properties of H⁺ secretion by vacuolar H⁺-ATPases in urinary epithelia were first elucidated in the turtle urinary bladder, a model epithelium that acidifies its lumen, and which has transport properties similar to those of the cortical collecting tubule.[13,31] Under short-circuit conditions, after sodium transport is inhibited by ouabain or amiloride, the bladder exhibits lumen-positive current that is identical (in meq/min) to the rate of luminal acidification measured with a pH stat,[32] a strong indication that the pump is electrogenic.

Membrane vesicle fractions from the turtle bladder were found to acidify their lumen upon the addition of ATP, indicating that they contained an ATP-driven proton pump.[33,34] The rate of acidification was increased by conditions that collapse a potential difference across the vesicle membrane, indicating that the pump was electrogenic. Proton transport was inhibited by DCCD and sulfhydryl reagents, but not by the F_0F_1 inhibitors oligomycin and azide, or the E_1E_2 inhibitor vanadate,[33] demonstrating that the pump had a unique inhibitor sensitivity profile, subsequently shown to be a generally characteristic of the vacuolar H⁺-ATPases. These experiments provided the first direct evidence that the urinary bladder proton pump was an ATPase.

H⁺/ATP Stoichiometry of the Turtle Bladder Proton Pump
Most of the properties of the vacuolar H⁺-ATPase of urinary epithelia have been determined from studies of the purified enzyme (see below). The apparent H⁺/ATP stoichiometry of the turtle bladder proton pump, however, has been examined in the intact tissue using two different approaches. Dixon and Al-Awqati[35] used measurements of the $\Delta G'_{ATP}$ and the apparent PMF of the pump, extrapolated from a plot of transport rate as a function of the luminal pH gradient, to arrive at

a stoichiometry of 3 H$^+$/ATP. Andersen and Steinmetz[31,36] have pointed out, however, that the apparent PMF obtained in this manner can be misleading if the shape of the pump I-V curve has an inflection near zero. Steinmetz et al[37] measured the production of lactate as a function of H$^+$ transport rates in bladders maintained under anaerobic conditions using glucose as a substrate. The measured lactate formation/H$^+$ transport ratios were constant at 0.55 to 0.58 under a variety of conditions, yielding a calculated stoichiometry of 2 H$^+$/ATP.

Transport Properties of Vacuolar H$^+$-ATPase in Mammalian Kidney Membranes

Renal Medullary Membranes

Membrane fractions from bovine kidney medulla were found to contain an ATP-driven proton pump with transport properties identical to those of the proton pump in the turtle urinary bladder.[38] Proton transport was resistant to the mitochondrial H$^+$-ATPase inhibitors oligomycin, efrapeptin, and rutamycin, and resistant to 500 µM vanadate, but inhibited completely by N-ethylmaleimide (NEM) with a K_i of 28 µM,[38] and by Zn^{2+} with a K_i of 170 µM.[39] The H$^+$ transport activity required ATP, and was not supported by any other nucleotide.[38] The membranes were found to contain NEM-sensitive ATPase activity that was resistant to oligomycin, azide, and vanadate, and which was used as an assay for isolation of the H$^+$-ATPase.[39,40] Proton transport activity with nearly identical properties was subsequently described in membranes from rat[41] and human renal medulla.[42]

Renal Cortex Endosomes

In rat kidney cortex, vacuolar H$^+$-ATPase-driven proton transporting activity was found in a membrane fraction that was enriched for endosomes.[43] ATP-induced acidification in these membranes exhibited properties typical of the vacuolar H$^+$-ATPases: sensitivity to NEM, DCCD, and Zn^{2+}, and resistance to oligomycin.[43] H$^+$ transport was highly selective for ATP, with a K_m = 77 µM; vesicle acidification required a permeant anion,[43,44] with a relative efficacy of SCN$^-$>Cl$^-$>Br$^-$>I$^-$>>HPO$_4^{2-}$>HCO$_3^-$>F$^-$, but showed no preference for monovalent cations.[44] These membranes also displayed ATPase activity that was stimulated by ionophores and whose ionic dependence and reaction to inhibitors were similar to those of ATP-driven acidification, consistent with the presence of an H$^+$-ATPase whose activity is affected by the proton electrochemical gradient across the vesicle membrane.[45] ATP-driven proton transport in membranes enriched for endosomes from rabbit kidney cortex had properties similar to those of endosome-enriched rat kidney cortex membranes.[46]

Renal Cortex Brush Border Membranes

Renal brush border membranes are normally oriented "right side out", preventing access of ATP to the cytoplasmic domain of the vacuolar H$^+$-ATPase. Alternative approaches have therefore been employed for studying vacuolar H$^+$-ATPase-driven H$^+$ transport in brush border. In early studies, brush border membranes were found to have intravesicular ATPase activity inside the sealed membranes. The activity was stimulated by protonophores and inhibited by DCCD, which was interpreted as representing an H⁺-ATPase in the membrane.[47] Subsequent experiments, however, showed that the protonophore-stimulated ATPase activity was insensitive to high concentrations of NEM[48] and was therefore unlikely to have arisen from vacuolar H⁺-ATPase activity.

Burckhardt and colleagues[49] obtained the first evidence for a vacuolar H⁺-ATPase in renal brush border. They found that rat kidney brush border membranes into which ATP and an ATP-regenerating system were introduced by freeze-thawing were able to alkalinize their interior by a DCCD and NEM-inhibitable mechanism; these membranes also had internal NEM-sensitive ATPase activity. Simon and Burckhardt[50] demonstrated that treatment of brush border membranes with cholate resulted in a reorientation of the membrane, such that addition of ATP to the vesicles resulted in intravesicular acidification. The acidification in these reoriented brush border membranes had inhibitor sensitivities typical of the vacuolar H⁺-ATPases: it was resistant to azide, oligomycin, and vanadate, but was inhibited by DCCD and Zn^{2+}, and had K_i's for NEM and NBD-Cl of 0.77 μM and 0.39 μM. The K_m for ATP was 93 μM, with a substrate preference of ATP>GTP≈ITP>UTP; ADP was a competitive inhibitor of acidification with a K_i = 24 μM. The same laboratory later showed that acidification by the reoriented brush border membranes was highly sensitive to bafilomycin A_1, with a K_i < 10 μM.[51] These properties are in good agreement with those of the purified bovine brush border vacuolar H⁺-ATPase[52] (see below).

Enzymatic Properties
of the Isolated Mammalian Kidney Vacuolar H⁺-ATPase

The kidney vacuolar H⁺-ATPase was initially isolated from solubilized bovine kidney medullary membranes by precipitation and ion exchange chromatography;[39] a monoclonal antibody was generated against the isolated bovine H⁺-ATPase,[53] which enabled the development of an immunoaffinity method for purifying the active enzyme from solubilized kidney membranes in a single step.[39]

Kidney vacuolar H⁺-ATPase has been isolated by this immunoaffinity technique from several renal membrane fractions[39,40,52,54] (Table 6.1). Like all other known vacuolar H⁺-ATPases,[55-57] the renal vacuolar H⁺-ATPase isolated from these membranes was electrogenic.[38-40] H⁺-ATPase isolated from proximal tubule brush border membranes and lysosomes from bovine renal cortex, and from microsomes from bovine renal outer medulla differed, however, in several enzymatic properties. The pH optimum of the brush border H⁺-ATPase was 7.2, close to the steady-state intracellular pH of the proximal tubule cell,[14] while that from the microsomes was 6.3, well below the intracellular pH of about 7.2 in intercalated cells.[22,23,58-65] These findings imply that brush border enzyme would transport optimally in a constitutive state, where cell pH remained at its usual value. In contrast, the medullary microsomal enzyme, which probably comes mostly from intercalated cells because it is enriched in the B1 subunit isoform [39,40,52,66] (see below), would increase its activity were cytosolic pH to fall, enhancing renal acid excretion. The magnitude of such changes would be small, however: only 20-25% over a physiologic pH range of 6.2 to 7.5.[39]

Vacuolar H⁺-ATPases isolated from different kidney membrane fractions differed in other properties (Table 6.1). The medullary microsomal enzyme activity was activated by several phospholipids;[39] in contrast, the activities of the brush border and lysosomal enzymes were not affected by added lipids.[52,54] As the apical membrane of epithelia maintains a different lipid composition from that of the basolateral membrane and intracellular membrane compartments,[67-69] changes in the lipid environment of the enzyme could conceivably be a means for physiologic regulation.

The H⁺-ATPases from different membrane compartments varied in their substrate specificity (Table 6.1), implying that they might differ in the structure of the catalytic site. Striking differences were also found in the effects of divalent and trivalent cations on ATPase activity. Divalent or trivalent cations differed in their ability to substitute for magnesium in supporting ATPase activity among the vacuolar H⁺-ATPases from different kidney membrane compartments.[39,52,54] Divalent or trivalent cations also varied in their ability to inhibit ATPase activity among the different enzyme preparations. The inhibition by Al^{3+} was competitive with ATP,[54] again suggesting that the differences in the effects of the multivalent cations occurs at the active site of these enzymes.

Chloride neither stimulated nor was required for either the ATPase or proton transporting activity of the purified H⁺-ATPase.[39] Both ATP-induced acidification and potential generation by the purified, reconstituted H⁺-ATPase occurred in the absence of chloride.[39] Although chloride, or other permeant anions, were required for ATP-induced acidification of kidney membrane vesicles,[38,41,43,45] this requirement most

Table 6.1. Properties of H⁺-ATPase in different kidney membrane fractions

	Membrane Fraction		
	Microsome[39,40]	**Brush Border[52]**	**Lysosome[54]**
pH Optimum	6.3	7.4	6.9
Substrate specificity	ATPase:GTPase = 4:1	ATPase:GTPase = 2:1	ATPase:GTPase = 3:1
Lipid effects	Activated by PG > PI >> PC > PS	No lipid activation	No lipid activation
Multivalent cation effects	Cu^{+2}: 100% inhib Al^{+3}: 0% inhib	Cu^{+2}: 50% inhib Al^{+3}:0% inhib	Cu^{+2}: 20% inhib Al^{+3}: 80% inhib
Anion effects	No Cl⁻ requirement Inhibition by NO$_3^-$	No Cl⁻ requirement Inhibition by NO$_3^-$	No Cl⁻ requirement Inhibition by NO$_3^-$
Structure	A subnit: single 70 kD B subunit: 58 kD (B1) > 56 kD (B2) E subunit: single 31 kD	A subunit:single 70 kD B subunit: 56 kD (B2) only E subunit: multiple ~31 kD	A subunit:single 70 kD B subunit: 56 kD (B1) only E subunit: single 31 kD

likely reflects the need for chloride entry as a counterion to collapse an inside-positive potential difference produced by electrogenic H^+ secretion.

The kidney vacuolar H^+-ATPases appear to have an oxyanion binding site that affects enzymatic activity. For example, both nitrate and sulfite (at concentrations >5 mM) inhibited the H^+-ATPase with uncompetitive kinetics.[52,54]

STRUCTURAL PROPERTIES OF H⁺-ATPASE IN KIDNEY

The vacuolar H^+-ATPase isolated from bovine kidney membranes by ion exchange chromatography or immunoaffinity chromatography contained polypeptides at M_r = 70,000, ~56,000 (as a cluster of three to four discrete polypeptides), 45,000, 42,000, 38,000, 33,000, 31,000, 15,000, 14,000, and 12,000.[39,40,52] This composition was similar to that reported for other mammalian vacuolar H^+-ATPases (reviewed in ref. 55), but notably lacked the M_r = 110,000 or 116,000 intrinsic membrane polypeptide observed in other H^+-ATPase preparations. A summary of the subunits of the mammalian kidney vacuolar H^+-ATPase is given in Table 6.2. The kidney H^+-ATPase isolated by ion exchange chromatography exhibited two peaks of ATPase activity.[39] These fractions were subjected to polyacrylamide gel electrophoresis under nondenaturing conditions, and two complexes with approximate M_r values of 551,000 and 523,000 were found.[40] The polypeptide composition in the two peaks of activity showed some differences. Most notably, two B subunit polypeptides at M_r = 58,000 and 56,000 were found to segregate on the column, each following one of the peaks of ATPase activity eluting from the column.[39] These findings were an initial indication that different structural forms of the H^+-ATPase might exist containing isoforms of a B subunit.

The possibility that H^+-ATPases with different structures were located in different membrane compartments in the kidney was subsequently investigated. H^+-ATPase was isolated by immunoaffinity chromatography from kidney microsomes and from brush border. The structure of the isolated enzymes was examined on two dimensional polyacrylamide gels. Both of the enzyme preparations showed a single A subunit polypeptide. In contrast, the enzymes differed substantially in the composition of the B and E subunits. In both preparations, the B subunit showed several polypeptides at M_r = 56,000, but the pattern from the two preparations was reproducibly different. Only the microsome-derived H^+-ATPase had a prominent polypeptide at M_r = 58,000, which had a pI slightly more acidic than that of the M_r = 56,000 polypeptides.[52] In the brush border H^+-ATPase, multiple E subunit polypeptides were observed, with M_rs in the 31,000–35,000 range; all of these exhibited immunoreactivity with a monoclonal antibody to the COOH-terminus of the E subunit.[70] In contrast, in the microsomal H^+-ATPase, only a single E subunit polypeptide was observed. These results suggest that structural differences in the B and/or E subunits might underlie differences in enzymatic properties or distribution of vacuolar H^+-ATPases.

cDNA clones for the A, B1, B2, and E subunits have been isolated from kidney cDNA libraries.[66,70-74] The structure of the A, B,

Table 6.2. Subunits of the mammalian kidney vacuolar H⁺-ATPase

Domain	Subunit M_r	Letter/Name(s)	Copies per H⁺-ATPase	Function	Structural Features
V_1	70,000	A subunit A1 isoform	3	Site of ATP hydrolysis during catalysis	"P-loop" nucleotide-binding motif ATP-synthase motif ; homologous to F_0F_1 β subunit
	58,000[1] 56,000	B subunit B1 "kidney" isoform B2 "brain" isoform	3	? Regulatory subunit Influence on enzymatic properties	ATP-synthase motif; homologous to F_0F_1 α subunit
	42,000	C subunit	1	Unknown function; probably resides in "stalk"	Not homologous to any F_0F_1 subunit
	33,000		1	Unknown function; probably resides in "stalk"	Structure unknown
	31,000	E subunit	1	Unknown function; probably resides in "stalk"	Not homologous to any F_0F_1 subunit
V_0	15,000	Proteolipid c subunit	6	Probably major structural component of "proton channel"	DCCD-reactive glutamate residue; homologous to F_0F_1 proteolipid
	14,000[2]	Name not designated	Unknown	Unknown	Unknown
	12,000[2]	Name not designated	Unknown	Unknown	Unknown

[1] Evidence suggests that only three copies of a B1 isoform or three copies of a B2 isoform are present in a single H⁺-ATPase complex (D.M. Underhill and S.L. Gluck, unpublished observations).

[2] The precise composition of the vacuolar H⁺-ATPase remains unresolved, and differences in subunit composition have been reported among enzyme purified from mammalian sources. The $M_r = 14,000$ subunit may be part of the V_1 domain, as reported for the H⁺-ATPase isolated from insect intestine.[293]

and E subunit genes is discussed elsewhere. The B1 and B2 subunits represent two isoforms of the B subunit that are encoded by different genes. The predicted protein products of the two cDNAs are 98% identical over most of the protein. They differ markedly, however, in the sequences of the amino terminal 28 amino acids, and the carboxyl terminal 12 amino acids.[66]

The cDNAs for the A and E subunit are expressed in all mammalian tissues[71,73,75] and appear identical to those in the constitutively expressed vacuolar H⁺-ATPase common to all cells. The B1 isoform is also expressed in all mammalian tissues,[66,76] and encodes a subunit probably residing in constitutive forms of the vacuolar H⁺-ATPase, such as lysosomes and endosomes.[54] The B2 isoform, however, is expressed at detectable levels only in the kidney[66] and placenta (R.D.N., S. Bae, and S.L.G, unpublished observations), and may encode a subunit with cell-specific functions.

DISTRIBUTION OF VACUOLAR H⁺-ATPASE IN KIDNEY

Distribution of Vacuolar H⁺-ATPase by Physiologic Criteria

Several studies have provided physiologic evidence for vacuolar H⁺-ATPase-driven H⁺ secretion in the proximal tubule luminal membrane.[6,9,15,77-80] Approximately one-third of proximal tubule bicarbonate reabsorption (H⁺ secretion) is independent of Na⁺-H⁺ exchange,[9,78,79] suggesting that another transporter, such as an H⁺-ATPase, has a significant role in proximal H⁺ secretion. Kurtz[78] demonstrated that the S3 portion of the rabbit proximal tubule has a sodium-independent mechanism for pH_i recovery following acidification with NH_4Cl; this recovery process was inhibited by luminal NEM and DCCD, suggesting that it arose from a vacuolar H⁺-ATPase. Bank et al[15] showed that the mid- to late portions of the proximal convoluted tubule in rat kidney (probably representing the S2 and part of the S1 segments), Na⁺-H⁺ exchange-independent bicarbonate reabsorption is inhibitable by luminal NEM and DCCD, suggesting that it occurs through a luminal vacuolar H⁺-ATPase. Zimolo,[6] studying intracellular pH regulation in fragments of rat proximal tubule immobilized on coverslips, found a sodium-independent mechanism for intracellular pH recovery from intracellular acidification induced by the NH_4Cl prepulse method.[81] The intracellular pH recovery was inhibited by inhibitors of ATP production, and inhibitable by bafilomycin A_1, indicating that it probably represented a vacuolar H⁺-ATPase.

In the rat early distal tubule, Wang et al[21] found that 20% of the bicarbonate reabsorption was resistant to amiloride and ethylisopropylamiloride (EIPA), inhibitors of Na⁺-H⁺, and that bafilomycin A_1 inhibited about 36% of bicarbonate reabsorption, consistent with a small contribution of a vacuolar H⁺-ATPase to overall H⁺ secretion. In the late distal tubule (corresponding the initial collecting tubule), amiloride and EIPA had no effect on bicarbonate reabsorption, whereas bafilomycin inhibited 47%, indicating that a vacuolar H⁺-ATPase may be the major transporter responsible for H⁺ in this segment.[21]

Numerous investigations have provided physiologic evidence for a vacuolar H⁺-ATPase in the cortical and outer medullary collecting tubule. The intact cortical collecting tubule, under suitable conditions, reabsorbs bicarbonate and generates a lumen-positive potential difference in the absence of sodium or presence of luminal amiloride, consistent with a model of luminal H⁺ secretion by an electrogenic H⁺-ATPase.[82-87] Evidence for a vacuolar H⁺-ATPase in the plasma membrane of collecting tubule cells has also come from studies on regulation of their intracellular pH. After pH_i of individual collecting tubule cells is acidified by the NH_4Cl prepulse technique,[81] pH_i rises partly by Na^+/H^+ exchange,[60-62] and partly by a sodium-independent amiloride-insensitive mechanism, that is resistant to the H^+/K^+-ATPase inhibitor SCH 28080,[61,63,88] but inhibitable by luminal NEM,[63,64,88] thought to represent the luminal vacuolar H⁺-ATPase.[3] Until recently, the distribution of the H⁺-ATPase in individual collecting tubule cells, as assessed by physiologic methods, was unclear. The problem of distinguishing H⁺ transport in intercalated cells from that in principal cells has largely been solved by an ingenious method of loading the pH probe BCECF selectively into intercalated cells developed by Weiner and Hamm.[89] Using this method, several investigators have found evidence for vacuolar H⁺-ATPase-driven pH_i recovery in intercalated cells of the cortical collecting tubule.[3,29,90] Limited electrophysiologic data on bicarbonate-secreting (β) intercalated cell is consistent with an electrogenic H⁺-ATPase in the basolateral membrane.[90,91]

The outer stripe of the outer medullary collecting tubule has functionally distinct H⁺-secreting intercalated cells, and Na⁺-transporting principal cells,[92,93] and reabsorbs bicarbonate by an amiloride-resistant sodium-independent mechanism,[92] consistent with H⁺ secretion by a vacuolar H⁺-ATPase. Studies of intracellular pH regulation in individual cells from outer stripe indicate that the intercalated cells, but not the principal cells, have a luminal plasma membrane vacuolar H⁺-ATPase.[63,64]

The inner stripe of the outer medullary collecting tubule (in the rabbit) is significantly different from the outer stripe. The inner stripe does not transport sodium, and normally has a lumen-positive potential difference thought to arise from electrogenic proton secretion.[26,30,86,87] This segment once was thought to exist of a single cell type,[61,94,95] but morphologic[12,96-98] and physiologic evidence[22,88] have shown that there are two functionally distinct cell types. In recent studies, vacuolar H⁺-ATPase-driven proton extrusion was measured in individual inner stripe intercalated cells from rates of sodium-independent recovery of pH_i following in vitro acid loading.[22,88] A minority cell population in the inner stripe, that probably represents intercalated cells, have higher

---

[3] Breyer and Jacobson[60] investigated the recovery of pH_i of outer medullary collecting tubules in CO_2-free buffers following acidification by exposure to NH_4Cl, and found that recovery was blocked by basolateral amiloride, and was absolutely dependent on basolateral sodium, indicating sole control by Na^+-H^+ exchange. Several subsequent studies[7,22,61] showed that, in CO_2-free buffers, the rate or time course for activation of sodium-independent recovery is slow, requiring as long as 25 minutes or more to reach baseline pH; Breyer and Jacobson may have failed to detect this H⁺-extrusion mechanism because their experiments were of short duration.

rates of vacuolar H⁺-ATPase-driven H⁺ extrusion than the neighboring majority (principal) cells,[22,88] although the latter also appear functionally to have apical membrane H⁺-ATPases.[22]

Distribution of Vacuolar H⁺-ATPase by Immunocytochemistry

The immunocytochemical distribution of vacuolar H⁺-ATPase in the rat nephron (see Table 6.3), performed using antibodies specific to the M_r 70,000, 56,000, and 31,000 subunits,[11,12,99] showed a distribution that agrees well with (and, in some instances, predicted) the distribution of the H⁺-ATPase assessed by physiologic methods.[5,15,17-20,22,61,63,88,100]

In the S1 and S2 segments of the proximal tubule, vacuolar H⁺-ATPase was found in the brush border microvilli, and in the invaginations and vesicles at the base of the microvilli.[12] H⁺-ATPase staining was absent from clathrin-coated pits in this latter region. In the S3 segment, H⁺-ATPase labeling was again found in the sub-villar invaginations of the apical membrane, but was absent from the microvilli.[12]

In the initial part of the thin descending limb, moderate H⁺-ATPase labeling was found in both the apical and basolateral membranes; staining

Table 6.3. Distribution of vacuolar H⁺-ATPase in mammalian kidney[1]

Segment	Distribution	Intensity
Proximal tubule: S1	Brush border microvilli	+++
	Sub-villar invaginations and endosomes	++
Proximal tubule: S2	Brush border microvilli	++
	Sub-villar invaginations and endosomes	++
Proximal tubule: S3	Brush border microvilli	0
	Sub-villar invaginations and endosomes	++
Loop: Initial thin descending limb	Apical membrane	+
	Basolateral membrane	+
Loop: thin limb	Apical membrane	0
	Basolateral membrane	0
Loop: thick ascending limb	Apical membrane and apical vesicles	+
Distal convoluted tubule	Apical membrane	+
Connecting tubule	Principal cells: Apical membrane	+
	Intercalated cells (plasma membrane and vesicles)	++++
Cortical collecting tubule	Principal cells:	0
	Intercalated cells (plasma membrane and vesicles)	++++
Outer medullary collecting tubule: outer stripe	Principal cells:	0
	Intercalated cells (plasma membrane and vesicles)	++++
Outer medullary collecting tubule: inner stripe	Principal cells: Apical membrane	0
	Intercalated cells (plasma membrane and vesicles)	++++
Inner medullary collecting tubule	Principal cells: Apical membrane	+
	Intercalated cells (1ˢᵗ 1/3 only)	++++

[1] The distribution is derived from immunocytochemical studies in rat kidney.[12] The distribution of H⁺-ATPase in rabbit kidney is largely similar, although differences are present.[98]

staining was absent from the inner part of the thin limb. Moderate H⁺-ATPase labeling was observed in the apical membrane and apical vesicles in the medullary and cortical thick ascending limb, and weak to moderate staining was present in the apical membrane of the distal convoluted tubule.[12]

In both the connecting tubule and cortical collecting tubule, intense H⁺-ATPase staining was observed in all of the intercalated cells, with variable distributions in apical membrane, basolateral membrane, or cytosolic vesicles among the cells.[12] Six different patterns of H⁺-ATPase staining were discernible in rat kidney:[99] well-polarized apical (WPA; 38%), poorly polarized apical (PPA; 25%), diffuse (D; 9%), bipolar (B-A; 6%), poorly-polarized basolateral (PPB; 21%), and well-polarized basolateral (WPB; 2%). As discussed below, the distribution of H⁺-ATPase in these cells is dynamic, responding to the acid-base status of the animal.

The principal cells in the connecting tubule had weak to moderate apical membrane H⁺-ATPase staining, resembling that in the distal convoluted tubule.[12] The principal cells in the cortical collecting tubule, however, were devoid of detectable H⁺-ATPase labeling.[12]

In the outer medullary collecting tubule, in both the outer and inner stripe, principal cells had no detectable H⁺-ATPase staining, and intercalated cells labeled intensively with H⁺-ATPase antibodies.[12] In contrast to the cortex, intercalated cells in the outer medullary segments showed only WPA or PPA H⁺-ATPase distributions,[99] consistent with physiologic evidence that no bicarbonate secretion occurs in these segments (reviewed in ref. 29).

In the inner medullary collecting duct, the intercalated cells (present only in the first third of the segment) had intense H⁺-ATPase labeling[12] with a WPA distribution.[99] The principal cells of the inner medulla showed weak apical staining with antibody to the B1 subunit isoform,[101] but no staining with antibodies to the A, E, or proteolipid subunits of the H⁺-ATPase,[101] indicating that functional vacuolar H⁺-ATPases probably do not reside in the plasma membrane.

Although vacuolar H⁺-ATPases are on endosomes, lysosomes, trans-Golgi network, and other acidic organelles, where they can be detected immunocytochemically on fixed intact cells,[53,102] H⁺-ATPase staining generally is not demonstrable on these intracellular organelles in sections of kidney tissue.[10-12] The reason for this apparent discrepancy is probably the relatively low density of H⁺-ATPases in intracellular organelles. It has been estimated that there are 22 H⁺-ATPases on a single chromaffin granule.[103] Using a value of 0.5 μm for the diameter of a chromaffin granule gives a density of 7 pumps/μm² in these organelles. In contrast, proton secreting urinary epithelial cells have a density of over 14,000 pumps/μm.[210] At these enormous densities, H⁺-ATPases are readily demonstrable on the plasma membrane in several cell types in the kidney.

Cell-Specific Expression of H⁺-ATPase Subunits

As indicated above, there are two isoforms of the B subunit that are encoded by different genes. Nelson et al[66] examined the distribution of the B1 and B2 subunit isoform in rat tissues. High levels of B1 subunit mRNA and protein expression were found only in kidney cortex and medulla, and not in any other tissues tested. The distribution of the B1 ("kidney") isoform protein was determined by immunocytochemistry in rat kidney and compared it with the distribution of the E (M_r 31,000) subunit. The B1 isoform was expressed at high levels in both A and B-type intercalated cells. In these cells, the distribution of B1 subunit staining was identical to that of E subunit staining, indicating no preferential targeting of the isoform to apical or basolateral poles. As indicated above, the B1 subunit antibody also labeled the apical membrane of inner medullary principal cells, but no other subunits of the vacuolar H⁺-ATPase were detectable in these cells. Physiologic studies also indicate that H⁺ secretion by the inner medullary principal cells is not from a vacuolar-type proton pump (reviewed in ref. 104), but may be from an H⁺/K⁺ ATPase.[105] No staining with B1 subunit isoform antibody was observed in the proximal tubule, loop of Henle, or distal convoluted tubules.

The B2 subunit isoform mRNA was found to be expressed in all rat tissues tested, with highest levels in brain and adrenal medulla.[66] In recent studies, our laboratory has generated polyclonal antibodies specific for the mammalian B2 subunit isoform, and we have examined its distribution in mammalian kidney.[106] Staining with antibodies to the B2 subunit isoform shows intense labeling of the proximal tubule brush border and subvillar invaginations in a distribution identical in the S1, S2, and S3 segments, to that obtained earlier with antibodies to the E subunit.[12,99] In the loop of Henle, moderate labeling was found in the apical membrane in thick ascending limb, and slight labeling of the apical membrane in distal convoluted tubule was observed. These staining patterns are also identical to those observed earlier with E subunit antibodies.[12,99] In marked contrast, no labeling of cortical or medullary collecting duct segments was found with the B2 isoform-specific antibody. No segments were detected which showed staining with both B1 and B2 antibodies.[106]

FACTORS THAT MODULATE VACUOLAR H⁺-ATPASE ACTIVITY

Kidney Cytosolic Proteins that Modulate Vacuolar H⁺-ATPase Activity

As discussed below, the kidney plasma membrane vacuolar H⁺-ATPase is subject to physiologic regulation that enables renal epithelial cells to control rates of transepithelial H⁺ transport. Cytosol from bovine kidney was found to contain factors that inhibit and stimulate the purified vacuolar H⁺-ATPase from kidney microsomes.[107,108] The properties of the inhibitor and activator are summarized in Table 6.4. The inhibitor was purified to a high degree using several precipitation, ion exchange, and size separation steps.[108] The inhibitor eluted from a gel filtration column with an apparent M_r = 14–20,000. The inhibitor activity was reduced by treatment with trypsin, but not by lipid and carbohydrate hydrolases or nucleases, indicating that it is a protein. Inhibition was concentration dependent, and required 20 minutes incubation time or longer to exhibit maximal inhibitory activity. The inhibition was irreversible, but neither various protease inhibitors nor addition of BSA had an effect on inhibitor activity, indicating that action by proteolysis was unlikely. The inhibitor reduced ATP-driven acidification in bovine kidney membranes by 90%, indicating that it is active against the H^+-ATPase in its native membrane state, and that it probably interacts with the H^+-ATPase on its cytoplasmic domain. The inhibitory effect on the purified H^+-ATPase was pH dependent, showing an increase in the maximum inhibition of the enzyme above pH 7.5. The inhibitor showed no preference for different kidney vacuolar H^+-ATPases; it was equally effective (at maximal concentrations) on H^+-ATPase purified from kidney microsomes, brush border membranes, and lysosomes.[108]

The activator was partially purified by precipitation, ion exchange, and gel filtration chromatography.[107] The activator eluted from a gel filtration column with an apparent M_r = 35,000. Activator activity was eliminated by treatment with trypsin, indicating that it is also a protein. Activation was concentration dependent and saturable. Upon addition of the activator, the stimulation of H^+-ATPase was nearly immediate, occurring with 1 minute or less of preincubation. The activation was reversible, and could be washed off. The pH of the buffer affected the ease of removal of the activator. At pH values of 6.5 and below, the activator required 10 minutes to dissociate from the H^+-ATPase, whereas at higher pH values the dissociation was immediate. Activation of the purified H^+-ATPase was highly pH dependent. In the pH range ≤ 6.6, the activator stimulated H^+-ATPase activity up to 1200 fold; at pH values ≥ 6.6, there was only a 1-2 fold stimulation of activity. The activator showed a significant selectivity in its effect on different kidney vacuolar H^+-ATPases. It showed a more than two-fold greater stimulation of microsomal and brush border

Table 6.4. Properties of inhibitor and activator of kidney vacuolar H^+-ATPase

	Inhibitor	Activator
Actions	Inhibits solubilized immunoaffinity-purified H^+-ATPase. Inhibits ATP-dependent proton transport	Activates solubilized immuno-affinitypurified H^+-ATPase
Mode of action	Saturable effect on H^+-ATPase Inhibition requires preincubation with the H^+-ATPase. Inhibition is irreversible; activity cannot be restored by washing enzyme at any pH	Saturable effect on H^+-ATPase Non-enzymatic effect. Probably works by binding. Activator readily washed off, but binds more tightly at pH < 7.5
pH effects	% Inhibition increases for pH > 7.5	Binding optimum at pH 6.5; % activation increases for pH < 6.5
Composition	Heat labile protein. Active M_r of 14 –20,000 by gel filtration	Heat-stable protein. Active M_r of ~35,000 by gel filtration
Specificity	Highly specific for vacuolar H^+-ATPases. Slight effect on F_0F_1 H^+-ATPase. Not selective for subclass of vacuolar H^+-ATPase	Highly specific for vacuolar H^+-ATPases. Slight effect on F_0F_1 H^+-ATPase. Selective for brush border > microsomal >> lysosomal V-ATPase

H^+-ATPase activity than it did on lysosomal H^+-ATPase activity. This property implies that the activator could exhibit selective regulatory effects on H^+-ATPases residing on the plasma membrane.[107]

Other Factors that Modulate ATP-Driven Proton Transport or Vacuolar H⁺-ATPase Activity

Gurich and Dubose[109] found that proton transport in endosome-enriched membranes from rabbit kidney cortex was influenced by in vitro addition of 8-Br-cAMP. In buffer containing 100 mM Cl^-, High concentrations of 8-Br-cAMP (250 µM) inhibited acidification in all endosome-enriched membrane fractions from a sucrose density gradient, but shoed a greater inhibition (up to 60%) in the higher density fractions. Low concentrations of 8-Br-cAMP (5 µM) slightly stimulated acidification in the light membrane fractions, and inhibited in the higher density fractions. The results suggest that the vacuolar H^+-ATPase in these preparations resides in a heterogenous population of vesicles which may be subject to different regulatory mechanisms. In buffers containing only 6 mM Cl^-, ATP-induced acidification was reduced; addition of 250 µM cAMP inhibited H^+ transport by about 90% in all gradient fractions. The authors interpreted this result to indicate a direct suppression of the H^+-ATPase, and suggested that this might occur through phosphorylation. Their results contrast with those of Mulberg et al,[110] who demonstrated that cAMP-dependent protein kinase (PKA) increased ATP-dependent acidification in bovine brain coated vesicles by increasing the chloride permeability. As Gurich and Dubose[109] used only 1 mM ATP and no regenerating system in their experiments, it is possible that the observed effects were due to competitive nucleotide substrate interactions.

Gurich et al[111] later found that both GTP and GTP-γ-S stimulated the initial rate of acidification in endosome-enriched membranes from rabbit renal cortex, and the effect was blocked by pretreatment of the membranes with pertussis toxin. They suggested that G proteins may be capable of stimulating the vacuolar H^+-ATPase in endosomes.

PHYSIOLOGIC REGULATION OF THE VACUOLAR H⁺-ATPase IN URINARY EPITHELIA⁴

The mammalian kidney has the ability to adjust net acid excretion as a response to changes in the physiologic environment of the animal. The physiologic factors that influence or control the rate of H^+ transport may be divided into those that produce rapid or *acute* responses, occurring in minutes to hours and usually readily reversible, and those that produce *chronic* responses, occurring in hours to days and often requiring major changes in cell architecture. Factors that affect H^+ transport may function, in principle, by one (or more) of three possible mechanisms (Figure 6.1): 1) a change in the kinetics of

---

⁴ This review will not discuss those aspects of the regulation of bicarbonate secretion directed at control of luminal Cl^-/HCO_3^- exchange rather than on the basolateral H^+-ATPase. Renal bicarbonate secretion has been the subject of recent excellent reviews.[29, 112, 115]

the H+-ATPase; 2) a change in the number of active H+-ATPases residing on the plasma membrane by redistribution to or from intracellular compartments; and 3) a change in the quantity of the H+-ATPase in the cell without any modification of the relative distribution in the cell.

H+ secretion by the vacuolar H+-ATPase in several nephron segments is subject to acute regulation by several factors, the most important of which are the luminal proton electrochemical gradient, CO_2, basolateral bicarbonate exit, and aldosterone. Many of the details by which acute factors influence the rate of H+ secretion by vacuolar H+-ATPases in urinary epithelia were first discovered in the turtle and toad urinary bladder, model epithelia with transport properties similar to those of the cortical collecting tubule.[13,31]

ACUTE REGULATION OF THE VACUOLAR H+-ATPASE IN TOAD AND TURTLE URINARY BLADDER

Proton transport in the bladder is performed by a subset of cells containing carbonic anhydrase, that are called CA cells in the turtle bladder (or mitochondria-rich cells in the toad bladder). The bladder CA cells are analogous to the cortical collecting tubule intercalated cells in many characteristics. There are two functional types of CA cells: α cells, that secrete H+, and β cells, that secrete bicarbonate.[116] The α cell has an apical membrane proton pump, belonging to the vacuolar class of H+-ATPases,[13,33,117] that acidifies its lumen,[32] and a Band 3-like Cl-/HCO_3^- exchanger on the basolateral membrane.[118-121] The β cell probably has a basolateral vacuolar H+-ATPase[13,122,123] and a

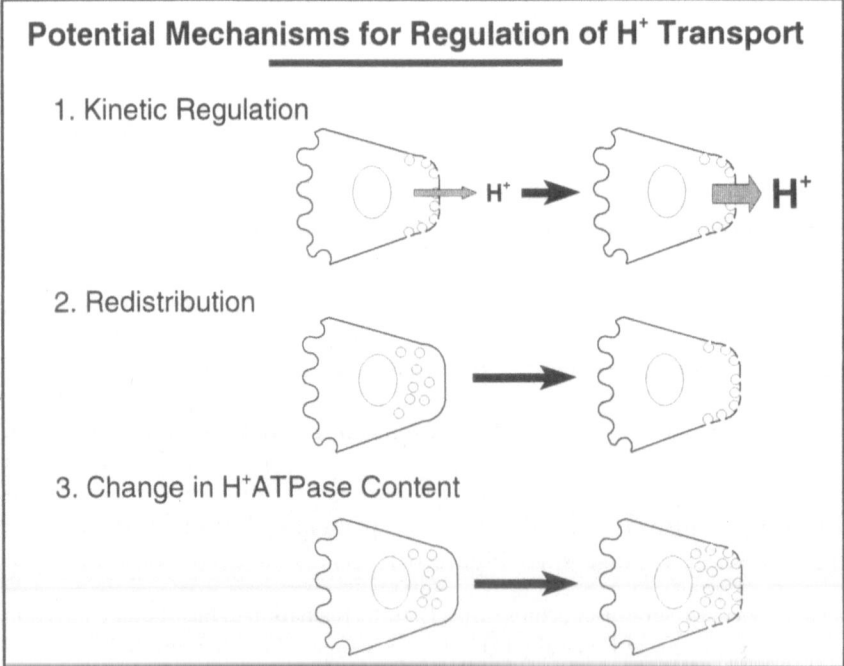

Fig. 6.1. General classes of mechanisms for regulation of epithelial H+ transport.

luminal Cl⁻/HCO₃⁻ exchanger[24,125] that differs from the basolateral α cell exchanger.[125]

Acute Regulation
by the Luminal Proton Electrochemical Gradient

The best studied factor producing acute changes in H^+ transport is the proton electrochemical gradient across the membrane in which the H^+-ATPase resides.[31] Lowering luminal pH of the bladder slows the rate of luminal H^+ secretion by imposing an adverse proton electrochemical gradient across the membrane.[36,126,127] Proton secretion by the bladder is electrogenic,[31,33,127,128] and does not require sodium transport or luminal sodium.[128-130] Under open-circuit conditions, however, sodium transport stimulates H^+ secretion by generating a lumen-negative transepithelial potential difference,[130,131] and by generating CO_2 (see below).

Application of a lumen-negative potential difference across the bladder epithelium stimulates H^+ secretion, and imposition of a lumen-positive potential difference across the epithelium inhibits H^+ secretion.[36,127,132] The effect of the proton electrochemical gradient across the luminal membrane on the rate of H^+ secretion is equivalent whether the proton-motive force is generated by pH gradients or electrical gradients.[36,127] The relationship between the active proton transport rate and the luminal proton electrochemical gradient is non-linear, saturating at a proton electrochemical gradient of near zero.[36] The observed current-voltage relationship of the bladder fits a mathematical model for the H^+-ATPase which assumes that it has a stoichiometry of 2 H^+/ATP, and a two-domain structure, with a catalytic "head" and a membrane-spanning proton channel, resembling that of the vacuolar and F-ATPases.[36] Studies on the bladder are generally consistent with a model in which the electrochemical gradient affects only the kinetics of the plasma membrane H^+-ATPase, without producing changes in the number of H^+-ATPases residing on the plasma membrane.[36]

Acute Regulation of H+ Secretion
by Intrinsic Regulatory Mechanisms: CO₂, Basolateral
Bicarbonate Exit, and the Possible Role of Cell pH

Like most epithelia, the turtle bladder has intrinsic or "homocellular" regulatory mechanisms that couple solute transport by the luminal and basolateral membrane.[133] In the bladder CA cell, luminal H^+ secretion is dependent on basolateral bicarbonate exit. Studies in the turtle bladder over several decades have led to the following overall paradigm for the effects of CO_2, basolateral HCO_3^-, and basolateral Cl⁻ on H^+ secretion (Figure 6.2).

Luminal H^+ secretion by an electrogenic pump (vacuolar H^+-ATPase) generates intracellular OH^-, which cannot exit the cell. In the absence of CO_2, the OH^- produced by the pump causes intracellular pH to rise, which in turn suppresses luminal H^+ secretion. In the presence of CO_2, carbonic anhydrase catalyzes the combination of OH^- with CO_2 to form bicarbonate, which can then exit the basolateral membrane through a Cl⁻/HCO₃⁻ exchanger. The chloride taken up by this exchange exits through a basolateral membrane chloride channel.

Many studies in the turtle bladder support individual aspects of this overall model. The rate of H^+ secretion in the bladder is critically dependent on the availability of CO_2. Limiting CO_2 by maintaining the bladder epithelium in CO_2-free buffers and bubbling them with nitrogen, inhibited over 90% of proton secretion.[134] Under CO_2-free conditions, bladder pH_i, measured using DMO, was 7.56 in air, and rose to 7.78 in nitrogen, compared with a pH_i of 7.42 in bladders in HCO_3^-/CO_2 buffers in air.[134] The rate of H^+ secretion in the bladder was greatly stimulated, within seconds, by raising the ambient PCO_2.[127,134-137] These findings were interpreted to indicate that CO_2 is required to buffer the OH^- generated by active H^+ secretion; in the absence of CO_2, cell pH alkalinizes, suppressing luminal H^+ transport.

Luminal H^+ secretion was reduced up to 90% by inhibitors of carbonic anhydrase such as acetazolamide.[134] The postulated action of these agents was to prevent the elimination of OH^- by inhibiting the enzymatic hydroxylation of CO_2,[134] consequently raising cell pH.

Steinmetz and colleagues demonstrated for several values of PCO_2 that raising the basolateral bicarbonate concentration at constant PCO_2 inhibits luminal H^+ secretion, and that the inhibition can be prevented by elevating the ambient PCO_2.[138] They proposed that H^+ secretion was responding to changes in intracellular pH. They obtained pH_i measurements, again using DMO, of the intact epithelium under these conditions, which suggested that the rate of H^+ secretion decreased sharply at cell H^+ concentrations below 40 nM (pH 7.40), with a half-

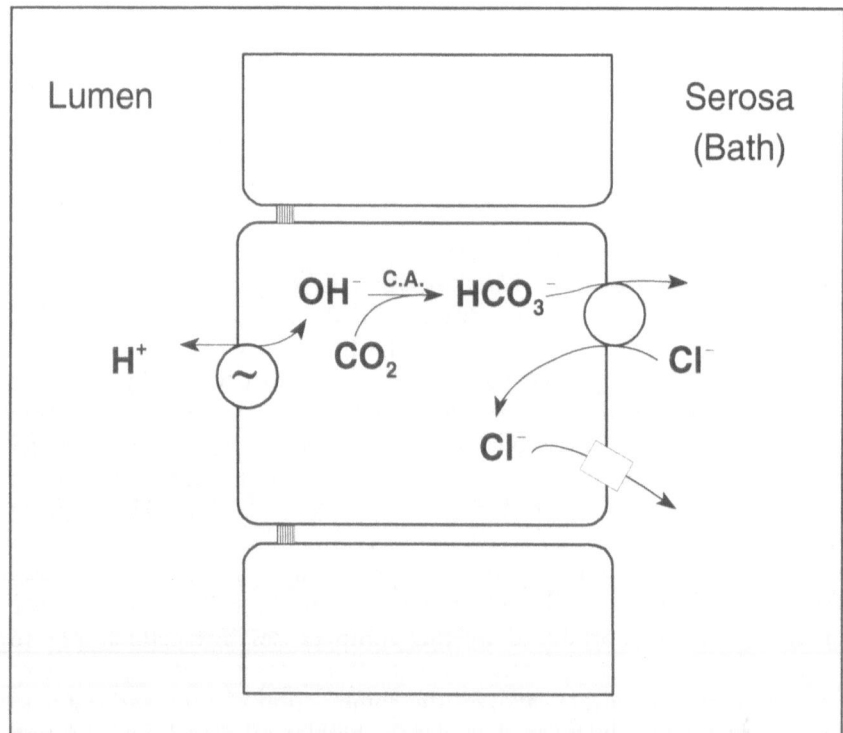

Fig. 6.2. Model for H^+ secretion in the turtle urinary bladder.

maximal effect at 25 nM (pH 7.30), and that H^+ secretion was near maximal at cell H^+ between 40 and 186 nM (pH 6.73).

When basolateral bicarbonate exit through the Cl^-/HCO_3^- exchanger was inhibited with disulfonic stilbenes[118] luminal H^+ secretion was inhibited. After application of the disulfonic stilbene SITS to the serosal side, H^+ secretion was inhibited 60%, and pH_i, measured with the DMO method,[139] increased from 7.48 to 7.61.[118] Exposure of SITS-treated bladders to 5% CO_2 transiently stimulated H^+ secretion, an effect attributed to an initial reduction in pH_i by CO_2, with a subsequent accumulation of bicarbonate and rise in pH_i.[118]

Removal of basolateral chloride inhibited H^+ secretion,[120] probably by inhibiting (rather than reversing) basolateral Cl^-/HCO_3^- exchange, since sub-millimolar chloride concentrations partially restored H^+ secretion. The inhibition of H^+ secretion was attributed to an effect on pH_i, rather than direct inhibition of the H^+-ATPase, as increasing the PCO_2 in low-chloride buffers transiently stimulated H^+ secretion.[120]

Although many aspects of the model have stood up to experimental scrutiny, the details of the intracellular mechanisms that control the luminal H^+-ATPase remain largely unresolved. A central issue is the hypothesis[31,138] that pH_i is the primary signal controlling the rate of luminal H^+ secretion. How does CO_2 stimulate H^+ secretion, and is its effect attributable to a fall in intracellular pH? How do removal of CO_2, elevation of basolateral bicarbonate, basolateral disulfonic stilbenes, and carbonic anhydrase inhibition suppress H^+ secretion, and are their actions the result of an elevation of intracellular pH?

MECHANISM OF THE STIMULATORY EFFECT OF CO_2 ON H^+ SECRETION

CO_2 and some maneuvers predicted to change the pH_i of CA cells exert at least part of their effect by changing the number of luminal H^+-ATPases through membrane fusion or endocytic removal.[136,137,140] Exposure of the basolateral membrane of bladders to the disulfonic stilbene SITS or with acetazolamide, both of which inhibit H^+ secretion, brought about a loss of microplicae (surface foldings) and decreased the luminal membrane surface area of a subpopulation of CA cells.[140] It was proposed that the inhibition of H^+ secretion was the result of a decrease in the number of luminal proton pumps achieved by reducing the area luminal membrane.[31] Gluck et al[136] showed that CA cells have a pool of cytoplasmic vesicles that are acidified by a proton pump. Fluorescent dextran in the luminal solution could be internalized into these vesicles, and was released back into the luminal medium within seconds when bladders were stimulated by CO_2, demonstrating that these vesicles fused rapidly with the luminal membrane in response to CO_2. The release of dextran and H^+ secretory response to CO_2 were both inhibited by the microtubule destabilizer colchicine. Stetson and Steinmetz[137] demonstrated that CA cells increased their apical membrane surface area and lost apical cytoplasmic vesicles, determined by electron microscopic (EM) measurements, when stimulated by CO_2. The cytoskeletal disruptive agents colchicine, cytochalasin B, and ^3H_2O inhibited the increase in H^+ secretion by CO_2, respectively, by 50%, 87%, and 88%, and ^3H_2O (the only compound that maintained adequate EM morphology) prevented the CO_2-induced loss of apical vesicles. Subsequently, Stetson and Steinmetz[141] found that CO_2 increased the luminal membrane area of α intercalated cells 3.9-fold, and increased the number of intramembranous particles, assessed by freeze fracture, by 9.8-fold, correlating well with an 8.8-fold increase in H^+ secretion.

Clausen and Dixon,[142] using impedance analysis, found that the turtle bladder apical membrane capacitance (a direct function of the membrane surface area) declined by 8.2% after treatment with 50 μM acetazolamide, and this returned to near baseline after exposure of the same bladders to CO_2. In a subsequent study, Dixon et al[143] found that 0.5 M acetazolamide inhibited H^+ secretion completely in 20 minutes, and that the agent reduced bladder apical membrane capacitance by 17%. They estimated from electron micrographs of CA cell horseradish peroxidase uptake that 15% of the CA cell luminal membrane is internalized each minute. Using endocytosis of FITC-dextran as an assay, they showed that 15-30 minutes after acetazolamide, the rate of dextran internalization increased 71%, and returned to baseline internalization rates by 75-90 minutes after acetazolamide. An estimate of the amount of membrane internalized, however, exceeded the measured change in membrane capacitance by five-fold, indicating that part of the increased endocytosis represented membrane recycling. Although the effect of acetazolamide on H^+ secretion was complete in 20 minutes, its effect on capacitance required over 45 minutes, suggesting that part of the inhibitory action of acetazolamide occurred through changes in pump kinetics rather than by altering membrane retrieval or fusion (see below). Dixon et al[144] subsequently found that azide prevented the reduction in H^+ secretion and luminal membrane capacitance induced by acetazolamide, although it did not affect constitutive dextran internalization rates. These interesting findings suggest that the vesicles involved in regulating H^+ transport by internalization and fusion, differ from those involved in constitutive membrane turnover. They also underscore the problems in relating rates of exocytosis and endocytosis to net changes in the number of apical H^+-ATPases.

It is possible that part of the effect of CO_2 is exerted through changes in the kinetics of the H^+-ATPase, since, in one investigation, colchicine was found to inhibit only partially the CO_2-stimulated increase in H^+ secretion,[137] even though the agent appeared to inhibit bladder membrane fusion events.[136,145]

ROLE OF INTRACELLULAR pH
IN REGULATING LUMINAL H^+ SECRETION

As indicated above, it was postulated that agents such as CO_2 and acetazolamide exerted their effects by changing pH_i of the CA cells. A major limitation of the DMO method used to measure pH_i in some of the studies above[118,134,138] is that it yields values for pH_i indicative of the entire epithelium rather than values specifically from the CA cells. The DMO method may also has an inherent inaccuracy, as the probe partitions into compartments other than the cytosol.

Several ensuing studies examined the hypothesis that CO_2 and acetazolamide act by changing pH_i of the CA cells using newer methods for measuring intracellular pH.[145-148] Cannon et al[146] measured the intracellular pH of individual CA cells in intact bladders with the pH-sensitive probe 6-carboxyfluorescein.[146] CO_2 produced a drop of pH_i in the CA cells of about 0.5 pH units over 5 minutes. Application of the weak acid butyrate to the bladder lumen under CO_2-free conditions also reduced pH_i by about 0.5 pH units over 5 minutes, and, in bladders allowed to internalize FITC-dextran,[136] was associated with release of FITC-dextran into the lumen. The authors interpreted these results to indicate that intracellular acidification, rather than CO_2 specifically, was the signal for membrane fusion. Several problems with the study undermine these conclusions. The authors found that the CA cells had a resting pH_i 0.5 pH units above the surrounding principal cells, indicating that principal cells have a resting pH_i of as low as 6.0, well out of the normal range, raising doubts about the validity of the measurements. The investigators did not exclude the possibility that butyrate enhanced endogenous CO_2 production, for example as a substrate, or by uncoupling mitochondrial oxidative phosphorylation. Most importantly, they did not determine whether raising the serosal bicarbonate concentration to levels that inhibit the H⁺ secretory response to CO_2 prevented the fall in pH_i or the release of FITC-dextran. In a following study, the same laboratory found that CO_2 acidified pH_i by only 0.25-0.35 pH units, and that pH_i returned to control levels within 4 minutes (in suspensions of bladder cells) or 7 minutes (in individual CA cells from an intact bladder), a response entirely different from their results reported earlier.[146] They interpreted these findings to indicate that insertion of luminal H⁺-ATPases was restoring intracellular pH. The principal experiment supporting this premise was that lowering the luminal pH to 5.5, which inhibits luminal H⁺ secretion (see above) prevented the restoration of pH_i over a period of 7 minutes following exposure to CO_2. It is unclear, however, why basolateral entry of bicarbonate entry did not restore intracellular pH, as later in the same article, the authors found that intracellular pH did recover under conditions where exocytosis was inhibited by calcium chelators (see below). Andersen and Steinmetz[36] showed earlier that lowering luminal pH sufficiently to inhibit H⁺ completely produced a transient "overshoot" in H⁺ secretion when the luminal pH was raised under CO_2-free conditions, but not in the presence of CO_2, presumably because basolateral entry of bicarbonate prevented intracellular acidification.

Graber, Dixon et al used 4-methylumbelliferone as a probe for measuring pHi in individual CA cells in the intact bladder. In contrast to the studies above,[146] they found that CA cells had a resting pH, of 0.1 pH units above that of principal cells. In a later study, Graber, Dixon et al[148] demonstrated that propionate, under appropriate conditions, acidified pH_i in CA cells without producing any increase in H⁺ secretion, roiling the evidence supporting pH_i as the immediate stimulus for H⁺ secretion.

As discussed above, it was postulated that acetazolamide inhibits H⁺ secretion by preventing the buffering of OH⁻ and raising pH_i.[134] Early evidence suggested that this might not be the case; acetazolamide, at concentrations that inhibited 73% of luminal H⁺ secretion, was found to produce no detectable change in cell pH by the DMO method.[134] van Adelsberg et al[145] found that acetazolamide raised pH_i in CA cells by 0.12 pH units, and proposed that this was the basis for its inhibition of H⁺ secretion. Graber, Dixon et al found that acetazolamide inhibition raised pHi by 0.25 pH units only in a small subpopulation of CA cells, yet inhibited H⁺ secretion 80%. In bladders in which pH_i was "clamped" at 7.0 with serosal DMO, acetazolamide still inhibited H⁺ transport 74%. They also found that in propionate-treated bladders, in which the pH_i of individual CA cells was acidified by 0.15 pH units, acetazolamide inhibited H⁺ transport 100%. As shown previously,[144] azide prevented completely the acetazolamide inhibition. These findings indicate that the inhibition of H⁺ secretion by acetazolamide is probably through an effect that is independent of cell pH.

If pH_i is an immediate signal controlling luminal H⁺ secretion, how does it act? It is unlikely that a fall in intracellular pH stimulates the H⁺-ATPase directly. The curve relating H⁺ secretion to apparent pH_i described by Cohen[138] is not at all similar to the pH dependence of ATPase activity of the purified renal enzyme.[39] Acidification of intracellular pH in CA cells has been reported to raise intracellular calcium activity; an elevation in intracellular calcium may be required to promote membrane fusion events, as calcium chelators inhibited CO_2 or butyrate-induced H⁺ secretion.[145,146,149,150] On the other hand, increasing the calcium permeability of the bladder with calcium ionophore inhibits H⁺ secretion.[151,152] In the mammalian intercalated cell (see below), some maneuvers that raise intracellular calcium, such as removal of basolateral sodium, have no effect on H⁺ secretion,[7] and the mild elevation of intracellular calcium produced by intracellular acidification appears to inhibit vacuolar H⁺-ATPase-driven H⁺ extrusion.[7]

In summary, the evidence that pH_i is the primary signal controlling luminal H⁺ secretion, and that CO_2 acts by changing intracellular pH, is at present inconclusive, and hampered by technical problems in obtaining reliable measurements of pH_i in individual CA cells in the intact bladder. Some of the more recent results are inconsistent with the hypothesis.

Acute Regulation of H⁺ Secretion by Intrinsic Regulatory Mechanisms: Coupling of H⁺ Secretion to Metabolism

Several studies have demonstrated that luminal H⁺ secretion is affected by the availability and type of metabolic substrate. Prolonged incubation of bladders in the absence of substrates results in a depletion of their endogenous substrate, and a decline in H⁺ secretion.[153-155] In these substrate depleted bladders, addition of glucose restored H⁺ transport to control rates, but addition of pyruvate had no effect.[153] Treatment of bladders with the glycolytic inhibitor 2-deoxyglucose has a similar effect: H⁺ secretion is inhibited, and is restored by addition of glucose, but not pyruvate.[154] These results suggest that the rate of glycolysis, rather than the rate of mitochondrial ATP production, has a strong influence on the rate of H+ secretion. The mechanism of this effect in the bladder is unknown, but it may indicate that glycolytic enzymes are bound in a close functional association with the luminal H+-ATPase, as has been demonstrated for the anion transporter Band 3 and other transport enzymes.[156-159] In other studies, substrate-depleted bladders did show an increase in H+ secretion;[155] the basis for the discrepant results has not been reported.

Under anaerobic conditions, the maximal rate of H+ secretion is achieved with 1% ambient P_{CO2}; under aerobic conditions, the maximal rate increases nearly four-fold, and is reached at 5% ambient P_{CO2}. The basis for the change in the response of the bladder to CO_2 with metabolic state is undetermined; it may reflect a requirement of aerobic metabolism for membrane insertion with CO_2, as discussed above.

Acute Regulation of H+ Secretion by Hormones

Aldosterone
Studies in the bladder have demonstrated that aldosterone stimulates H+ secretion independent of effects on sodium transport.[160-162] The effect occurs within one hour, and is due to a direct stimulation of the pump, and is not secondary to stimulation of the overall metabolic rate.[153] Whether the effect is the result of an increase in the number or kinetics of luminal H+-ATPases is unknown. In one study, the RNA transcription inhibitor actinomycin D prevented the response to aldosterone,[162] suggesting that new protein synthesis is required for the response, but does not exclude any of the mechanisms shown in Figure 6.1.

β-Adrenergic Agonists and Cyclic AMP
In the turtle bladder, addition of serosal cAMP, treatment with β-adrenergic agonists and other agents that elevate cAMP,[163,164] or treatment with carbachol[165,166] all stimulate bicarbonate secretion. Evidence suggests that part or most of this effect arises from an increase in the number of functional Cl-/HCO3- exchangers and chloride channels in the luminal membrane (presumably of the β CA cells)[116,167] that may occur through membrane fusion events.[168,169] There is at present no information for or against a direct effect of these agents on the basolateral membrane vacuolar H+-ATPase.

ACUTE REGULATION OF THE VACUOLAR H+-ATPASE IN KIDNEY

Acute Regulation
by the Luminal Proton Electrochemical Gradient
Information on the effect of electrochemical gradients on proton transport in the kidney collecting duct is limited, since the tubules cannot be studied under voltage-clamped conditions. As indicated above, the isolated renal vacuolar H+-ATPase is electrogenic.[38-40] Physiologic evidence in intact cortical collecting tubules and in the outer stripe of

the outer medullary collecting tubule is consistent with a model of luminal H^+ secretion by an electrogenic H^+-ATPase that does not require luminal sodium, in which the transepithelial potential difference, which is determined partly by sodium reabsorption in the cortical collecting tubule and outer stripe, influences the rate of H^+ secretion.[82-87] Limited electrophysiologic data on bicarbonate-secreting (β) intercalated cell is consistent with an electrogenic H^+-ATPase in the basolateral membrane.[90,91]

The inner stripe of the outer medullary collecting tubule does not transport sodium, and normally has a lumen-positive potential difference thought to arise from electrogenic proton secretion.[26,30,86,87] Lowering the luminal pH to <5 in this segment reduced the intracellular pH of intercalated cells from 7.27 to 7.03, consistent with inhibition of luminal H^+ secretion.[23] Depolarizing the tubule cells with K^+ and valinomycin produced a rise in pH_i, consistent with a stimulation of the H^+-ATPase.[62] Removing basolateral chloride suppressed H^+ secretion, partly by reducing basolateral bicarbonate exit through Cl^-/HCO_3^- exchange (see below), and partly by preventing the normal charge counterbalance provided by chloride secretion.[26] Under chloride-free conditions, the tubule maintained the lumen-positive potential despite reduced H^+ secretion.[26,85] These observations are consistent with a model similar to the turtle bladder, in which the luminal proton electrochemical gradient acutely regulates tubular H^+ secretion.

In the proximal tubule, it has been demonstrated that luminal pH has a significant effect on net rates of H^+ secretion,[14,170,171] but the component of total H^+ secretion attributable to the vacuolar H^+-ATPase in response to changes in the proton electrochemical gradient under these conditions has not been determined.

Acute Regulation of H^+ Secretion by Intrinsic Regulatory Mechanisms: CO_2, Basolateral Bicarbonate Exit, and the Possible Role of Cell pH

Outer Medullary Collecting Tubule

The outer medullary collecting tubule secretes H^+ without co-existing bicarbonate secretion.[26,30] Consequently, studies on the acute regulation of vacuolar H^+-ATPase-driven proton transport from this segment are simpler to interpret than those from the cortical collecting tubule (see below).

H^+ secretion in the outer stripe of the outer medullary collecting tubule resembles electrogenic H^+ secretion in the turtle bladder, with functionally distinct H^+-secreting intercalated cells, and Na^+-transporting principal cells.[92,93] In contrast, the inner stripe of the outer medullary collecting tubule (in the rabbit) is significantly different. This inner stripe once was thought to exist of a single cell type,[61,94,95] but morphologic[12,96-98] and physiologic evidence[22,88] has shown that there are two functionally distinct cell types. A minority cell population in the inner stripe, probably representing intercalated cells, has higher rates of vacuolar H^+-ATPase-driven H^+ extrusion than the neighboring majority cells, although the latter also appear functionally to have apical membrane H⁺-ATPases.[22] In both the outer and inner stripe, the problem of distinguishing H⁺ transport in intercalated cells from that in principal cells has largely been solved by the method of loading the pH probe BCECF selectively into intercalated cells.[89]

Many properties of H⁺ secretion by α intercalated cells in the outer medullary collecting tubule fit the turtle bladder paradigm for acute control of H⁺ secretion by CA cells discussed above. As in the bladder, lowering or raising basolateral bicarbonate at constant PCO_2 respectively stimulates or inhibits H⁺ secretion,[27] and lowering[27] or raising[172] the PCO_2 at a constant basolateral bicarbonate concentration respectively inhibits or stimulates H⁺ secretion. The α intercalated cell has a basolateral Cl^-/HCO_3^- exchanger[22-24,60,96,173-175] and chloride conductance.[176] Acetazolamide,[30,92] basolateral disulfonic stilbenes,[23,26,92] and removal of basolateral chloride[26] all inhibit H⁺ secretion as they do in the turtle bladder.

Some properties of the intercalated cells in the outer medullary collecting tubule differ from those reported for the CA cells of the turtle bladder. The intercalated cells have basolateral Na^+-H^+ exchange[7,22,60,61,64,88,175] and $Na^+-HCO_3^-$ cotransport, which have not been found in turtle bladder cells.[145] There is also some evidence for a luminal H⁺/K⁺-ATPase in intercalated cells, particularly during potassium depletion (reviewed in refs. 28 and 177), although 90-95% or more of outer medullary H⁺ secretion is accompanied by chloride secretion rather than K⁺ reabsorption,[26,27] and studies of H⁺ transport by individual intercalated cells have not found evidence for an H⁺/K⁺-ATPase.[61,64] Ouabain-sensitive active reabsorption of luminal potassium has been described in turtle bladder[178] that has about 14% of the flux of active H⁺ secretion, but there is no evidence for coupling of H⁺ secretion to K⁺ reabsorption;[128] hence, at present, there is no direct evidence for an H⁺/K⁺-ATPase in turtle bladder.

MECHANISMS FOR ALTERING RATES OF LUMINAL H⁺ SECRETION

A number of studies have provided information on the intrinsic regulatory mechanisms that control vacuolar H⁺-ATPase-driven H⁺ secretion by α intercalated cells. Some of the available evidence suggests that changes in the number of luminal H⁺-ATPases by membrane fusion or internalization is an important mechanism controlling rates of H⁺ secretion in the α intercalated cell. Madsen and Tisher[179] found that acute respiratory acidosis of 5 to 6 hours duration in rats produced an increase in apical surface density and loss of tubulovesicular structures in intercalated cells of the outer medullary collecting tubule, consistent with recruitment of luminal membrane by fusion events. This length of exposure to an elevated PCO_2 is probably too short for chronic changes in H⁺ transport to have occurred[180] (see below). Schwartz and Al-Awqati[181] found that intercalated cells in rabbit medullary collecting tubules internalized FITC-dextran into acidic compartments, and that a change from CO_2-free to HCO_3^-/CO_2 buffers produced a 24-27% reduction of the fluorescent intensity in the cells in each segment. This was interpreted to represent CO_2-induced release of the dextran by exocytosis because it was inhibited by colchicine, although they did not demonstrate that the dextran was released into the lumen, nor did they exclude other possible causes for loss of fluorescent intensity, such as increased scatter, cell volume changes, or a change in the focal plane of the vesicles. They also did not determine whether tubular H^+ transport was altered with the change in CO_2. In contrast, McKinney and Davidson[172] showed that elevating the PCO_2 stimulated H^+ secretion 46-52%, an effect that was inhibited by colchicine, but which did not produce any significant change in intercalated cell apical membrane surface density assessed by electron microscopy. Thus at present, the extent to which changes in pump number by membrane insertion and internalization are responsible for intrinsic regulation of vacuolar H^+-ATPase in intercalated cells remains unresolved.

Another possible mechanism for intrinsic regulation of the luminal vacuolar H^+-ATPase is changes in enzyme kinetics (i.e. the intrinsic rate of ATP hydrolysis in the absence of imposed electrochemical gradients) rather than changes in enzyme number. There is some evidence for accessory proteins that may regulate H^+-ATPase kinetics. Fanestil and colleagues first uncovered indirect evidence for H^+-ATPase inhibitory proteins when they showed that inhibitors of protein and RNA synthesis stimulated H^+ secretion by the toad bladder.[161] More recently, partially purified fractions from bovine kidney cytosol were found to have activities that activated[107] or inhibited[108] the purified bovine kidney vacuolar H^+-ATPase (discussed above). The effect of these putative regulatory proteins was found to be highly pH dependent, suggesting that they might participate in the intrinsic regulation of the H^+-ATPase: The activator increased activity about 100% over a pH range of 8.0 to 6.6; below pH 6.6, it increased activity up to 1200%. The inhibitor reduced the activity of the purified H^+-ATPase by 45% over a pH range of 6.0 to 7.5; above pH 7.5, the inhibition increased to 70%. The effects of these regulatory proteins could therefore serve to stimulate the H^+-ATPase during cytosolic acidification, and stimulate the H^+-ATPase during cytosolic alkalinization. As discussed above, Gurich and DuBose have found evidence that GTP-binding proteins may be involved in regulating renal endosomal acidification,[109,111] but there is at present no indication that these proteins regulate transepithelial H^+ secretion.

Intracellular Signals Controlling Rates of Luminal H^+ Secretion

Some of available information for the outer medullary collecting tubule supports the turtle bladder paradigm of intracellular pH as the primary regulator of luminal H^+ secretion. For example, in one study, changing from HCO_3^-/CO_2 buffers to CO_2-free conditions, which inhibits H^+ secretion, raised intracellular pH.[22] Lowering basolateral bicarbonate, which stimulates H^+ secretion, lowered cell pH.[62,182] Both addition of basolateral DIDS[23,60,182] or removal of basolateral Cl^-,[23] which inhibit basolateral Cl^-/HCO_3^- exchange, alkalinized the cytosol and suppressed H^+ secretion presumably by elevating pH_i.

Other results, however, indicate that factors other than intracellular pH control the luminal H^+-ATPase in intrinsic regulation. In the turtle bladder paradigm, raising the basolateral bicarbonate concentration suppresses H^+ secretion by changing intracellular pH. Yet, in two studies there were no significant differences in the intracellular pH of α intercalated cells in CO_2-free, normal bicarbonate, or high bicarbonate buffers.[63,88]

Removal of basolateral sodium did not change the rate of outer medullary collecting tubule H^+ secretion (bicarbonate reabsorption)[85,183] even though, in several studies, removal of sodium or addition of basolateral amiloride caused an intracellular acidification.[7,60,62]

Studies of intracellular pH regulation in intercalated cells have provided information on the intrinsic regulation of the luminal H^+-ATPase. As indicated above, α intercalated cells have a sodium-independent amiloride-insensitive mechanism for intracellular pH recovery that is thought to represent the luminal vacuolar H^+-ATPase.[61,63,64,88] It is still uncertain how accurately the rate of pH_i recovery reflects tubular secretion rates, as the rates of H^+ secretion calculated from these studies are 4 to 5 times the measured rate of bicarbonate reabsorption in the outer medullary collecting tubule, assuming that the inner stripe consists of a single cell type.[7,175] Weiner et al[22] identified two distinct populations of cells in this segment, presumably intercalated and principal cells, on the basis of pH_i regulation; they showed that both the minority (intercalated) and majority (principal) cells have sodium-independent pH_i recovery, but the rate of this process is slower in principal cells.

Both Hays and Alpern[7] and Weiner et al[22] found that the rate of recovery of pH_i following intracellular acidification was greater in the presence of HCO_3^-/CO_2 than in its absence. Weiner et al found that the rate of sodium-independent pH_i recovery after acidification to pH 6.25 in intercalated cells from the inner stripe of outer medulla was 0.034 pH units/min in CO_2/HCO_3^--free buffers. Using CO_2/HCO_3^- buffers, Kuwahara et al[63] found that the rate of sodium-independent pH_i recovery after acidification to pH 6.40 was 0.234 pH units/min, nearly 10-fold higher for the same value of intracellular pH. Since Weiner and Hamm[184] have shown in β intercalated cells that NH_4Cl-induced acidification does not provoke entry of HCO_3^- by reversal of Cl^-/HCO_3^- exchange, these results strongly suggest that CO_2 directly increases vacuolar H^+-ATPase-driven H^+ extrusion independent of its effects on pH_i.

Other results also support an effect of CO_2/HCO_3^- on intercalated cell H^+ secretion independent of intracellular pH. Kuwahara et al[63] found that the rate of sodium-independent pH_i recovery in α intercalated cells was no different in the presence or absence of bicarbonate buffers, but the rate of vacuolar H^+-driven proton extrusion again was higher for any value of pH_i in bicarbonate buffer since the buffering capacity of the cells is higher in bicarbonate buffers. Hayashi et al[88] showed that elevation of basolateral HCO_3^- in vitro for 3 hours, which lowers tubular H^+ secretion, reduced the rate of pH_i recovery from an NH_4Cl prepulse. This effect was not attributable to a change in buffer capacity or to the degree of initial acidification. This result indicates that bicarbonate suppresses the intrinsic rate of luminal H^+ secretion independent of its effect on intracellular pH.

Hays and Alpern,[62] using CO_2-free conditions, found that changing to Na^+ and Cl^--free buffers, which prevents basolateral base exit, produced a lower steady-state value for pH_i in α intercalated cells. The value of this steady state pH_i presumably was a result of inhibition of luminal H^+ secretion rather than basolateral base exit; yet following cell acidification with an NH_4Cl prepulse, cell pH rose above the steady-state value,[7] indicating that cell pH is not the primary regulator of the H^+-ATPase.[7] The same investigators also demonstrated that exposure of the tubule to propionate produced a transient intracellular acidification, followed by a rise in pH_i to a steady-state level above the initial resting level.[7]

These results suggest a model in which H^+ secretion by the luminal vacuolar H^+-ATPase responds to relative changes in intracellular pH, but in which the "set point" (or magnitude of proton flux at a given pH) is controlled by other factors. In the intercalated cell, it is unlikely that the luminal H^+-ATPase has a primary function in regulating or restoring intracellular pH, as the rate of pH_i recovery from basolateral Na^+-H^+ exchange exceeds that from the H^+-ATPase.[22,60-65]

Whether intracellular calcium (Ca_i) has a role in regulating the luminal H^+-ATPase is unclear. Hays and Alpern[7] found that removal of basolateral sodium produced a transient "spike" increase in Ca_i, arising by release from intracellular stores, followed by a sustained increase in Ca_i, due to entry from the extracellular medium, that was reduced or elevated slightly by intracellular alkalinization or acidification, respectively. The sustained elevation in Ca_i inhibited the initial rate of sodium-independent pH_i recovery from intracellular acidification, and the later recovery of pH_i to baseline did not require any elevation of pH_i. In the presence of HCO_3^-/CO_2 buffers, the late recovery of pH_i was accelerated, although intracellular calcium levels were unaltered. The authors concluded that Ca_i elevation may inhibit the H^+-ATPase. Furuya et al,[65] however, demonstrated basolateral Ca^{+2}-H^+ exchange in outer medullary collecting tubule, likely arising from a Ca^{+2}-ATPase, which may have been activated by the elevated Ca_i, producing an increase in cellular H^+ entry. The results of Hays and Alpern[7] suggest that changes in Ca_i are not required for activation of the H^+-ATPase following a fall in intracellular pH or in response to CO_2.[5]

Cortical Collecting Tubule

Information on the intrinsic regulation of the vacuolar H+-ATPase in the cortical collecting tubule is more difficult to interpret, because of the co-existing processes of H^+ secretion (bicarbonate reabsorption)

---

[5] In this study, the authors proposed that calcium was required for the late recovery of pH_i following intracellular acidification, since the intracellular calcium chelator BAPTA "inhibited" the late recovery of pH_i. However, the resting pH_i level in the presence of BAPTA was 7.6, far above normal, and recovered to the more normal level of pH 7.1 following acidification. Intracellular calcium levels also dropped nearly to undetectable in the experiments, and this may have produced some non-specific inhibitory effects.

and bicarbonate secretion, and the failure, in most studies, to distinguish between the two mechanisms. Some evidence is consistent with coupling of the basolateral vacuolar H^+-ATPase to luminal bicarbonate entry through the Cl^-/HCO_3^- exchanger. Raising or lowering basolateral bicarbonate was found to suppress or stimulate, respectively, net bicarbonate reabsorption in cortical collecting tubule.[185] Removal of basolateral chloride, which stimulates bicarbonate secretion,[186] acidified the intracellular pH of the β intercalated cell.[58,90,100]

In contrast, lowering the ambient P_{CO2} in vitro had no effect on net bicarbonate transport in the cortical collecting tubule[185] even though it elevated the intracellular pH of β intercalated cells,[184] suggesting that the H^+-ATPase was not suppressed by intracellular alkalinization. Acute elevation of the P_{CO2} in vitro lowered intracellular pH,[184] but had no effect on net bicarbonate transport,[185] and in vivo elevation of P_{CO2} for 5-6 hours produced no discernible morphologic changes in β intercalated cells.[187] From the information presently available, it appears likely that the intrinsic regulation of the basolateral membrane H^+-ATPase differs substantially from that controlling the apical H^+-ATPase. In the β intercalated cell, basolateral Na^+/H^+ exchange has an even greater role in maintaining pH_i than it does in the α intercalated cell.[90,100,188]

Proximal Tubule

Although there are many factors that acutely affect rates of H^+ secretion by the proximal tubule,[14] most of the investigations to date have not determined whether these factors affect the luminal vacuolar H^+-ATPase. There are a few studies, however, that provide evidence for participation of the vacuolar H^+-ATPase in acute regulation of proximal H^+ secretion.

Kurtz[78] demonstrated that the S3 portion of the proximal tubule has a sodium independent mechanism for pH_i recovery following acidification with NH_4Cl; this recovery process was inhibited by luminal NEM and DCCD, suggesting that it was a vacuolar H^+-ATPase. Nakhoul et al[80] found, in the S3 part of proximal tubule, that basolateral CO_2/HCO_3^-, but not luminal, induced an intracellular alkalinization that was not attributable to bicarbonate entry, and appeared to arise from a luminal proton pump. The alkalinization did not occur in response to a fall in intracellular pH induced by luminal CO_2/HCO_3^-, nor was the alkalinization inhibited by acetazolamide, which prevented the fall in pH_i induced by basolateral CO_2/HCO_3^-; these results indicate that CO_2/HCO_3^- stimulated the activity independently of its effects on intracellular pH.

Schwartz and Al-Awqati[181] found that the S2 portion of the proximal tubule internalized FITC-dextran into acidic vacuoles, and that CO_2 produced a colchicine-inhibitable decrease in the fluorescent signal (as described above), which they interpreted to represent release of dextran by exocytosis and a presumptive insertion of H^+-ATPases into the luminal membrane. The decrease in the fluorescent signal occurred within one minute. Kurtz[78] found sodium-independent pH_i recovery in the S2 proximal tubule, although its activity was only 11% of that in the S3 segment. The sodium-independent pH_i recovery mechanism described by Kurtz in the S3 proximal tubule was inhibited by pre-treatment of the tubule with colchicine; this is interpreted to indicate that the response might be dependent on membrane fusion. It is also possible, however, that colchicine caused a loss of H^+-ATPase from the brush border membrane by membrane traffic, as this was found to occur in the proximal tubule of rats injected with colchicine.[189]

In summary, the proximal tubule S3 segment luminal H^+-ATPase is activated by basolateral CO_2/HCO_3^-, and it appears unlikely that this effect is exerted through changes in intracellular pH. As discussed below, removal of bicarbonate from the proximal tubule reduces intracellular ATP levels and has other metabolic effects that might affect the vacuolar H^+-ATPase. The mechanism of activation of the proximal tubule H^+-ATPase by CO_2/HCO_3^- remains unclear; insertion of H^+-ATPases by membrane fusion remains an unproven possibility.

Acute Regulation of H^+ Secretion by Intrinsic Regulatory Mechanisms: Coupling of H^+ Secretion to Metabolism

Very little information is available on the influence of metabolism on the plasma membrane vacuolar H^+-ATPase in the kidney. In the studies by Kurtz[78] discussed above, the sodium-independent mechanism for pH_i recovery in the S3 portion of the proximal tubule, thought to represent a vacuolar H^+-ATPase, was not affected by the mitochondrial cytochrome inhibitor cyanide, but was markedly reduced by the glycolysis inhibitor iodoacetamide. The activity was restored by ATP or AMP plus iodoacetamide in the absence, but not in the presence, of cyanide, indicating that mitochondrion-derived ATP could be used, if sufficient AMP were available. This suggests that the kidney proximal tubule vacuolar H^+-ATPase, like that in the turtle bladder, is coupled preferentially to glycolysis.

Vinay and colleagues[190] examined the metabolic cost of the vacuolar H^+-ATPase in suspensions of rabbit proximal tubule. Of the total phosphorylative (i.e. oligomycin-inhibitable) respiration, only 4.5% was attributed to the vacuolar H^+-ATPase (inhibitable by bafilomycin), as compared to the 58.8% attributed to the Na^+/K^+-ATPase (inhibitable by ouabain).

Dickman and Mandel[191] examined the effect of CO_2/HCO_3^- removal on metabolic oxygen consumption in suspensions of rabbit proximal tubule. Either removal of CO_2/HCO_3^- or addition of bafilomycin A_1 reduced the ouabain-insensitive respiration by 20%. The reduction in oxygen consumption under CO_2/HCO_3^--free conditions originated from a limitation in mitochondrial uptake of metabolic substrates. Removal of CO_2/HCO_3^- also reduced cellular ATP content by 50%, and increased cellular lactate production about 10-fold. These impressive metabolic effects of CO_2/HCO_3^- are candidate mechanisms which might influence vacuolar H^+-ATPase activity independent of effects on cell pH.

Acute Regulation of H⁺ Secretion by Hormones

Aldosterone

Aldosterone stimulates H⁺ secretion in the isolated renal collecting tubule.[175,183] At least part of the acute action of mineralocorticoids is a direct stimulation, not attributable changes in the collecting tubule potential difference, and independent of effects on sodium transport.[183] The mechanism of the direct effect remains uncertain. As indicated above (Figure 6.1), increases in H⁺ secretion could be the result of increases in the quantity of H⁺-ATPase protein, recruitment of preexisting enzyme from an intracellular or inactive pool, or kinetic activation of the H⁺-ATPase. Preliminary studies suggest that mineralocorticoids induce a modest overall increase in H⁺-ATPase in both cortex and medulla, and a modest increase in H⁺-ATPase in the luminal membrane of outer medullary α intercalated cells by redistribution;[192] an effect on the kinetics of the H⁺-ATPase was not excluded.

β-*Adrenergic Agonists and Cyclic AMP*

β-adrenergic agonists and other agents that elevate cAMP stimulate H⁺ and bicarbonate secretion. Isoproterenol elevates cAMP levels in isolated intercalated cells from cortical[193] and outer medullary collecting tubule.[97,194]

In the cortical collecting tubule, isoproterenol and cAMP enhance bicarbonate secretion[195-197] by increasing luminal Cl^-/HCO_3^- exchange,[198] and possibly by opening a luminal anion channel.[196] Since these agents decrease the lumen-negative potential in this segment, it has been suggested that they may stimulate the luminal H⁺-ATPase directly in α intercalated cells.[29,197] In the rabbit outer medullary collecting tubule (inner stripe), cAMP and forskolin increased H⁺ secretion 34.5% and 39%, respectively.[199] The mechanism of these effects (see Figure 6.1) is unknown.

Other Hormones

Several other hormones have modest effects on H⁺ transport of uncertain physiologic significance. The peptide hormones vasopressin[200,201] and glucagon[202,203] inhibit bicarbonate secretion (or enhanced H⁺ secretion) in the rat cortical collecting tubule and distal convoluted tubule.[201] Calcitonin produces an apparent stimulation of H⁺ secretion in rat cortical collecting tubule.[204] The prostaglandin PGE_2 produced a 12% inhibition of H⁺ secretion in the outer stripe of rabbit outer medullary collecting tubule.[199]

Protein kinase C activators have no effect on net bicarbonate reabsorption in the cortical collecting tubule.[205]

CHRONIC REGULATION OF H⁺ SECRETION BY THE VACUOLAR H⁺-ATPASE IN RENAL EPITHELIA

Prolonged acid or alkali administration, mineralocorticoid administration, and sustained elevation of the P_{CO_2} are the three most important stimuli that produce chronic changes in renal acid excretion involving the kidney vacuolar H⁺-ATPase.

Pitfalls in Assays Used to Assess Changes in Vacuolar H⁺-ATPase During Chronic Regulation

Enzymatic Assays: NEM-Sensitive ATPase Activity

Physiologic and biochemical studies have attempted to determine to what extent chronic changes in tubular H^+ secretion originate from the vacuolar H^+-ATPase. The paradigm underlying this type of investigation are the numerous studies of changes in Na^+/K^+-ATPase activity that occur during chronic adaptation in various physiologic settings.[206] Assays of whole cell Na^+/K^+-ATPase activity in different nephron segments have provided useful information because: 1) there is a highly specific inhibitor for the Na^+/K^+-ATPase, ouabain, which can be used in intact cells, and that can provide measurements of enzymatic activity (as ouabain-sensitive ATPase or Rb^+ uptake) and enzyme numbers (as ^3H-ouabain binding sites); and 2) nearly all of the cellular Na^+/K^+-ATPase resides on the plasma membrane.

In contrast, there are several problems with studies that have attempted to measure changes in vacuolar H^+-ATPase activity. First, is that there is no proven specific inhibitor of the vacuolar H^+-ATPases. A number of studies have used N-ethylmaleimide (NEM) inhibitable ATPase as an assay,[207-221] and more recently studies have examined bafilomycin A_1 inhibitable ATPase[222] as an assay for vacuolar H^+-ATPase activity. The validity of these assays in whole cell homogenates or even membrane fractions remains unproven. For example, no studies have attempted to determine if immunodepletion of vacuolar H^+-ATPase from these preparations removes all of the NEM-sensitive or bafilomycin-sensitive ATPase activity. NEM is a sulfhydryl reactive reagent that inhibits a variety of transport ATPases. Its presumed specificity relies on assay conditions that reduce the activity of ATPases other than the vacuolar H^+-ATPase. Bafilomycin is a high affinity inhibitor of the vacuolar H^+-ATPases with an IC_{50} in the low nanomolar range, orders of magnitude below its effects on other ATPases.[51, 223] When used at concentrations needed to inhibit 100% of the vacuolar H^+-ATPase activity, however, it does have effects on other enzymes. Bafilomycin has not been studied for a sufficient length of time for any other possible effects to have emerged.

The second problem with assaying vacuolar H^+-ATPase activity in intact tubule segments is that much, or even most of the enzyme resides in intracellular membrane compartments instead of on the plasma membrane.[12,93,224-226] None of the investigators using the assays of total enzyme activity (as NEM-sensitive ATPase activity) in intact cells have attempted to distinguish whether the activity arises from the plasma membrane. It has been proposed that, in permeabilized cells, the activity measured represents the total enzyme content the cell,[207,216,227] but some experiments cast serious doubts on this assumption. For example, colchicine inhibits nearly completely the NEM-sensitive ATPase activity in permeabilized kidney cells.[215] As colchicine acts, presumably, by preventing microtubule-mediated insertion of the H^+-ATPase in the plasma membrane, the inhibition of ATPase activity suggests that the cellular location of the enzyme has a powerful influence on the measured activity, and that NEM-sensitive ATPase activity is not a simple function of cellular ATPase content.

The third problem is that measurements of total enzymatic activity in tubules provide very little insight into adaptational mechanisms. It cannot be determined whether an increase in activity represents an increase in the number of H⁺-ATPases per cell, a kinetic activation of extant H⁺-ATPases, or a redistribution of pre-existing H⁺-ATPases.

As an example of the pitfalls of using whole tubule NEM-sensitive ATPase assays, Sabatini et al[216] showed that the pH optimum and ATP concentration dependence of NEM-sensitive ATPase in cortical and medullary collecting tubules were entirely different from those of the purified kidney H⁺-ATPase.[39,40,52] The purified enzyme from medulla had a pH optimum of 6.5 and a K_m for ATP of 150 µM,[39] a value in good agreement with that for other purified vacuolar H⁺-ATPases[55,57] and with proton transport in kidney membrane vesicles.[50] In contrast, medullary collecting tubule had a pH optimum of 7.4, and a K_m for ATP of 2.43 mM. On the basis of enzymatic properties alone, it is unlikely that NEM-sensitive ATPase activity is the equivalent of H⁺-ATPase activity in the intact tubule.

Additionally, four studies have found that NEM-sensitive ATPase activity in the thick ascending limb exceeds the activity in all segments of the collecting duct[207,212,214,227] despite immunocytochemical studies showing far less vacuolar H⁺-ATPase in thick ascending limb than in collecting tubule.[12] Chronic acid administration increases H⁺ secretion in the thick ascending limb, and excellent physiologic studies demonstrate that essentially all of the increase in acid secretion is the result of increased luminal Na⁺/H⁺ exchange activity.[20,228,229] Yet, in two studies,[214,218] chronic acid administration produced NEM-sensitive ATPase activity values in thick limb that were 192% and 222%, respectively, of control, and these increases exceeded those in any other nephron segments.

Vesicular H⁺ Transport Assays

Several studies have attempted to quantify changes in H⁺-ATPase under different physiologic conditions by measuring acidification kinetics in membrane fractions. These assays are susceptible to a variety of problems, including: 1) the membranes are impure and include endosomes, lysosomes, and other acidic compartments; 2) factors other than the H⁺-ATPase affect the acidification kinetics, such as the vesicle volume, the membrane potential of the vesicle, the internal buffering capacity, and the passive proton conductance; consequently, the acidification kinetics may not be a linear function of the number of H⁺-ATPases in the membrane; 3) the experimental physiologic condition may affect the membrane fractionation procedure.

Chronic H⁺-ATPase Regulation in Response to Prolonged Acid and Alkali Administration

Physiologic studies have shown that chronic acid administration increases net acid excretion in the proximal tubule,[14,230-234] thick ascending limb,[20,228] distal convoluted tubule,[21,235,236] cortical collecting tubule,[3,64,83,112,114,237-243] and inner medullary collecting duct.[104,244-247] The studies discussed below indicate that the vacuolar H⁺-ATPase participates in this adaptational increase in H⁺ secretion in the cortical and medullary collecting tubules, and possibly in the proximal tubule.

Response to Prolonged Acid Administration

Cortical Collecting Tubule

As indicated above, the cortical collecting tubule has the capacity to secrete protons, through the α intercalated cells[3,11,29,90,248] and to secrete bicarbonate[112,237,249] by chloride-bicarbonate exchange[186,196,250] in the luminal membrane of β intercalated cells[90,100,184,198] and γ intercalated cells.[25]

The net acid intake and generation of the animal is the major determinant of whether the cortical collecting tubule manifests net acid or bicarbonate secretion. Chronic administration of acid in vivo suppresses cortical collecting tubule bicarbonate secretion, usually producing a tubule with net acid secretion,[30,58,83,87,196,237,241,251,252] with little or no stimulation of H⁺ secretion (bicarbonate reabsorption).[241,251,252] In vitro incubation of rabbit cortical collecting tubules in acidic medium also suppresses bicarbonate secretion,[243,253] without affecting H⁺ secretion.[243] Although suppression of bicarbonate secretion is thought to involve inactivation or internalization of luminal Cl⁻/HCO₃⁻ exchange,[243,253] at least part of the response to in vivo administration of acid may be attributable to loss of vacuolar H⁺-ATPase from the basolateral membrane of bicarbonate-secreting type β intercalated cells. Chronic administration of acid to rats for 14 days produced a reduction in basolateral membrane H⁺-ATPase polarization and an increase of cytoplasmic vesicular staining for H⁺-ATPase in β intercalated cells of the cortical collecting tubule.[99] A progressive and significant increase in the number of cells with apical H⁺-ATPase staining was observed, suggesting that H⁺ secretion might have increased even though, as discussed above, physiologic studies have not demonstrated an increase. These immunocytochemical observations are supported by the earlier morphologic findings of Dørup[254] who found that acute acid administration to rats increased the luminal membrane surface and decreased the number of intermediate size apical vesicles in intercalated cells of the cortical collecting tubule, consistent with recruitment of H⁺-ATPase to the luminal membrane by vesicle fusion.

In the study by Bastani et al,[99] no change in H⁺-ATPase content per mg of protein from cortical microsomes (assessed by an anti-H⁺-ATPase E (31 kD) subunit monoclonal antibody binding assay) was observed in kidneys from the acid-loaded rats.[99] It is possible, however, that the failure to detect a change in H⁺-ATPase content may have resulted from H⁺-ATPase-rich proximal tubule brush border membrane contaminating the microsome preparation, obscuring changes originating in the collecting duct; the monoclonal antibody used also detects vacuolar H⁺-ATPase in the proximal tubule,[66,70,99] which contains the majority of the H⁺-ATPase in the cortex.[17] The recent finding that the H⁺-ATPase B2 (56 kD "kidney") subunit isoform is expressed in intercalated cells but not in proximal tubule or thick limb[66] may make it possible to determine if changes in H⁺-ATPase content occur in the cortical collecting tubule during acid administration.

Without resolving the assay problems discussed above, several investigators have examined changes in NEM-sensitive ATPase activity in different segments of the nephron in response to chronic acid administration and other physiologic conditions. Table 6.5 provides a summary from three different laboratories of baseline NEM-sensitive ATPase activity (expressed as pmol/hr/mm tubule length) in different microdissected segments of the rat nephron. At baseline, some of the values vary more than three-fold, indicative of the problems with the assay. Table 6.2 shows the effects of chronic acid administration on NEM-sensitive ATPase activity in different nephron segments. In investigations originating from three different laboratories, only two found a significant increase in NEM-sensitive ATPase activity in the cortical collecting tubule.

Outer Medullary Collecting Tubule

In several studies, no change in in vitro H⁺ secretion (bicarbonate reabsorption) was found in outer medullary collecting tubules obtained from animals subjected to chronic acid administration.[30,87,92,241] In recent studies by Hayashi et al,[88] vacuolar H⁺-ATPase-driven proton extrusion was measured in individual intercalated cells from rates of sodium-independent recovery of pH_i following in vitro acid loading. In outer medullary collecting tubules from rabbits subjected to 14 days of high acid intake, there was no significant change in sodium-independent pH recovery in the intercalated cells.

In contrast, several studies have demonstrated marked morphologic alterations and a redistribution of H⁺-ATPase with chronic acid administration. Madsen and Tisher[255] found that administration of acid

---

Table 6.5. ATPase activity in individual microdissected nephron segments (pmol/hr/mm)

Segment[1]	NEM-sensitive ATPase Activity		
	Refs. 207, 212	Ref. 214	Ref. 227
PT-early (PCT)	–	440	550
PT-late (PST)	–	–	350
mTAL	720	220	360
cTAL	1080	340	420
DCT	2160	–	–
CNT	–	–	–
CCT	924	280	350
OMCT	714	220	250
IMCD	804	320	–

[1] Abbreviations: PT, proximal tubule; PCT, proximal convoluted tubule; PST, proximal straight tubule; mTAL, medullary thick ascending limb; cTAL, cortical thick ascending limb; DCT, distal convoluted tubule; CNT, connecting tubule; CCT, cortical collecting tubule; OMCT, outer medullary collecting tubule; IMCD, inner medullary collecting duct

to rats for 15 days produced an increase in apical membrane area, and loss of tubulovesicular structures in the apical cytoplasm of intercalated cells in the inner and outer stripe of outer medulla; they suggested that the changes might signify recruitment of H+-ATPases from the tubulovesicular structures to the apical membrane.

Bastani et al[99] used a monoclonal antibody to the E (31 kD) subunit of the vacuolar H+-ATPase to examine H+-ATPase distribution during chronic acid administration to rats at several time points over a 14 day period. They found that chronic acid administration produced a profound redistribution of the vacuolar H+-ATPase from cytoplasmic vesicles to the plasma membrane in the intercalated cells. The shift in H+-ATPase was most marked in the inner stripe of the outer medulla; only 7% of the intercalated cells exhibited plasma membrane H+-ATPase staining in control rats, whereas 63% of the cells showed plasma membrane staining after one day of acid administration, and this increased progressively to 89% at 14 days. The outer stripe of the outer medulla showed a similar response, but with a slower time course.

Bastani et al[99] also used a monoclonal antibody and cDNA probe to the E subunit of the vacuolar H+-ATPase to examine whether changes in H+-ATPase content occurred during chronic acid administration. They found that there was no significant change in the quantity of H+-ATPase (measured in medullary microsomes by quantitative immunobinding assay) or in levels of E subunit mRNA at any time point over a 2-week period. These observations must be viewed with

Table 6.6. Effect of acid and alkali administration on NEM–sensitive ATPase in microdissected nephron segments. Values expressed as a percent of control activity

Segment	Acid Administration			Alkali Administration		
	Ref. 207	Ref. 214	Ref. 218	Ref. 212	Ref. 214	Ref. 218
PT–early (PCT)	–	62	90	–	96	88
PT–late (PST)	–	–	96	–	–	96
mTAL	–	192	222	108	96	34
cTAL	147	94	–	–	71	94
DCT	97	–	–	97	–	–
CCT	173	186	114	41	107	92
OMCT	183	140	169	43	115	53
IMCD	169	100	–	53	76	–

the caveat, discussed above, that H⁺-ATPase-rich proximal tubule membrane contaminating the microsomes could have concealed a change.

In three studies, a modest increase in NEM-sensitive ATPase activity was found in outer medullary collecting tubule with chronic acid administration (Table 6.6); Chang et al found no change in H⁺ transport rates in medullary membranes from acid-loaded rabbits.[256]

The evidence indicates, therefore, that redistribution of the H⁺-ATPase to the plasma membrane by vesicular traffic occurs during chronic acid administration, and suggests that an increase in H⁺ secretion may occur in the outer medullary collecting tubule. The basis for the discrepancy between the transport studies and morphologic/immunocyto-chemical studies is unknown at the present time. As discussed below, studies on chronic regulation of H⁺ transport in response to alkali administration and hypercapnia also demonstrate the response of the tubule in vitro differs from that in vivo.

Distal Tubule

The "distal tubule" examined in physiologic studies over several decades consists of several distinct tubular epithelia: the distal convoluted tubule (early distal tubule), which has a single cell type, and the connecting tubule and initial collecting tubule (late distal tubule), which are composed of a mixture of principal cells and intercalated cells.[257] Studies suggesting the participation of a vacuolar H⁺-ATPase in distal tubular H⁺ secretion demonstrated acidification that was not inhibited by amiloride, and that was affected by the tubular potential difference.[258] Recent evidence[21] indicates that a vacuolar H⁺-ATPase accounts for 20-35% of early distal tubular H⁺-ATPase H⁺ secretion (bicarbonate reabsorption), and 84% or more₆ of late distal tubular H⁺ secretion.

The rat distal tubule shows either no bicarbonate reabsorption,[235] net bicarbonate reabsorption[258,261-263] or net bicarbonate secretion[201,249,263] depending on the flow rate,[263,264] the luminal bicarbonate concentration,[21,264] the diet of the animal,[249,258] the luminal chloride concentration,[21,265] and the presence or absence of potassium depletion.[266]

Chronic administration of acid[235,236,267] or an acid-ash diet[258] increases distal tubular H⁺ secretion, an response that requires longer than 2 hours.[235] The role of the vacuolar H⁺-ATPase in this response is at present unknown.

Proximal Tubule

During states of metabolic acidosis and chronic acid loading, proximal tubular bicarbonate reabsorption rises, primarily by an increase in the activity of the luminal Na⁺/H⁺ antiporter.[14] Whether any of the acid-induced increase in proximal tubule H⁺ secretion arises from an increase in vacuolar H⁺-ATPase activity is unclear. As indicated above,

---

₆Part of the H⁺ secretion in the late distal tubule was attributed to an H⁺/K⁺-ATPase on the basis of partial inhibition by SCH 28080, a putative specific H⁺/K⁺-ATPase inhibitor. Recent evidence, however, demonstrates that SCH 28080 also inhibits H⁺ secretion by the electrogenic vacuolar H⁺-ATPase in intact turtle bladder,[259,260] probably by a mechanism other than direct inhibition of the enzyme.

Bastani et al[99] found no change in quantity of immunoreactive H⁺-ATPase E (31 kD) subunit in cortex microsomes (which contain brush border membrane) at any time point during 14 days of acid administration to rats. Recently, Chambrey et al[222] found that acid administration to rats for 4 days produced a 26% increase in NEM-sensitive ATPase activity and a 33% increase in bafilomycin-sensitive ATPase activity in brush border membranes from acidotic kidneys, as compared to brush border membranes from control kidneys, but no difference in either activity between the acidotic and control homogenates. Using a cholate treatment and dilution method to reorient the brush border membranes, they showed that membranes from acidotic animals had 25% higher initial rates of acidification than did cholate-reoriented membranes from control kidneys. Since acidosis increased the apparent H⁺-ATPase activity in brush border without increasing total bafilomycin A_1-sensitive ATPase activity, they proposed that acidosis produced a redistribution of vacuolar H⁺-ATPase from intracellular compartments to the plasma membrane. One significant problem with the study is that the authors used 200 nM bafilomycin A_1, a concentration far in excess of the 15 nM needed to inhibit completely H⁺ in cholate-treated brush border membranes;[51] at this higher concentration of bafilomycin, inhibition of other ATPases has been observed.[223] In two studies on intact microdissected proximal tubules from acid-loaded rats, one found no change in NEM-sensitive ATPase activity, and the other observed a decrease (Table 6.2).

Response to Prolonged Alkali Administration

Cortical Collecting Tubule

Chronic alkali administration and induction of a sustained metabolic alkalosis both increase bicarbonate secretion in the cortical collecting tubule[30,87,186,196,237,268,269] (although Hamm et al observed no change in bicarbonate secretion.[241] Bastani et al[99] examined the effect of chronic administration of bicarbonate on the distribution of H⁺-ATPase in the intercalated cells of rat cortical and medullary collecting tubules between 3 and 14 days following bicarbonate administration, and found significant increases in the number of intercalated cells with bipolar and basolateral H⁺-ATPase polarization. The magnitude of the response may have been blunted as a result of the failure of the animals to develop a metabolic alkalosis.[99] No change in H⁺-ATPase content of cortical microsomes was found by the anti-E subunit immuno-binding assay discussed above.

More recently, Verlander et al[270] examined the distribution of vacuolar H⁺-ATPase in rat intercalated cells in a model of chloride depletion metabolic alkalosis induced by peritoneal dialysis of rats with 150 mM sodium bicarbonate.[271] In the alkalotic rats, antibody labeling for the H⁺-ATPase A (70 kD) subunit, assayed by immunoelectron microscopy, was lost from the apical membrane and increased in intracellular vacuoles of intercalated cells in the cortical and outer medullary collecting tubules. H⁺-ATPase labeling was increased in the basolateral plasma membrane of β intercalated cells of the cortical collecting tubule.

These findings suggest that in chloride depletion alkalosis, H+ secretion is inhibited, and bicarbonate secretion is enhanced, and that both adaptational processes result, at least in part, from regulated changes in H+-ATPase distribution. The change in bicarbonate secretion occurred within 45 minutes, and the change in H+-ATPase distribution appeared within 2 hours in the intact rat.[269,270] In contrast, rat cortical collecting tubules that were incubated in alkaline solutions in vitro failed to exhibit an increase in bicarbonate secretion,[269] indicating that as yet unidentified factors in vivo are responsible for modifying bicarbonate secretion, and possibly for changing the H+-ATPase distribution. It is also possible that the redistribution of H+-ATPase occurred as a secondary response to increased luminal Cl-/HCO3- exchange; Kim et al[272] demonstrated that chronic bumetanide administration to rats, which increases Cl- delivery to the collecting tubule, produced nearly identical changes in H+-ATPase distribution.

Of three studies on the effect of alkali administration on NEM-sensitive ATPase activity in cortical collecting tubule, two found no change, and one found a decrease (Table 6.6).

Distal Tubule

In a number of investigations, it has been demonstrated that distal tubule bicarbonate secretion increases during chronic alkali administration or metabolic alkalosis.[235,249,263,264,273-275] Distal tubule bicarbonate secretion probably contributes to the correction of metabolic alkalosis,[275] although the collecting duct may have a greater role.[276] Bicarbonate secretion in the distal tubule is an active process[263] that is stimulated by luminal chloride,[21,265,274,275] and probably occurs in connecting tubule and initial collecting tubule intercalated cells by a basolateral vacuolar H+-ATPase[12] and luminal Cl-/HCO3- exchange.[114] Chronic metabolic alkalosis induced by diuretic administration also increases distal tubule H+ secretion;[277,278] this response may be an effect of chronically increased aldosterone levels (see below).

Chronic H+-ATPase Regulation
from Prolonged Mineralocorticoid Administration

Chronic mineralocorticoid administration increases renal net acid excretion.[279-283] The effect of chronic mineralocorticoid administration on H+ transport in the cortical and outer medullary collecting tubules has been examined in several studies.

Cortical Collecting Tubule

Chronic administration of the mineralocorticoid deoxycorticosterone (DOC) to rabbits produced an increase in the lumen-positive potential of the cortical collecting tubules observed in the absence of sodium, an index of apical H+-ATPase activity.[85] This suggests that mineralocorticoids stimulate H+ secretion in cortical α intercalated cells. It is unlikely, however, that aldosterone has direct effects on the vacuolar H+-ATPase in β (or γ) intercalated cells (for an excellent review, see ref. 29). Immunocytochemical studies indicate that the mineralocorticoid receptor is not detectable in most of the cortical collecting tubule intercalated cells,[284-286] probably indicating absence of the receptor in β (or γ) intercalated cells, since recent studies show that as few as 4% of cortical intercalated cell function as α cells.[25] Cortical collecting tubules from rabbits treated chronically with DOC, which produces a slight alkalosis, showed an increase in unidirectional bicarbonate secretion.[251] The bicarbonate secretion was eliminated, however, by concomitant administration of acid and DOC to the animals, suggesting that the acid-base status of the animal, rather than a direct effect of DOC, was the primary factor regulating the response to mineralocorticoid;[251] evidence that alkalosis increases bicarbonate secretion partly by causing a redistribution of the vacuolar H⁺-ATPase in the β intercalated cells is discussed above. Mineralocorticoid administration produced no change in apical or basolateral membrane surface density of cortical intercalated cells.[287,288] Yamaji et al[188] found that the β intercalated cells in cortical collecting tubules of rabbits treated chronically with DOC had an increase in basolateral Na⁺/H⁺ exchange activity, but no change in the amiloride-resistant sodium-independent rate of pH$_i$ recovery (an assay for H⁺-ATPase-mediated H⁺ extrusion).

Outer Medullary Collecting Tubule

Prolonged in vivo administration of mineralocorticoids increases H⁺ secretion 65-84% in outer medullary collecting tubules studied in vitro.[92,183] Using sodium-independent pH$_i$ recovery from an in vitro acid load (delivered by NH$_4$Cl prepulse) as an assay for vacuolar H⁺-ATPase-driven proton extrusion, Hays[175] found that pH$_i$ recovery activity increased 76% in outer medullary tubules from DOC-treated

Table 6.7. Effect of mineralocorticoid administration and adrenalectomy on NEM-sensitive ATPase activity in microdissected nephron segments. Values expressed as a percent of control activity.

| Segment | Mineralocorticoid Administration | | | | Adrenalectomy |
	Ref. 289	Ref. 221	Ref. 211	Ref. 210	Ref. 208
PT-early (PCT)	–	–	–	–	94
PT-late (PST)	–	–	–	–	113
mTAL	–	–	–	–	95
cTAL	–	–	–	125	107
DCT	–	–	–	127	ND
CCT	100	161	200	188	38
OMCT	208	191	300	213	41
IMCD	–	–	–	371	–

rabbits, and decreased 56% in tubules from adrenalectomized animals. In this study, no attempt was made to distinguish intercalated cells from principal cells. In studies of individual intercalated cells of outer medullary collecting tubules from mineralocorticoid-treated rabbits, Kuwahara et al[64] found that sodium-independent in vitro recovery of pH_i, an assay for vacuolar H⁺-ATPase-driven H⁺ extrusion, increased 111% in intercalated cells from the outer stripe, studied in HCO_3^-/CO_2 buffers. Weiner et al[22] found that DOC administration to the animal increased sodium-independent pH_i recovery 306% in intercalated (minority) cells from the inner stripe, studied under CO_2-free conditions. The baseline intracellular pH in the intercalated (minority) cells in sodium-containing buffers increased from 7.24 to 7.51 with DOC treatment. This probably indicates a primary increase in H⁺-ATPase activity induced by mineralocorticoid, since DOC had no effect on α intercalated cell basolateral Na^+-H^+ exchange, and did not decrease basolateral Cl^-/HCO_3^- activity.[64]

The results of several studies examining the effect of mineralocorticoids on NEM-sensitive ATPase activity in different nephron segments[208,210,211,221,289] are listed in Table 6.7. Three of the four studies found an increase in activity in the cortical and outer medullary collecting tubules, though, as discussed above, it is doubtful that this assay is specific for the vacuolar H⁺-ATPase. In addition, the changes in activity noted do not provide a distinction between changes in number, kinetics, or distribution of the enzyme.

Chronic H⁺-ATPase Regulation in Response to Prolonged Elevation of P_{CO2}

Prolonged hypercapnia (rise of the P_{CO2}) increases renal net acid excretion producing a sustained elevation in extracellular fluid bicarbonate concentration.[290-292] Laski and Kurtzman[180] studied in vitro bicarbonate reabsorption in collecting tubules from rabbits subjected to hypercapnia for varying lengths of time. Cortical collecting tubules began to show an increase in net bicarbonate reabsorption by 6 hours, that was significant by 24 hours; after 48 hours of in vivo hypercapnia, cortical collecting tubule net bicarbonate reabsorption was 650% of control rates. As unidirectional rates of bicarbonate transport were not measured, it is unknown whether this response represented a suppression of bicarbonate secretion, or a stimulation of H⁺ secretion. Laski and Kurtzman also found that there was no change in bicarbonate reabsorption in outer medullary collecting tubules,[180] suggesting that the changes in the cortex most likely were from suppression of bicarbonate secretion. The basis of this adaptational response is presently unknown. Verlander et al[187] found a 73-95% increase in the apical membrane surface density of α intercalated cells of the cortical collecting tubule and connecting tubule in rats after 4-5 hours of hypercapnia, but no significant change in the β cells. On the basis of the time course for the physiologic adaptation, however, the time course of their studies were probably too short for the adaptive changes to chronic CO_2 elevation to have occurred.

As discussed above, Madsen and Tisher[179] found a significant increase in apical surface density of intercalated cells with a concomitant loss of apical tubulovesicles in the outer medullary collecting tubule in rats after 4-5 hours of hypercapnia, which may again represent only an acute regulatory response. McKinney and Davidson,[172] however, found that outer medullary collecting tubules increased H^+ secretion 146-252% in response to an in vitro elevation of P_{CO2}, but when the same tubules were subjected to electron microscopic morphometric analysis, there was no change from controls in the apical membrane surface density or any other indices. The discrepancies between the in vitro physiologic and morphologic responses and the in vivo morphologic observations suggests that an additional unidentified factor(s) participates in vivo in the redistribution of vacuolar H^+-ATPase. The basis for the adaptive increase in H^+ secretion with prolonged hypercapnia at present remains unsolved.

Chang[256] found no difference in H^+ transport rates in renal medullary vesicles obtained from rabbits subjected to hypercapnia for two days, and concluded that no changes in cellular H^+-ATPase content occurred, but there are too many problems with variation in fractionation purity and efficiency in such studies to obtain reliable conclusions.

THE RENAL VACUOLAR H^+-ATPASE: SUMMARY AND CONCLUSIONS

In most eukaryotic cells, vacuolar H^+-ATPases reside predominantly in intracellular compartments, where they are responsible for the acidification of the vacuolar system required for constitutive endocytic and secretory processes. In some specialized cells, such as proton-transporting renal epithelial cells, most of the vacuolar H^+-ATPase resides on the plasma membrane where it functions in the bicarbonate reabsorption and regeneration needed for acid-base homeostasis. Acute and chronic regulation of vacuolar H^+-ATPase-driven H^+ transport in renal tubular epithelial cells may occur, in principle, through the three general mechanisms shown in Figure 6.1: kinetic regulation, redistribution, and change in enzyme content. The experimental evidence discussed above demonstrates that both kinetic regulation and redistribution are employed by the kidney. How then does the kidney accomplish the physiologic regulation of plasma membrane vacuolar H^+-ATPases while allowing constitutive acidification of intracellular compartments to proceed?

The studies of the enzyme to date have provided a broad overview of how such regulation may occur: 1) Vacuolar H^+-ATPases from different compartments of the kidney have intrinsic differences in their enzymatic properties, such as pH optima and response to changes in the lipid and ionic environment. These enzymes also appear distinct in the degree to which their activity is affected by regulatory factors. The basis for these differences in properties remains unclear; the observation that the B and E subunits are dissimilar among these enzymes provides a possible clue. 2) Renal epithelial cells have the capacity to express different structural forms of vacuolar H^+-ATPase. This may provide a basis for changing the properties and regulation of the vacuolar H^+-ATPase in different segments of the nephron. The cell-specific expression of different structural forms of the H⁺-ATPase may provide a means for sequestering the enzyme into specialized membrane compartments capable of inserting the enzyme into the plasma membrane in a regulated manner. 3) The kidney produces cytosolic regulatory proteins that interact directly with the vacuolar H⁺-ATPase to modify its activity. Future investigations may determine where these proteins are expressed, and whether they participate in the acute and chronic regulatory responses of renal H⁺ transport discussed above.

ACKNOWLEDGEMENT

This work was supported in part by NIH grants DK38848, DK09976, AR32087, and DK45181, and by a grant from the Monsanto-Washington University Biomedical Agreement. SLG was a Sandoz Pharmaceutical Corporation Established Investigator of the American Heart Association.

REFERENCES

1. DuBose, T., Jr., Reclamation of filtered bicarbonate.Review. Kidney Int, 1990. 38(4): p. 584-9.

2. Rector, F., Jr., Renal regulation of acid-base balance. Aust N Z J Med, 1981. 11(Suppl 1): p. 1-5.

3. Hamm, L.L. and K.S. Hering-Smith, Acid-base transport in the collecting duct.Review. Semin Nephrol, 1993. 13(2): p. 246-55.

4. Ulate, G., R. Fernandez, and G. Malnic, Effect of bafilomycin on proximal bicarbonate absorption in the rat. Braz J Med Biol Res, 1993. 26(7): p. 773-7.

5. Preisig, P.A. and R.J. Alpern, Pathways for apical and basolateral membrane NH3 and NH4⁺ movement in rat proximal tubule. Am J Physiol, 1990. 259(4 Pt 2): p. F587-93.

6. Zimolo, Z., M.H. Montrose, and H. Murer, H+ extrusion by an apical vacuolar-type H(+)-ATPase in rat renal proximal tubules. J Membr Biol, 1992. 126(1): p. 19-26.

7. Hays, S.R. and R.J. Alpern, Inhibition of Na(⁺)-independent H⁺ pump by Na(+)-induced changes in cell Ca2⁺. J Gen Physiol, 1991. 98(4): p. 791-813.

8. Moe, O.W., P.A. Preisig, and R.J. Alpern, Cellular model of proximal tubule NaCl and NaHCO3 absorption.Review. Kidney Int, 1990. 38(4): p. 605-11.

9. Preisig, P.A., et al., Role of the Na⁺/H⁺ antiporter in rat proximal tubule bicarbonate absorption. J Clin Invest, 1987. 80(4): p. 970-8.

10. Brown, D., S. Gluck, and J. Hartwig, Structure of the novel membrane-coating material in proton-secreting epithelial cells and identification as an H⁺-ATPase. J Cell Biol, 1987. 105(4): p. 1637-48.

11. Brown, D., S. Hirsch, and S. Gluck, An H⁺-ATPase in opposite plasma membrane domains in kidney epithelial cell subpopulations. Nature, 1988. 331(6157): p. 622-4.

12. Brown, D., S. Hirsch, and S. Gluck, Localization of a proton-pumping ATPase in rat kidney. J Clin Invest, 1988. 82(6): p. 2114-26.

13. Steinmetz, P.R., Cellular organization of urinary acidification.Review. Am J Physiol, 1986. 251(2 Pt 2): p. F173-87.

14. Alpern, R.J., Cell mechanisms of proximal tubule acidification.Review. Physiol Rev, 1990. 70(1): p. 79-114.

15. Bank, N., H.S. Aynedjian, and B.F. Mutz, Proximal bicarbonate absorption independent of Na$^+$-H$^+$ exchange: effect of bicarbonate load. Am J Physiol, 1989. 256(4 Pt 2): p. F577-82.

16. Good, D.W., M.A. Knepper, and M.B. Burg, Ammonia and bicarbonate transport by thick ascending limb of rat kidney. Am J Physiol, 1984. 247(1 Pt 2): p. F35-44.

17. Capasso, G., R. Unwin, and G. Giebisch, Role of the loop of Henle in urinary acidification. Kidney Int Suppl, 1991. 33(5): p. S33-5.

18. Capasso, G., et al., Bicarbonate transport along the loop of Henle. I. Microperfusion studies of load and inhibitor sensitivity. J Clin Invest, 1991. 88(2): p. 430-7.

19. Froissart, M., et al., Plasma membrane Na($^+$)-H$^+$ antiporter and H($^+$)-ATPase in the medullary thick ascending limb of rat kidney. Am J Physiol, 1992. 262(4 Pt 1): p. C963-70.

20. Good, D.W., The thick ascending limb as a site of renal bicarbonate reabsorption.Review. Semin Nephrol, 1993. 13(2): p. 225-35.

21. Wang, T., et al., Renal bicarbonate reabsorption in the rat. IV. Bicarbonate transport mechanisms in the early and late distal tubule. J Clin Invest, 1993. 91(6): p. 2776-84.

22. Weiner, I.D., C.S. Wingo, and L.L. Hamm, Regulation of intracellular pH in two cell populations of inner stripe of rabbit outer medullary collecting duct. Am J Physiol, 1993. 265(3 Pt 2): p. F406-15.

23. Schwartz, G.J., L.M. Satlin, and J.E. Bergmann, Fluorescent characterization of collecting duct cells: a second H$^+$-secreting type. Am J Physiol, 1988. 255(5 Pt 2): p. F1003-14.

24. Alper, S.L., et al., Subtypes of intercalated cells in rat kidney collecting duct defined by antibodies against erythroid band 3 and renal vacuolar H+-ATPase. Proc Natl Acad Sci U S A, 1989. 86(14): p. 5429-33.

25. Emmons, C. and I. Kurtz, Functional characterization of three intercalated cell subtypes in the rabbit outer cortical collecting duct. J Clin Invest, 1994. 93(1): p. 417-23.

26. Stone, D.K., et al., Anion dependence of rabbit medullary collecting duct acidification. J Clin Invest, 1983. 71(5): p. 1505-8.

27. Jacobson, H.R., Medullary collecting duct acidification. Effects of potassium, HCO3 concentration, and pCO2. J Clin Invest, 1984. 74(6): p. 2107-14.

28. Wingo, C.S. and B.D. Cain, The renal H-K-ATPase: physiological significance and role in potassium homeostasis.Review. Annu Rev Physiol, 1993. 55: p. 323-47.

29. Schuster, V.L., Function and regulation of collecting duct intercalated cells.Review. Annu Rev Physiol, 1993. 55: p. 267-88.

30. Lombard, W.E., J.P. Kokko, and H.R. Jacobson, Bicarbonate transport in cortical and outer medullary collecting tubules. Am J Physiol, 1983. 244(3): p. F289-96.

31. Steinmetz, P.R. and O.S. Andersen, Electrogenic proton transport in epithelial membranes.Review. J Membr Biol, 1982. 65(3): p. 155-74.

32. Steinmetz, P.R., Cellular mechanisms of urinary acidification.Review. Physiol Rev, 1974. 54(4): p. 890-956.

33. Gluck, S., S. Kelly, and Q. Al-Awqati, The proton translocating ATPase responsible for urinary acidification. J Biol Chem, 1982. 257(16): p. 9230-3.

34. Youmans, S.J., H.J. Worman, and W.A. Brodsky, ATPase activity and ATP-dependent proton translocation in plasma membrane vesicles of turtle bladder epithelial cells. Biochim Biophys Acta, 1983. 730(1): p. 173-7.

35. Dixon, T.E. and Q. Al-Awqati, H*/ATP stoichiometry of proton pump of turtle urinary bladder. J Biol Chem, 1980. 255(8): p. 3237-9.

36. Andersen, O.S., J.E. Silveira, and P.R. Steinmetz, Intrinsic characteristics of the proton pump in the luminal membrane of a tight urinary epithelium. The relation between transport rate and delta mu H. J Gen Physiol, 1985. 86(2): p. 215-34.

37. Steinmetz, P.R., et al., Coupling between H* transport and anaerobic glycolysis in turtle urinary bladder: effect of inhibitors of H* ATPase. J Membr Biol, 1981. 59(1): p. 27-34.

38. Gluck, S. and Q. Al-Awqati, An electrogenic proton-translocating adenosine triphosphatase from bovine kidney medulla. J Clin Invest, 1984. 73(6): p. 1704-10.

39. Gluck, S. and J. Caldwell, Immunoaffinity purification and characterization of vacuolar H*-ATPase from bovine kidney. J Biol Chem, 1987. 262(32): p. 15780-9.

40. Gluck, S. and J. Caldwell, Proton-translocating ATPase from bovine kidney medulla: partial purification and reconstitution. Am J Physiol, 1988. 254(1 Pt 2): p. F71-9.

41. Kaunitz, J.D., R.D. Gunther, and G. Sachs, Characterization of an electrogenic ATP and chloride-dependent proton translocating pump from rat renal medulla. J Biol Chem, 1985. 260(21): p. 11567-73.

42. Diaz-Diaz, F.D., et al., ATP-dependent proton transport in human renal medulla. Am J Physiol, 1986. 251(2 Pt 2): p. F297-302.

43. Sabolic, I., W. Haase, and G. Burckhardt, ATP-dependent H* pump in membrane vesicles from rat kidney cortex. Am J Physiol, 1985. 248(6 Pt 2): p. F835-44.

44. Sabolic, I. and G. Burckhardt, Characteristics of the proton pump in rat renal cortical endocytotic vesicles. Am J Physiol, 1986. 250(5 Pt 2): p. F817-26.

45. Sabolic, I. and G. Burckhardt, Proton ATPase in rat renal cortical endocytotic vesicles. Biochim Biophys Acta, 1988. 937(2): p. 398-410.

46. Hilden, S.A., C.A. Johns, and N.E. Madias, Cl(-)-dependent ATP-driven H* transport in rabbit renal cortical endosomes. Am J Physiol, 1988. 255(5 Pt 2): p. F885-97.

47. Kinne-Saffran, E., R. Beauwens, and R. Kinne, An ATP-driven proton pump in brush-border membranes from rat renal cortex. J Membr Biol, 1982. 64(1-2): p. 67-76.

48. Kinne-Saffran, E. and R. Kinne, Proton pump activity and Mg-ATPase activity in rat kidney cortex brushborder membranes: effect of 'proton ATPase' inhibitors. Pflugers Arch, 1986. 407(2): p. S180-5.

49. Turrini, F., et al., Relation of ATPases in rat renal brush-border membranes to ATP-driven H* secretion. J Membr Biol, 1989. 107(1): p. 1-12.

50. Simon, B.J. and G. Burckhardt, Characterization of inside-out oriented H(*)-ATPases in cholate-pretreated renal brush-border membrane vesicles.

J Membr Biol, 1990. 117(2): p. 141-51.

51. Jehmlich, K., et al., Biochemical aspects of H(+)-ATPase in renal proximal tubules: inhibition by N,N'-dicyclohexylcarbodiimide, N-ethylmaleimide, and bafilomycin. Kidney Int Suppl, 1991. 33(70): p. S64-70.

52. Wang, Z.Q. and S. Gluck, Isolation and properties of bovine kidney brush border vacuolar H(+)-ATPase. A proton pump with enzymatic and structural differences from kidney microsomal H(+)-ATPase. J Biol Chem, 1990. 265(35): p. 21957-65.

53. Yurko, M.A. and S. Gluck, Production and characterization of a monoclonal antibody to vacuolar H⁺-ATPase of renal epithelia. J Biol Chem, 1987. 262(32): p. 15770-9.

54. Gluck, S.L., Z.-Q. Wang, and K. Zhang, Properties of the lysosomal vacuolar H⁺-ATPase isolated from bovine kidney cortex. In preparation, 1994.

55. Forgac, M., Structure and function of vacuolar class of ATP-driven proton pumps.Review. Physiol Rev, 1989. 69(3): p. 765-96.

56. Nelson, N., Structural conservation and functional diversity of V-ATPases. Review. J Bioenerg Biomembr, 1992. 24(4): p. 407-14.

57. Gluck, S.L., The vacuolar H(+)-ATPases: versatile proton pumps participating in constitutive and specialized functions of eukaryotic cells.Review. Int Rev Cytol, 1993: p. 105-37.

58. Schwartz, G.J., J. Barasch, and Q. Al-Awqati, Plasticity of functional epithelial polarity. Nature, 1985. 318(6044): p. 368-71.

59. Satlin, L.M. and G.J. Schwartz, Postnatal maturation of rabbit renal collecting duct: intercalated cell function. Am J Physiol, 1987. 253(4 Pt 2): p. F622-35.

60. Breyer, M.D. and H.R. Jacobson, Regulation of rabbit medullary collecting duct cell pH by basolateral Na⁺/H⁺ and Cl-/base exchange. J Clin Invest, 1989. 84(3): p. 996-1004.

61. Hays, S.R. and R.J. Alpern, Apical and basolateral membrane H⁺ extrusion mechanisms in inner stripe of rabbit outer medullary collecting duct. Am J Physiol, 1990. 259(4 Pt 2): p. F628-35.

62. Hays, S.R. and R.J. Alpern, Basolateral membrane Na(⁺)-independent Cl⁻/HCO3- exchange in the inner stripe of the rabbit outer medullary collecting tubule. J Gen Physiol, 1990. 95(2): p. 347-67.

63. Kuwahara, M., S. Sasaki, and F. Marumo, Cell pH regulation in rabbit outer medullary collecting duct cells: mechanisms of HCO3(-)-independent processes. Am J Physiol, 1990. 259(6 Pt 2): p. F902-9.

64. Kuwahara, M., S. Sasaki, and F. Marumo, Mineralocorticoids and acidosis regulate H+/HCO3- transport of intercalated cells. J Clin Invest, 1992. 89(5): p. 1388-94.

65. Furuya, H., H.R. Jacobson, and M.D. Breyer, Evidence for basolateral membrane Ca2⁺/H⁺ exchange in outer medullary collecting duct. Am J Physiol, 1993. 264(1 Pt 2): p. F88-93.

66. Nelson, R.D., et al., Selectively amplified expression of an isoform of the vacuolar H(+)-ATPase 56-kilodalton subunit in renal intercalated cells. Proc Natl Acad Sci U S A, 1992. 89(8): p. 3541-5.

67. van Meer, G. and K. Simons, Lipid polarity and sorting in epithelial cells.Review. J Cell Biochem, 1988. 36(1): p. 51-8.

68. Simons, K. and A. Wandinger-Ness, Polarized sorting in epithelia.Review. Cell, 1990. 62(2): p. 207-10.

69. Simons, K., et al., Biogenesis of cell-surface polarity in epithelial cells and neurons.Review. Cold Spring Harb Symp Quant Biol, 1992. 57: p. 611-9.

70. Hemken, P., et al., Immunologic evidence that vacuolar H⁺ ATPases with heterogeneous forms of Mr = 31,000 subunit have different membrane distributions in mammalian kidney. J Biol Chem, 1992. 267(14): p. 9948-57.

71. Hirsch, S., et al., Isolation and sequence of a cDNA clone encoding the 31-kDa subunit of bovine kidney vacuolar H⁺-ATPase. Proc Natl Acad Sci U S A, 1988. 85(9): p. 3004-8.

72. Sudhof, T.C., et al., Human endomembrane H⁺ pump strongly resembles the ATP-synthetase of Archaebacteria. Proc Natl Acad Sci U S A, 1989. 86(16): p. 6067-71.

73. Marushack, M.M., et al., cDNA sequence and tissue expression of bovine vacuolar H(+)-ATPase M(r) 70,000 subunit. Am J Physiol, 1992. 263(1 Pt 2): p. F171-4.

74. Sander, I., et al., Sequence analysis of the catalytic subunit of H(+)-ATPase from porcine renal brush-border membranes. Biochim Biophys Acta, 1992. 1112(1): p. 129-41.

75. Puopolo, K., et al., A single gene encodes the catalytic "A" subunit of the bovine vacuolar H(+)-ATPase. J Biol Chem, 1991. 266(36): p. 24564-72.

76. Puopolo, K., et al., Differential expression of the "B" subunit of the vacuolar H(+)-ATPase in bovine tissues. J Biol Chem, 1992. 267(6): p. 3696-706.

77. Chan, Y.L. and G. Giebisch, Relationship between sodium and bicarbonate transport in the rat proximal convoluted tubule. Am J Physiol, 1981. 240(3): p. F222-30.

78. Kurtz, I., Apical Na⁺/H⁺ antiporter and glycolysis-dependent H⁺-ATPase regulate intracellular pH in the rabbit S3 proximal tubule. J Clin Invest, 1987. 80(4): p. 928-35.

79. Preisig, P.A., Luminal flow rate regulates proximal tubule H-HCO3 transporters. Am J Physiol, 1992. 262(1 Pt 2): p. F47-54.

80. Nakhoul, N.L., L.K. Chen, and W.F. Boron, Effect of basolateral CO2/HCO3- on intracellular pH regulation in the rabbit S3 proximal tubule. J Gen Physiol, 1993. 102(6): p. 1171-205.

81. Roos, A. and W.F. Boron, Intracellular pH.Review. Physiol Rev, 1981. 61(2): p. 296-434.

82. Stoner, L.C., M.B. Burg, and J. Orloff, Ion transport in cortical collecting tubule; effect of amiloride. Am J Physiol, 1974. 227(2): p. 453-9.

83. McKinney, T.D. and M.B. Burg, Bicarbonate absorption by rabbit cortical collecting tubules in vitro. Am J Physiol, 1978. 234(2): p. F141-5.

84. Hanley, M.J., et al., Electrophysiologic study of the cortical collecting tubule of the rabbit. Kidney Int, 1980. 17(1): p. 74-81.

85. Koeppen, B.M. and S.I. Helman, Acidification of luminal fluid by the rabbit cortical collecting tubule perfused in vitro. Am J Physiol, 1982. 242(5): p. F521-31.

86. Laski, M.E. and N.A. Kurtzman, Characterization of acidification in the cortical and medullary collecting tubule of the rabbit. J Clin Invest, 1983. 72(6): p. 2050-9.

87. Atkins, J.L. and M.B. Burg, Bicarbonate transport by isolated perfused rat collecting ducts. Am J Physiol, 1985. 249(4 Pt 2): p. F485-9.

88. Hayashi, M., et al., Effects of in vivo and in vitro alkali treatment on intracellular pH regulation of OMCDis cells. Am J Physiol, 1993. 265(5 Pt 2): p. F729-35.

89. Weiner, I.D. and L.L. Hamm, Use of fluorescent dye BCECF to measure intracellular pH in cortical collecting tubule. Am J Physiol, 1989. 256(5 Pt 2): p. F957-64.

90. Furuya, H., M.D. Breyer, and H.R. Jacobson, Functional characterization of alpha- and beta-intercalated cell types in rabbit cortical collecting duct. Am J Physiol, 1991. 261(3 Pt 2): p. F377-85.

91. Muto, S., et al., Electrophysiological identification of alpha- and beta-intercalated cells and their distribution along the rabbit distal nephron segments. J Clin Invest, 1990. 86: p. 1829-1839.

92. McKinney, T.D. and K.K. Davidson, Bicarbonate transport in collecting tubules from outer stripe of outer medulla of rabbit kidneys. Am J Physiol, 1987. 253(5 Pt 2): p. F816-22.

93. Verlander, J.W., K.M. Madsen, and C.C. Tisher, Structural and functional features of proton and bicarbonate transport in the rat collecting duct.Review. Semin Nephrol, 1991. 11(4): p. 465-77.

94. Koeppen, B.M., Conductive properties of the rabbit outer medullary collecting duct: inner stripe. Am J Physiol, 1985. 248(4 Pt 2): p. F500-6.

95. Ridderstrale, Y., et al., Morphological heterogeneity of the rabbit collecting duct. Kidney Int, 1988. 34(5): p. 655-70.

96. Schuster, V.L., S.M. Bonsib, and M.L. Jennings, Two types of collecting duct mitochondria-rich (intercalated) cells: lectin and band 3 cytochemistry. Am J Physiol, 1986. 251(3 Pt 1): p. C347-55.

97. Burnatowska-Hledin, M.A. and W.S. Spielman, Immunodissection of mitochondria-rich cells from rabbit outer medullary collecting tubule. Am J Physiol, 1988. 254(6 Pt 2): p. F907-11.

98. Schuster, V.L., et al., Colocalization of H(+)-ATPase and band 3 anion exchanger in rabbit collecting duct intercalated cells. Am J Physiol, 1991. 260(4 Pt 2): p. F506-17.

99. Bastani, B., et al., Expression and distribution of renal vacuolar proton-translocating adenosine triphosphatase in response to chronic acid and alkali loads in the rat. J Clin Invest, 1991. 88(1): p. 126-36.

100. Weiner, I.D. and L.L. Hamm, Regulation of intracellular pH in the rabbit cortical collecting tubule. J Clin Invest, 1990. 85(1): p. 274-81.

101. Sabolic, I., et al., Apical endosomes isolated from kidney collecting duct principal cells lack subunits of the proton pumping ATPase. J Cell Biol, 1992. 119(1): p. 111-22.

102. Rodman, J.S., P.D. Stahl, and S. Gluck, Distribution and structure of the vacuolar H' ATPase in endosomes and lysosomes from LLC-PK1 cells. Exp Cell Res, 1991. 192(2): p. 445-52.

103. Schmidt, W., H. Winkler, and H. Plattner, Adrenal chromaffin granules: evidence for an ultrastructural equivalent of the proton-pumping ATPase. Eur J Cell Biol, 1982. 27(1): p. 96-104.

104. Wall, S.M. and M.A. Knepper, Acid-base transport in the inner medullary collecting duct.Review. Semin Nephrol, 1990. 10(2): p. 148-58.

105. Kleinman, J.G., P. Tipnis, and R. Pscheidt, H(+)-K(+)-ATPase of rat inner medullary collecting duct in primary culture. Am J Physiol, 1993. 265(5 Pt 2): p. F698-704.

106. Nelson, R.D., et al., Cell-specific amplification of vacuolar H⁺-ATPase B subunit expression in mammalian kidney. In preparation, 1994.

107. Zhang, K., Z.Q. Wang, and S. Gluck, Identification and partial purification of a cytosolic activator of vacuolar H(+)-ATPases from mammalian kidney. J Biol Chem, 1992. 267(14): p. 9701-5.

108. Zhang, K., Z.Q. Wang, and S. Gluck, A cytosolic inhibitor of vacuolar H(+)-ATPases from mammalian kidney. J Biol Chem, 1992. 267(21): p. 14539-42.

109. Gurich, R.W. and T. DuBose Jr., Heterogeneity of cAMP effect on endosomal proton transport. Am J Physiol, 1989. 257(5 Pt 2): p. F777-84.

110. Mulberg, A.E., B.M. Tulk, and M. Forgac, Modulation of coated vesicle chloride channel activity and acidification by reversible protein kinase A-dependent phosphorylation. J Biol Chem, 1991. 266(31): p. 20590-3.

111. Gurich, R.W., J. Codina, and T. DuBose Jr., A potential role for guanine nucleotide-binding protein in the regulation of endosomal proton transport. J Clin Invest, 1991. 87(5): p. 1547-52.

112. Schuster, V.L., Cortical collecting duct bicarbonate secretion.Review. Kidney Int Suppl, 1991. 33(50): p. S47-50.

113. Schuster, V.L., Organization of collecting duct intercalated cells.Review. Kidney Int, 1990. 38(4): p. 668-72.

114. Schuster, V.L., Bicarbonate reabsorption and secretion in the cortical and outer medullary collecting tubule.Review. Semin Nephrol, 1990. 10(2): p. 139-47.

115. Schuster, V.L., Physiology and cell biology update: control mechanisms for bicarbonate secretion.Review. Am J Kidney Dis, 1989. 13(4): p. 348-52.

116. Stetson, D.L., et al., A double-membrane model for urinary bicarbonate secretion. Am J Physiol, 1985. 249(4 Pt 2): p. F546-52.

117. Al-Awqati, Q., Proton-translocating ATPases Review. Annu Rev Cell Biol, 1986. 2: p. 179-99.

118. Cohen, L.H., A. Mueller, and P.R. Steinmetz, Inhibition of the bicarbonate exit step in urinary acidification by a disulfonic stilbene. J Clin Invest, 1978. 61(4): p. 981-6.

119. Husted, R.F., L.H. Cohen, and P.R. Steinmetz, Pathways for bicarbonate transfer across the serosal membrane of turtle urinary bladder: studies with a disulfonic stilbene. J Membr Biol, 1979. 47(1): p. 27-37.

120. Fischer, J.L., R.F. Husted, and P.R. Steinmetz, Chloride dependence of the HCO3 exit step in urinary acidification by the turtle bladder. Am J Physiol, 1983. 245(5 Pt 1): p. F564-8.

121. Drenckhahn, D., et al., Band 3 is the basolateral anion exchanger of dark epithelial cells of turtle urinary bladder. Am J Physiol, 1987. 252(5 Pt 1): p. C570-4.

122. Oliver, J.A., S. Himmelstein, and P.R. Steinmetz, Energy dependence of urinary bicarbonate secretion in turtle bladder. J Clin Invest, 1975. 55(5): p. 1003-8.

123. Fritsche, C., et al., HCO3- secretion in mitochondria-rich cells is linked to an H⁺-ATPase. Am J Physiol, 1989. 256(5 Pt 2): p. F869-74.

124. Leslie, B.R., J.H. Schwartz, and P.R. Steinmetz, Coupling between Cl- absorption and HCO3- secretion in turtle urinary bladder. Am J Physiol, 1973. 225(3): p. 610-7.

125. Kohn, O.F., P.P. Mitchell, and P.R. Steinmetz, Characteristics of apical Cl-HCO3 exchanger of bicarbonate-secreting cells in turtle bladder. Am J Physiol, 1990. 258(1 Pt 2): p. F9-14.

126. Beauwens, R. and Q. Al-Awqati, Active H⁺ transport in the turtle urinary bladder. Coupling of transport to glucose oxidation. J Gen Physiol, 1976. 68(4): p. 421-39.

127. Al-awqati, Q., A. Mueller, and P.R. Steinmetz, Transport of H⁺ against electrochemical gradients in turtle urinary bladder. Am J Physiol, 1977. 233(6): p. F502-8.

128. Steinmetz, P.R., R.S. Omachi, and H.S. Frazier, Independence of hydrogen ion secretion and transport of other electrolytes in turtle bladder. J Clin Invest, 1967. 46(10): p. 1541-8.

129. Ludens, J.H. and D.D. Fanestil, Acidification of urine by the isolated urinary bladder of the toad. Am J Physiol, 1972. 223(6): p. 1338-44.

130. Husted, R.F. and P.R. Steinmetz, The effects of amiloride and ouabain on urinary acidification by turtle bladder. J Pharmacol Exp Ther, 1979. 210(2): p. 264-8.

131. Steinmetz, P.R., Characteristics of hydrogen ion transport in urinary bladder of water turtle. J Clin Invest, 1967. 46(10): p. 1531-40.

132. Ziegler, T.W., D.D. Fanestil, and J.H. Ludens, Influence of transepithelial potential difference on acidification in the toad urinary bladder. Kidney Int, 1976. 10(4): p. 279-86.

133. Schultz, S.G., Homocellular regulatory mechanisms in sodium-transporting epithelia: avoidance of extinction by "flush-through".Review. Am J Physiol, 1981. 241(6): p. F579-90.

134. Steinmetz, P.R., Acid-base relations in epithelium of turtle bladder: site of active step in acidification and role of metabolic CO_2. J Clin Invest, 1969. 48(7): p. 1258-65.

135. Schwartz, J.H. and P.R. Steinmetz, CO_2 requirements for H⁺ secretion by the isolated turtle bladder. Am J Physiol, 1971. 220(6): p. 2051-7.

136. Gluck, S., C. Cannon, and Q. Al-Awqati, Exocytosis regulates urinary acidification in turtle bladder by rapid insertion of H⁺ pumps into the luminal membrane. Proc Natl Acad Sci U S A, 1982. 79(14): p. 4327-31.

137. Stetson, D.L. and P.R. Steinmetz, Role of membrane fusion in CO2 stimulation of proton secretion by turtle bladder. Am J Physiol, 1983. 245(1): p. C113-20.

138. Cohen, L.H. and P.R. Steinmetz, Control of active proton transport in turtle urinary bladder by cell pH. J Gen Physiol, 1980. 76(3): p. 381-93.

139. Waddell, W.J. and T.C. Butler, Calculation of intracellular pH from the distribution of 5,5'-dimethyl-2,4-oxazolidinedione (DMO): application to skeletal muscle of the dog. J Clin Invest, 1959. 38: p. 720-729.

140. Husted, R.F., et al., Surface characteristics of carbonic-anhydrase-rich cells in turtle urinary bladder. Kidney Int, 1981. 19(4): p. 491-502.

141. Stetson, D.L. and P.R. Steinmetz, Correlation between apical intramembrane particles and H⁺ secretion rates during CO2 stimulation in turtle bladder. Pflugers Arch, 1986. 407(2): p. S80-4.

142. Clausen, C. and T.E. Dixon, Membrane electrical parameters in turtle bladder measured using impedance-analysis techniques. J Membr Biol, 1986. 92(1): p. 9-19.

143. Dixon, T.E., et al., Proton transport and membrane shuttling in turtle

bladder epithelium. J Membr Biol, 1986. 94(3): p. 233-43.

144. Dixon, T.E., C. Clausen, and D. Coachman, Constitutive and transport-related endocytotic pathways in turtle bladder epithelium. J Membr Biol, 1988. 102(1): p. 49-58.

145. van Adelsberg, J. and Q. Al-Awqati, Regulation of cell pH by Ca·2-mediated exocytotic insertion of H⁺-ATPases. J Cell Biol, 1986. 102(5): p. 1638-45.

146. Cannon, C., et al., Carbon-dioxide-induced exocytotic insertion of H⁺ pumps in turtle-bladder luminal membrane: role of cell pH and calcium. Nature, 1985. 314(6010): p. 443-6.

147. Graber, M.L., et al., Fluorescence identifies an alkaline cell in turtle urinary bladder. Am J Physiol, 1986. 250(1 Pt 2): p. F159-68.

148. Graber, M., et al., Acetazolamide inhibits acidification by the turtle bladder independent of cell pH. Am J Physiol, 1989. 256(5 Pt 2): p. F923-31.

149. Arruda, J.A., Z. Talor, and C. Dytko, Effect of agents that alter cell calcium and microfilaments on CO2 stimulated H⁺ secretion in the turtle bladder. Arch Int Pharmacodyn Ther, 1988. 293: p. 273-83.

150. Arruda, J.A., G. Dytko, and Z. Talor, Stimulation of H⁺ secretion by CO2 in turtle bladder: role of intracellular pH, exocytosis, and calcium. Am J Physiol, 1990. 258(1 Pt 2): p. R222-31.

151. Arruda, J.A., Calcium inhibits urinary acidification: effect of the ionophore A23187 on the turtle bladder. Pflugers Arch, 1979. 381(2): p. 107-11.

152. Ehrenspeck, G., Effect of calcium ionophore A23187 on electrogenic acid-base transport in turtle bladder. Inhibition of acidification and stimulation of alkalinization. Biochim Biophys Acta, 1983. 732(1): p. 146-53.

153. Al-Awqati, Q., Effect of aldosterone on the coupling between H⁺ transport and glucose oxidation. J Clin Invest, 1977. 60(6): p. 1240-7.

154. Schwartz, J.H. and P.R. Steinmetz, Metabolic energy and PCO2 as determinants of H⁺ secretion by turtle urinary bladder. Am J Physiol, 1977. 233(2): p. F145-9.

155. Kelly, S., T.E. Dixon, and Q. Al-Awqati, Metabolic pathways coupled to H+ transport in turtle urinary bladder. J Membr Biol, 1980. 54(3): p. 237-43.

156. Choate, G.L., L. Lan, and T.E. Mansour, Heart 6-phosphofructo-1-kinase. Subcellular distribution and binding to myofibrils. J Biol Chem, 1985. 260(8): p. 4815-22.

157. Harrison, M.L., et al., Role of band 3 tyrosine phosphorylation in the regulation of erythrocyte glycolysis. J Biol Chem, 1991. 266(7): p. 4106-11.

158. Beitner, R., Control of glycolytic enzymes through binding to cell structures and by glucose-1,6-bisphosphate under different conditions. The role of Ca2⁺ and calmodulin.Review. Int J Biochem, 1993. 25(3): p. 297-305.

159. Low, P.S., P. Rathinavelu, and M.L. Harrison, Regulation of glycolysis via reversible enzyme binding to the membrane protein, band 3. J Biol Chem, 1993. 268(20): p. 14627-31.

160. Ludens, J.H. and D.D. Fanestil, Aldosterone stimulation of acidification of urine by isolated urinary bladder of the Colombian toad. Am J Physiol, 1974. 226(6): p. 1321-6.

161. Ludens, J.H., D.A. Vaughn, and D.D. Fanestil, Stimulation of urinary

acidification by aldosterone and inhibitors of RNA and protein synthesis. J Membr Biol, 1978. : p. 199-211.

162. Al-Awqati, Q., et al., Characteristics of stimulation of H' transport by aldosterone in turtle urinary bladder. J Clin Invest, 1976. 58(2): p. 351-8.

163. Ehrenspeck, G., Effect of 3-isobutyl-1-methylxanthine on HCO3- transport in turtle bladder. Evidence for electrogenic HCO3- secretion. Biochim Biophys Acta, 1982. 684(2): p. 219-27.

164. Satake, N., et al., Active electrogenic mechanisms for alkali and acid transport in turtle bladders. Am J Physiol, 1983. 244(3): p. C259-69.

165. Arruda, J.A. and S. Sabatini, Cholinergic inhibition of urinary acidification by the turtle bladder. Kidney Int, 1980. 17(5): p. 622-30.

166. Schneider, E.S., et al., Alkali secretion in the turtle bladder: up-regulation by the phospho-inositol cascade and inhibition by diphenylamine carboxylate (DPC). Prog Clin Biol Res, 1988. 258: p. 81-92.

167. Stetson, D.L. and P.R. Steinmetz, Alpha and beta types of carbonic anhydrase-rich cells in turtle bladder. Am J Physiol, 1985. 249(4 Pt 2): p. F553-65.

168. Rich, A., T.E. Dixon, and C. Clausen, Changes in membrane conductances and areas associated with bicarbonate secretion in turtle bladder. J Membr Biol, 1990. 113(3): p. 211-9.

169. Rich, A., T.E. Dixon, and C. Clausen, Electrogenic bicarbonate secretion in the turtle bladder: apical membrane conductance characteristics. J Membr Biol, 1991. 119(3): p. 241-52.

170. Alpern, R.J., M.G. Cogan, and F. Rector Jr., Effect of luminal bicarbonate concentration on proximal acidification in the rat. Am J Physiol, 1982. 243(1): p. F53-9.

171. Alpern, R.J. and M. Chambers, Cell pH in the rat proximal convoluted tubule. Regulation by luminal and peritubular pH and sodium concentration. J Clin Invest, 1986. 78(2): p. 502-10.

172. McKinney, T.D. and K.K. Davidson, Effects of respiratory acidosis on HCO3- transport by rabbit collecting tubules. Am J Physiol, 1988. 255(4 Pt 2): p. F656-65.

173. Drenckhahn, D., et al., Colocalization of band 3 with ankyrin and spectrin at the basal membrane of intercalated cells in the rat kidney. Science, 1985. 230(4731): p. 1287-9.

174. Verlander, J.W., et al., Immunocytochemical localization of band 3 protein in the rat collecting duct. Am J Physiol, 1988. 255(1 Pt 2): p. F115-25.

175. Hays, S.R., Mineralocorticoid modulation of apical and basolateral membrane H'/OH-/HCO3- transport processes in the rabbit inner stripe of outer medullary collecting duct. J Clin Invest, 1992. 90(1): p. 180-7.

176. Koeppen, B.M., Electrophysiological identification of principal and intercalated cells in the rabbit outer medullary collecting duct. Pflugers Arch, 1987. 409(1-2): p. 138-41.

177. Wingo, C.S. and F.E. Armitage, Potassium transport in the kidney: regulation and physiological relevance of H', K(')-ATPase.Review. Semin Nephrol, 1993. 13(2): p. 213-24.

178. Husted, R.F. and P.R. Steinmetz, Potassium absorptive pump at the luminal membrane of turtle urinary bladder. Am J Physiol, 1981. 241(3): p. F315-21.

179. Madsen, K.M. and C.C. Tisher, Cellular response to acute respiratory acidosis in rat medullary collecting duct. Am J Physiol, 1983. 245(6): p. F670-9.

180. Laski, M.E. and N.A. Kurtzman, Collecting tubule adaptation to respiratory acidosis induced in vivo. Am J Physiol, 1990. 258(1 Pt 2): p. F15-20.

181. Schwartz, G.J. and Q. Al-Awqati, Carbon dioxide causes exocytosis of vesicles containing H⁺ pumps in isolated perfused proximal and collecting tubules. J Clin Invest, 1985. 75(5): p. 1638-44.

182. Kuwahara, M., S. Sasaki, and F. Marumo, Cl-HCO3 exchange and Na-HCO3 symport in rabbit outer medullary collecting duct cells. Am J Physiol, 1991. 260(5 Pt 2): p. F635-42.

183. Stone, D.K., et al., Mineralocorticoid modulation of rabbit medullary collecting duct acidification. A sodium-independent effect. J Clin Invest, 1983. 72(1): p. 77-83.

184. Weiner, I.D. and L.L. Hamm, Regulation of Cl⁻/HCO3⁻ exchange in the rabbit cortical collecting tubule. J Clin Invest, 1991. 87(5): p. 1553-8.

185. Breyer, M.D., J.P. Kokko, and H.R. Jacobson, Regulation of net bicarbonate transport in rabbit cortical collecting tubule by peritubular pH, carbon dioxide tension, and bicarbonate concentration. J Clin Invest, 1986. 77(5): p. 1650-60.

186. Star, R.A., M.B. Burg, and M.A. Knepper, Bicarbonate secretion and chloride absorption by rabbit cortical collecting ducts. Role of chloride/bicarbonate exchange. J Clin Invest, 1985. 76(3): p. 1123-30.

187. Verlander, J.W., K.M. Madsen, and C.C. Tisher, Effect of acute respiratory acidosis on two populations of intercalated cells in rat cortical collecting duct. Am J Physiol, 1987. 253(6 Pt 2): p. F1142-56.

188. Yamaji, Y., et al., Chronic DOC treatment enhances Na(⁺)-H⁺ exchanger activity of beta-intercalated cells in rabbit CCD. Am J Physiol, 1992. 262(5 Pt 2): p. F712-7.

189. Brown, D., I. Sabolic, and S. Gluck, Colchicine-induced redistribution of proton pumps in kidney epithelial cells. Kidney Int Suppl, 1991. 33(83): p. S79-83.

190. Noel, J., et al., Metabolic cost of bafilomycin-sensitive H⁺ pump in intact dog, rabbit, and hamster proximal tubules. Am J Physiol, 1993. 264(4 Pt 2): p. F655-61.

191. Dickman, K.G. and L.J. Mandel, Relationship between HCO3- transport and oxidative metabolism in rabbit proximal tubule. Am J Physiol, 1992. 263(2 Pt 2): p. F342-51.

192. Bastani, B., M. Kalkbrenner, and S. Gluck, Chronic DOCA administration increases polarization of H⁺-ATPase in medullary intercalated cells. J Am Soc Nephrol, 1991. 2: p. 694 (Abstr.).

193. Fejes-Toth, G. and A. Naray-Fejes-Toth, Isolated principal and intercalated cells: hormone responsiveness and Na⁺-K⁺-ATPase activity. Am J Physiol, 1989. 256(4 Pt 2): p. F742-50.

194. Koseki, C., et al., Isolation by monoclonal antibody of intercalated cells of rabbit kidney. Kidney Int, 1988. 33(2): p. 543-54.

195. Star, R.A., et al., Calcium and cyclic adenosine monophosphate as second messengers for vasopressin in the rat inner medullary collecting duct. J Clin Invest, 1988. 81(6): p. 1879-88.

196. Schuster, V.L., Cyclic adenosine monophosphate-stimulated bicarbonate

secretion in rabbit cortical collecting tubules. J Clin Invest, 1985. 75(6): p. 2056-64.

197. Schuster, V.L., Cyclic adenosine monophosphate-stimulated anion transport in rabbit cortical collecting duct. Kinetics, stoichiometry, and conductive pathways. J Clin Invest, 1986. 78(6): p. 1621-30.

198. Hayashi, M., et al., Effect of isoproterenol on intracellular pH of the intercalated cells in the rabbit cortical collecting ducts. J Clin Invest, 1991. 87(4): p. 1153-7.

199. Hays, S., J.P. Kokko, and H.R. Jacobson, Hormonal regulation of proton secretion in rabbit medullary collecting duct. J Clin Invest, 1986. 78(5): p. 1279-86.

200. Tomita, K., et al., Effects of vasopressin and bradykinin on anion transport by the rat cortical collecting duct. Evidence for an electroneutral sodium chloride transport pathway. J Clin Invest, 1986. 77(1): p. 136-41.

201. Bichara, M., et al., Effects of antidiuretic hormone on urinary acidification and on tubular handling of bicarbonate in the rat. J Clin Invest, 1987. 80(3): p. 621-30.

202. Delahousse, M., et al., Glucagon inhibits urinary acidification in the rat. Am J Physiol, 1988. 254(5 Pt 2): p. F762-9.

203. Mercier, O., et al., Effects of glucagon on $H(+)$-HCO_3- transport in Henle's loop, distal tubule, and collecting ducts in the rat. Am J Physiol, 1989. 257(6 Pt 2): p. F1003-14.

204. Siga, E., et al., Effects of calcitonin on function of intercalated cells of rat cortical collecting duct. Am J Physiol, 1993. 264(2 Pt 2): p. F221-7.

205. Hays, S.R., M. Baum, and J.P. Kokko, Effects of protein kinase C activation on sodium, potassium, chloride, and total CO_2 transport in the rabbit cortical collecting tubule. J Clin Invest, 1987. 80(6): p. 1561-70.

206. Bertorello, A.M. and A.I. Katz, Short-term regulation of renal Na-K-ATPase activity: physiological relevance and cellular mechanisms.Review. Am J Physiol, 1993. 265(6 Pt 2): p. F743-55.

207. Garg, L.C. and N. Narang, Stimulation of an N-ethylmaleimide-sensitive ATPase in the collecting duct segments of the rat nephron by metabolic acidosis. Can J Physiol Pharmacol, 1985. 63(10): p. 1291-6.

208. Khadouri, C., et al., Effect of adrenalectomy on NEM-sensitive ATPase along rat nephron and on urinary acidification. Am J Physiol, 1987. 253(3 Pt 2): p. F495-9.

209. Garg, L.C. and N. Narang, Effects of potassium bicarbonate on distal nephron Na-K-ATPase in adrenalectomized rabbits. Pflugers Arch, 1987. 409(1-2): p. 126-31.

210. Garg, L.C. and N. Narang, Effects of aldosterone on NEM-sensitive ATPase in rabbit nephron segments. Kidney Int, 1988. 34(1): p. 13-7.

211. Khadouri, C., et al., Short-term effect of aldosterone on NEM-sensitive ATPase in rat collecting tubule. Am J Physiol, 1989. 257(2 Pt 2): p. F177-81.

212. Garg, L.C. and N. Narang, Decrease in N-ethylmaleimide-sensitive AT-Pase activity in collecting duct by metabolic alkalosis. Can J Physiol Pharmacol, 1990. 68(8): p. 1119-23.

213. Garg, L.C. and N. Narang, Effects of low-potassium diet on N-ethylmaleimide-sensitive ATPase in the distal nephron segments. Renal Physiol Biochem, 1990. 13(3): p. 129-36.

214. Sabatini, S., M.E. Laski, and N.A. Kurtzman, NEM-sensitive ATPase activity in rat nephron: effect of metabolic acidosis and alkalosis. Am J Physiol, 1990. 258(2 Pt 2): p. F297-304.

215. Khadouri, C., et al., Characterization and control of proton-ATPase along the nephron. Kidney Int Suppl, 1991. 33(8): p. S71-8.

216. Sabatini, S., et al., Characterization of the N-ethylmaleimide-sensitive ATPase in rat cortical and medullary collecting tubule. Miner Electrolyte Metab, 1991. 17(5): p. 324-30.

217. Garg, L.C. and N. Narang, Changes in H-ATPase activity in the distal nephron segments of the rat during metabolic acidosis and alkalosis. Contrib Nephrol, 1991. 92: p. 39-45.

218. Khadouri, C., et al., Effect of metabolic acidosis and alkalosis on NEM-sensitive ATPase in rat nephron segments. Am J Physiol, 1992. 262(4 Pt 2): p. F583-90.

219. Eiam-Ong, S., et al., The biochemical basis of hypokalemic metabolic alkalosis. Trans Assoc Am Physicians, 1992. 105: p. 157-64.

220. Eiam-Ong, S., et al., H-K-ATPase in distal renal tubular acidosis: urinary tract obstruction, lithium, and amiloride. Am J Physiol, 1993. 265(6 Pt 2): p. F875-80.

221. Eiam-Ong, S., N.A. Kurtzman, and S. Sabatini, Regulation of collecting tubule adenosine triphosphatases by aldosterone and potassium. J Clin Invest, 1993. 91(6): p. 2385-92.

222. Chambrey, R., M. Paillard, and R.A. Podevin, Enzymatic and functional evidence for adaptation of the vacuolar H(+)-ATPase in proximal tubule apical membranes from rats with chronic metabolic acidosis. J Biol Chem, 1994. 269(5): p. 3243-50.

223. Bowman, E.J., A. Siebers, and K. Altendorf, Bafilomycins:A class of inhibitors of membrane ATPases from microorganisms, animal cells, and plant cells. ProcNatl AcadSci, 1988. 85: p. 7972-7976.

224. Verlander, J.W., et al., Immunocytochemical localization of intracellular acidic compartments: rat proximal nephron. Am J Physiol, 1989. 257(3 Pt 2): p. F454-62.

225. Tisher, C.C., K.M. Madsen, and J.W. Verlander, Structural adaptation of the collecting duct to acid-base disturbances.Review. Contrib Nephrol, 1991. 95: p. 168-77.

226. Madsen, K.M., et al., Morphological adaptation of the collecting duct to acid-base disturbances. Kidney Int Suppl, 1991. 33(63): p. S57-63.

227. Ait-Mohamed, A.K., et al., Characterization of N-ethylmaleimide-sensitive proton pump in the rat kidney. Localization along the nephron. J Biol Chem, 1986. 261(27): p. 12526-33.

228. Good, D.W., Adaptation of HCO₃⁻ and NH₄⁺ transport in rat MTAL: effects of chronic metabolic acidosis and Na⁺ intake. Am J Physiol, 1990. 258(5 Pt 2): p. F1345-53.

229. Good, D.W., Bicarbonate absorption by the thick ascending limb of Henle's loop.Review. Semin Nephrol, 1990. 10(2): p. 132-8.

230. Glassman, V.P., R. Safirstein, and V.A. DiScala, Effects of metabolic acidosis on proximal tubule ion reabsorption in dog kidney. Am J Physiol, 1974. 227(4): p. 759-65.

231. Cogan, M.G., et al., Control of proximal bicarbonate reabsorption in normal and acidotic rats. J Clin Invest, 1979. 64(5): p. 1168-80.

232. Kunau, R., Jr., J.I. Hart, and K.A. Walker, Effect of metabolic acidosis on proximal tubular total CO_2 absorption. Am J Physiol, 1985. 249(1 Pt 2): p. F62-8.

233. Preisig, P.A. and R.J. Alpern, Chronic metabolic acidosis causes an adaptation in the apical membrane Na/H antiporter and basolateral membrane $Na(HCO_3)_3$ symporter in the rat proximal convoluted tubule. J Clin Invest, 1988. 82(4): p. 1445-53.

234. Santella, R.N., F.J. Gennari, and D.A. Maddox, Metabolic acidosis stimulates bicarbonate reabsorption in the early proximal tubule. Am J Physiol, 1989. 257(1 Pt 2): p. F35-42.

235. Lucci, M.S., et al., Evaluation of bicarbonate transport in rat distal tubule: effects of acid-base status. Am J Physiol, 1982. 243(4): p. F335-41.

236. Vandorpe, D.H. and D.Z. Levine, Distal tubule bicarbonate reabsorption in NH4Cl acidotic rats. Clin Invest Med, 1989. 12(4): p. 224-9.

237. McKinney, T.D. and M.B. Burg, Bicarbonate transport by rabbit cortical collecting tubules. Effect of acid and alkali loads in vivo on transport in vitro. J Clin Invest, 1977. 60(3): p. 766-8.

238. McKinney, T.D. and M.B. Burg, Bicarbonate secretion by rabbit cortical collecting tubules in vitro. J Clin Invest, 1978. 61(6): p. 1421-7.

239. Hamm, L.L., C. Gillespie, and S. Klahr, NH_4Cl inhibition of transport in the rabbit cortical collecting tubule. Am J Physiol, 1985. 248(5 Pt 2): p. F631-7.

240. Levine, D.Z. and H.R. Jacobson, The regulation of renal acid secretion: new observations from studies of distal nephron segments.Review. Kidney Int, 1986. 29(6): p. 1099-109.

241. Hamm, L.L., K.S. Hering-Smith, and V.M. Vehaskari, Control of bicarbonate transport in collecting tubules from normal and remnant kidneys. Am J Physiol, 1989. 256(4 Pt 2): p. F680-7.

242. Hamm, L.L., I.D. Weiner, and V.M. Vehaskari, Structural-functional characteristics of acid-base transport in the rabbit collecting duct.Review. Semin Nephrol, 1991. 11(4): p. 453-64.

243. Yasoshima, K., L.M. Satlin, and G.J. Schwartz, Adaptation of rabbit cortical collecting duct to in vitro acid incubation. Am J Physiol, 1992. 263(4 Pt 2): p. F749-56.

244. Graber, M.L., et al., Acute metabolic acidosis augments collecting duct acidification rate in the rat. Am J Physiol, 1981. 241(6): p. F669-76.

245. Bengele, H.H., et al., Chronic metabolic acidosis augments acidification along the inner medullary collecting duct. Am J Physiol, 1986. 250(4 Pt 2): p. F690-4.

246. Bengele, H.H., et al., Inner medullary collecting duct function during rebound alkalemia. Am J Physiol, 1987. 252(4 Pt 2): p. F712-6.

247. Alexander, E.A. and J.H. Schwartz, Regulation of acidification in the rat inner medullary collecting duct.Review. Am J Kidney Dis, 1991. 18(5): p. 612-8.

248. Jacobson, H.R., H. Furuya, and M.D. Breyer, Mechanism and regulation of proton transport in the outer medullary collecting duct.Review. Kidney Int Suppl, 1991. 33(6): p. S51-6.

249. Levine, D.Z., et al., Secretion of bicarbonate by rat distal tubules in vivo. Modulation by overnight fasting. J Clin Invest, 1988. 81(6): p. 1873-8.

250. Tago, K., et al., Effects of inhibitors of Cl conductance on Cl self-exchange

in rabbit cortical collecting tubule. Am J Physiol, 1986. 251(6 Pt 2): p. F1009-17.

251. Garcia-Austt, J., et al., Deoxycorticosterone-stimulated bicarbonate secretion in rabbit cortical collecting ducts: effects of luminal chloride removal and in vivo acid loading. Am J Physiol, 1985. 249(2 Pt 2): p. F205-12.

252. Satlin, L.M. and G.J. Schwartz, Cellular remodeling of HCO$_3$(-)-secreting cells in rabbit renal collecting duct in response to an acidic environment. J Cell Biol, 1989. 109(3): p. 1279-88.

253. Satlin, L.M., T. Matsumoto, and G.J. Schwartz, Postnatal maturation of rabbit renal collecting duct. III. Peanut lectin-binding intercalated cells. Am J Physiol, 1992. 262(2 Pt 2): p. F199-208.

254. Dørup, J., Structural adaptation of intercalated cells in rat renal cortex to acute metabolic acidosis and alkalosis. J Ultrastruct Res, 1985. 92(1-2): p. 119-31.

255. Madsen, K.M. and C.C. Tisher, Response of intercalated cells of rat outer medullary collecting duct to chronic metabolic acidosis. Lab Invest, 1984. 51(3): p. 268-76.

256. Chang, C.S., Z. Talor, and J.A. Arruda, Effect of metabolic or respiratory acidosis on rabbit renal medullary proton-ATPase. Biochem Cell Biol, 1988. 66(1): p. 20-4.

257. Kriz, W. and L. Bankir, A standard nomenclature for structures of the kidney. The Renal Commission of the International Union of Physiological Sciences (IUPS). Kidney Int, 1988. 33(1): p. 1-7.

258. Kunau, R., Jr. and K.A. Walker, Total CO2 absorption in the distal tubule of the rat. Am J Physiol, 1987. 252(3 Pt 2): p. F468-73.

259. Kohn, O.F., P.P. Mitchell, and P.R. Steinmetz, Sch-28080 inhibits bafilomycin-sensitive H⁺ secretion in turtle bladder independently of luminalK⁺. Am J Physiol, 1993. 265(2 Pt 2): p. F174-9.

260. Graber, M.L. and P. Devine, Omeprazole and SCH 28080 inhibit acid secretion by the turtle urinary bladder. Renal Physiol Biochem, 1993. 16(5): p. 257-67.

261. Malnic, G., M. De Mello Aires, and G. Giebisch, Micropuncture study of renal tubular hydrogen ion transport in the rat. Am J Physiol, 1972. 222(1): p. 147-58.

262. Capasso, G., et al., Renal bicarbonate reabsorption in the rat. I. Effects of hypokalemia and carbonic anhydrase. J Clin Invest, 1986. 78(6): p. 1558-67.

263. Iacovitti, M., et al., Distal tubule bicarbonate accumulation in vivo. Effect of flow and transtubular bicarbonate gradients. J Clin Invest, 1986. 78(6): p. 1658-65.

264. Chan, Y.L., G. Malnic, and G. Giebisch, Renal bicarbonate reabsorption in the rat. III. Distal tubule perfusion study of load dependence and bicarbonate permeability. J Clin Invest, 1989. 84(3): p. 931-8.

265. Levine, D.Z., D. Vandorpe, and M. Iacovitti, Luminal chloride modulates rat distal tubule bidirectional bicarbonate flux in vivo. J Clin Invest, 1990. 85(6): p. 1793-8.

266. Capasso, G., et al., Renal bicarbonate reabsorption in the rat. II. Distal tubule load dependence and effect of hypokalemia. J Clin Invest, 1987. 80(2): p. 409-14.

267. Levine, D.Z., An in vivo microperfusion study of distal tubule bicarbon-

ate reabsorption in normal and ammonium chloride rats. J Clin Invest, 1985. 75(2): p. 588-95.

268. Gifford, J.D., et al., Total CO_2 transport in rat cortical collecting duct in chloride-depletion alkalosis. Am J Physiol, 1990. 258(4 Pt 2): p. F848-53.

269. Gifford, J.D., et al., HCO_3- transport in rat CCD: rapid adaptation by in vivo but not in vitro alkalosis. Am J Physiol, 1993. 264(3 Pt 2): p. F435-40.

270. Verlander, J.W., et al., Response of intercalated cells to chloride depletion metabolic alkalosis. Am J Physiol, 1992. 262(2 Pt 2): p. F309-19.

271. Galla, J.H., et al., Segmental chloride and fluid handling during correction of chloride-depletion alkalosis without volume expansion in the rat. J Clin Invest, 1984. 73(1): p. 96-106.

272. Kim, J., et al., Immunocytochemical response of type A and type B intercalated cells to increased sodium chloride delivery. Am J Physiol, 1992. 262(2 Pt 2): p. F288-302.

273. Wesson, D.E. and G.M. Dolson, Augmented bidirectional HCO_3 transport by rat distal tubules in chronic alkalosis. Am J Physiol, 1991. 261(2 Pt 2): p. F308-17.

274. Levine, D.Z., M. Iacovitti, and V. Harrison, Bicarbonate secretion in vivo by rat distal tubules during alkalosis induced by dietary chloride restriction and alkali loading. J Clin Invest, 1991. 87(5): p. 1513-8.

275. Wesson, D.E. and G.M. Dolson, Enhanced HCO_3 secretion by distal tubule contributes to NaCl-induced correction of chronic alkalosis. Am J Physiol, 1993. 264(5 Pt 2): p. F899-906.

276. Galla, J.H., D.N. Bonduris, and R.G. Luke, Superficial distal and deep nephrons in correction of metabolic alkalosis. Am J Physiol, 1989. 257(1 Pt 2): p. F107-13.

277. Wesson, D.E., Augmented bicarbonate reabsorption by both the proximal and distal nephron maintains chloride-deplete metabolic alkalosis in rats. J Clin Invest, 1989. 84(5): p. 1460-9.

278. Wesson, D.E. and G.M. Dolson, Maximal proton secretory rate of rat distal tubules is higher during chronic metabolic alkalosis. Am J Physiol, 1991. 261(5 Pt 2): p. F753-9.

279. Mills, J.N., S. Thomas, and K.S. Williamson, The acute effect of hydrocortisone, deoxycorticosterone, and aldosterone upon the excretion of sodium, potassium, and acid by the human kidney. J Physiol (Lond), 1960. 151: p. 312-331.

280. Hulter, H.N., et al., Impaired renal H' secretion and NH_3 production in mineralocorticoid-deficient glucocorticoid-replete dogs. Am J Physiol, 1977. 232(2): p. F136-46.

281. Sebastian, A., et al., Effect of mineralocorticoid replacement therapy on renal acid-base homeostasis in adrenalectomized patients. Kidney Int, 1980. 18(6): p. 762-73.

282. Dubrovsky, A.H., et al., Renal net acid excretion in the adrenalectomized rat. Kidney Int, 1981. 19(4): p. 516-28.

283. Wilcox, C.S., D.A. Cemerikic, and G. Giebisch, Differential effects of acute mineralo- and glucocorticosteroid administration on renal acid elimination. Kidney Int, 1982. 21(4): p. 546-56.

284. Lombes, M., et al., Immunohistochemical localization of renal mineralocorticoid receptor by using an anti-idiotypic antibody that is an internal

image of aldosterone. Proc Natl Acad Sci U S A, 1990. 87(3): p. 1086-8.

285. Farman, N., et al., Immunolocalization of gluco- and mineralocorticoid receptors in rabbit kidney. Am J Physiol, 1991. 260(2 Pt 1): p. C226-33.

286. Farman, N., Steroid receptors: distribution along the nephron.Review. Semin Nephrol, 1992. 12(1): p. 12-7.

287. Wade, J.B., et al., Modulation of cell membrane area in renal collecting tubules by corticosteroid hormones. J Cell Biol, 1979. 81(2): p. 439-45.

288. Stanton, B., et al., Ultrastructure of rat initial collecting tubule. Effect of adrenal corticosteroid treatment. J Clin Invest, 1985. 75(4): p. 1327-34.

289. Mujais, S.K., Effects of aldosterone on rat collecting tubule N-ethyl-maleimide-sensitive adenosine triphosphatase. J Lab Clin Med, 1987. 109(1): p. 34-9.

290. Relman, A.S., B. Esten, and W.B. Schwartz, The regulation of renal bicarbonate reabsorption by plasma carbon dioxide. J Clin Invest, 1953. 32: p. 972-978.

291. Polak, A., et al., Effects of chronic hypercapnia on electrolyte and acid-base equilibrium. 1. Adaptation. J Clin Invest, 1961. 40: p. 1223-1237.

292. Schwartz, W.B. and J.J. Cohen, The nature of the renal response to chronic disorders of acid-base equilibrium.Review. Am J Med, 1978. 64(3): p. 417-28.

293. Graf, R., et al., A novel 14-kDa V-ATPase subunit in the tobacco hornworm midgut. J Biol Chem, 1994. 269(5): p. 3767-74.

INDEX

QUESTIONNAIRE

Receive a FREE BOOK of your choice

Please help us out—Just answer the questions below, then select the book of your choice from the list on the back and return this card.

R.G. Landes Company publishes five book series: *Medical Intelligence Unit, Molecular Biology Intelligence Unit, Neuroscience Intelligence Unit, Tissue Engineering Intelligence Unit* and *Biotechnology Intelligence Unit*. We also publish comprehensive, shorter than book-length reports on well-circumscribed topics in molecular biology and medicine. The authors of our books and reports are acknowledged leaders in their fields and the topics are unique. Almost without exception, there are no other comprehensive publications on these topics.

Our goal is to publish material in important and rapidly changing areas of bioscience for sophisticated scientists. To achieve this goal, we have accelerated our publishing program to conform to the fast pace in which information grows in bioscience. Most of our books and reports are published within 90 to 120 days of receipt of the manuscript.

Please circle your response to the questions below.

1. We would like to sell our *books* to scientists and students at a deep discount. But we can only do this as part of a prepaid subscription program. The retail price range for our books is $59-$99. Would you pay $196 to select four *books* per year from any of our Intelligence Units–$49 per book–as part of a prepaid program?

   Yes   No

2. We would like to sell our *reports* to scientists and students at a deep discount. But we can only do this as part of a prepaid subscription program. The retail price range for our reports is $39-$59. Would you pay $145 to select five *reports* per year–$29 per report–as part of a prepaid program?

   Yes   No

3. Would you pay $39–the retail price range of our books is $59-$99–to receive any single book in our Intelligence Units if it is spiral bound, but in every other way identical to the more expensive hardcover version?

   Yes   No

To receive your free book, please fill out the shipping information below, select your free book choice from the list on the back of this survey and mail this card to:

R.G. Landes Company, 909 S. Pine Street, Georgetown, Texas 78626   U.S.A.

Your Name _____

Address _____

City _____ State/Province: _____

Country: _____ Postal Code: _____

My computer type is Macintosh_____ ; IBM-compatible _____ ; Other _____

Do you own ____ or plan to purchase ___ a CD-ROM drive?

AVAILABLE FREE TITLES

Please check three titles in order of preference.
Your request will be filled based on availability. Thank you.

❑ Water Channels
Alan Verkman,
University of California-San Francisco

❑ The Na,K-ATPase:
Structure-Function Relationship
J.-D. Horisberger, University of Lausanne

❑ Intrathymic Development of T Cells
J. Nikolic-Zugic,
Memorial Sloan-Kettering Cancer Center

❑ Cyclic GMP
Thomas Lincoln, University of Alabama

❑ Primordial VRM System and the Evolution
of Vertebrate Immunity
John Stewart, Institut Pasteur-Paris

❑ Thyroid Hormone Regulation
of Gene Expression
Graham R. Williams, University of Birmingham

❑ Mechanisms of Immunological Self Tolerance
Guido Kroemer, CNRS Génétique Moléculaire et
Biologie du Développement-Villejuif

❑ The Costimulatory Pathway
for T Cell Responses
Yang Liu, New York University

❑ Molecular Genetics of Drosophila Oogenesis
Paul F. Lasko, McGill University

❑ Mechanism of Steroid Hormone Regulation
of Gene Transcription
M.-J. Tsai & Bert W. O'Malley, Baylor University

❑ Liver Gene Expression
François Tronche & Moshe Yaniv,
Institut Pasteur-Paris

❑ RNA Polymerase III Transcription
R.J. White, University of Cambridge

❑ src Family of Tyrosine Kinases in Leukocytes
Tomas Mustelin, La Jolla Institute

❑ MHC Antigens and NK Cells
Rafael Solana & Jose Peña,
University of Córdoba

❑ Kinetic Modeling of Gene Expression
James L. Hargrove, University of Georgia

❑ PCR and the Analysis of the T Cell Receptor
Repertoire
Jorge Oksenberg, Michael Panzara & Lawrence
Steinman, Stanford University

❑ Myointimal Hyperplasia
Philip Dobrin, Loyola University

❑ Transgenic Mice as an In Vivo Model
of Self-Reactivity
David Ferrick & Lisa DiMolfetto-Landon,
University of California-Davis and Pamela Ohashi,
Ontario Cancer Institute

❑ Cytogenetics of Bone and Soft Tissue Tumors
Avery A. Sandberg, Genetrix & Julia A. Bridge ,
University of Nebraska

❑ The Th1-Th2 Paradigm and Transplantation
Robin Lowry, Emory University

❑ Phagocyte Production and Function Following
Thermal Injury
Verlyn Peterson & Daniel R. Ambruso,
University of Colorado

❑ Human T Lymphocyte Activation Deficiencies
José Regueiro, Carlos Rodríguez-Gallego
and Antonio Arnaiz-Villena,
Hospital 12 de Octubre-Madrid

❑ Monoclonal Antibody in Detection and
Treatment of Colon Cancer
Edward W. Martin, Jr., Ohio State University

❑ Enteric Physiology of the Transplanted Intestine
Michael Sarr & Nadey S. Hakim, Mayo Clinic

❑ Artificial Chordae in Mitral Valve Surgery
Claudio Zussa, S. Maria dei Battuti Hospital-Treviso

❑ Injury and Tumor Implantation
Satya Murthy & Edward Scanlon,
Northwestern University

❑ Support of the Acutely Failing Liver
A.A. Demetriou, Cedars-Sinai

❑ Reactive Metabolites of Oxygen and Nitrogen
in Biology and Medicine
Matthew Grisham, Louisiana State-Shreveport

❑ Biology of Lung Cancer
Adi Gazdar & Paul Carbone,
Southwestern Medical Center

❑ Quantitative Measurement
of Venous Incompetence
Paul S. van Bemmelen, Southern Illinois University
and John J. Bergan, Scripps Memorial Hospital

❑ Adhesion Molecules in Organ Transplants
Gustav Steinhoff, University of Kiel

❑ Purging in Bone Marrow Transplantation
Subhash C. Gulati,
Memorial Sloan-Kettering Cancer Center

❑ Trauma 2000: Strategies for the New Millennium
David J. Dries & Richard L. Gamelli,
Loyola University